# Climate Change Management

**Series editor**

Walter Leal Filho, Hamburg, Germany

More information about this series at http://www.springer.com/series/8740

Walter Leal Filho · Evangelos Manolas
Anabela Marisa Azul · Ulisses M. Azeiteiro
Henry McGhie
Editors

# Handbook of Climate Change Communication: Vol. 1

## Theory of Climate Change Communication

Springer

*Editors*
Walter Leal Filho
HAW Hamburg
Hamburg
Germany

Ulisses M. Azeiteiro
Department of Biology and CESAM
University of Aveiro
Aveiro
Portugal

Evangelos Manolas
Department of Forestry and Management of the Environment and Natural Resources
Democritus University of Thrace
Orestiada
Greece

Henry McGhie
University of Manchester
Manchester
UK

Anabela Marisa Azul
Center for Neuroscience and Cell Biology
University of Coimbra
Coimbra
Portugal

ISSN 1610-2010 ISSN 1610-2002 (electronic)
Climate Change Management
ISBN 978-3-319-69837-3 ISBN 978-3-319-69838-0 (eBook)
https://doi.org/10.1007/978-3-319-69838-0

Library of Congress Control Number: 2017957177



Printed on acid-free paper

This Springer imprint is published by Springer Nature
The registered company is Springer International Publishing AG
The registered company address is: Gewerbestrasse 11, 6330 Cham, Switzerland

# Contents

# Africa's Dilemmas in Climate Change Communication: Universalistic Science Versus Indigenous Technical Knowledge

**Innocent Chirisa, Abraham Matamanda and John Mutambwa**

## 1 Introduction

Climate change communication is increasingly becoming a topical issue on discussions relating to climate change (Ward 2008; Ockwell et al. 2009). In this effort to communicate issues relating to climate change, information narrowly focuses on scientific findings, synthesis reports and at times proceedings from high level conferences and policy meetings (Weart 2003). This is so because the task of communicating climate change information is often perceived as a technical and scientific process that requires experts to interpret and pass on information relating to climate variability and trends in a given locality (Hiwasaki et al. 2014; Chanza and de Wit 2015). As a result of these scientific observations and projections, some strides have been made in addressing climate change when the findings are communicated to the general public or when they inform policy related to climate change mitigation and adaptation. However, science is universalistic while the impacts of climate change are localised. Therefore, using scientific methods and adaptation strategies alone proves to be inadequate in efforts aimed at addressing certain local conditions which are specific to certain areas and are not generalised (Ziervogel and Opere 2010; Hiwasaki et al. 2014). Furthermore, science is currently failing to predict reliably either the duration or distribution of seasonal rainfall which results in forecast being probabilistic. It is at this juncture that ITK comes in and integrates with the scientific methods and ultimately contributing in the realm of climate change communication yet its utility seems to be overlooked by policy makers (Ajani et al. 2013).

I. Chirisa (✉) · J. Mutambwa
Department of Rural and Urban Planning, University of Zimbabwe, Harare, Zimbabwe
e-mail: chirisa.innocent@gmail.com

I. Chirisa · A. Matamanda
University of Free State, Bloemfontein, South Africa



W. Leal Filho et al. (eds.), *Handbook of Climate Change Communication: Vol. 1*,
Climate Change Management, https://doi.org/10.1007/978-3-319-69838-0_1

Roncoli et al. (2002) have argued that the integration of scientific forecasts with indigenous knowledge has the potential to allow for inferences which will contribute greatly in rainfall forecasting. As a result, there is a dichotomy between science and ITK which complicates the success of climate change communication since a great deal of information is side-lined and eventually left out when national and regional climate change policies are being formulated due to their scientific orientation (Moser 2010). Furthermore, despite the over emphasis on scientific knowledge, there are some factors that contribute to the failure to tap from ITK chief among these being generation gap where the youth are not aware of some of the ITK (Chanza and de Wit 2015). In this way ITK is threatened as it may fail to be inter-generational. Religious and cultural beliefs also affect the ways in which ITK and science is perceived thus affecting the communication of climate change issues. It is observed that as climate change worsens and causes massive destruction to property and people's livelihoods especially in the developing countries, there is need to embrace communication approaches that integrate ITK and universalistic science. The integration of the two follows a study in Nepal and other regions of the world where it has been found that indigenous knowledge practices on their own will not help overcome climate change adaptation deficit (Chanza and de Wit 2015). Rather, the integration of ITK and science is the only way to go if climate change information is to be disseminated effectively. The aim of the paper is to analyse the opportunities for integrating ITK and universalistic science for effective climate change communication in Africa. The overarching objective thus is to explore on how best the fissures in communication if any between the two world views may be closed with the view to enhance the effectiveness and efficiency of climate change communication and adaption in Africa. The objectives of the paper thus are to:

- Identify the ITK systems that are used in different localities around the world with a view to contextualise some of these approaches that may be applicable in the African context.
- Explore the factors that stifle the successful adoption and integration of ITK systems with universalistic science systems in relation to climate change adaptation and communication.
- Assess how ITK systems and universalistic science may be synthesised so as to promote effective climate change communication and adaptation.

## 2 Context of the Study

Climate change is no longer a myth in Africa (Eriksen et al. 2008; Brown et al. 2007). It is increasingly becoming a priority policy agenda as governments are realising the severity of climate change related disasters which require immediate action. Events on the ground bear testimony to the occurrence of various climate change related disasters in Africa. Eastern Africa is plagued by prolonged and intensified droughts (Lisk 2009); floods are common in western and parts of southern Africa (Christensen et al. 2007); the rainforests in equatorial Africa are being depleted due to climate change; increased water scarcity is also increasing in Africa and creating ground for conflicts (De Wit and Jacek 2006; Brown et al. 2007) while the ocean acidity is increasing around Africa's southern coast (Lisk 2009). As a result of the foregoing effects of climate change in Africa, it emerges from the Global Humanitarian Forum Human Impact Report (2009) 15 out of the 20 countries in the world that are most vulnerable to climate change are from Africa.

Considering that Africa is haunted by a host of other socio-economic and environmental problems, climate change is likely to exacerbate these challenges as it impacts on the livelihoods of communities thus reducing their well-being and standard of living (IPCC 2014). Adaptation and mitigation to climate change then becomes the panacea to address the climate change debacle in Africa. However, efforts in this regard are often stifled by the fact that most discussions together with scholarship on climate change are biased towards scientific knowledge with little input from the ITK. This situation complicates the successful implementation of climate change responses since science is universal yet climate change effects tends to be experienced at the local level hence the need to incorporate the concerns, beliefs, values and attitudes of the indigenous communities when focusing on climate change. According to BBC World Service Trust (2010) there is a critical shortage of relevant, useful information for the African audiences which does little serve complicate Africans citizens' response to climate change. Tapping information from the locals and their experiences proves to yield positive results because the emerging policies will be in line with locals' experiences and beliefs with regards climate change. Therefore, climate change seem to be complicated in Africa due to the fact that there is much reliance on scientific information at the expense of local beliefs and cultures that prove to yield better results in adapting and mitigating climate change when integrated with ITK (Chanza and de Wit 2015).

Adaptation is a process through which individuals and communities formulate mechanisms aimed at coping with the undesired effects of a particular event such as flooding or earthquake. For climate change, adaption is also critical since it enables societies to be better placed in efforts to cope with the negative impacts of climate change and uncertain future. The rationale of climate change adaptation is to reduce the negative effects of climate change by making appropriate adjustments and changes in ecological, social or economic systems in response to climate change (Smith et al. 2003; Lim and Spranger-Siegffried 2004). These adjustments are meant to reduce harm posed by climate change or maximise the benefits emerging from the climate variability. Climate change adaptation is thus context bound since climate change presents varying impacts on different areas across the world. The adaptation strategies for climate change are then best formulated through the integration of ITK systems because both are localised initiatives. Effective climate change adaptation responses for Africa are those that are heavily informed by ITK rather than solely focusing on scientific methods that do not adequately address the climate change problems from a local (African) perspective. Adaptation comes in different ways and the options can either be scientific or informed by ITK. Options include increased sea defences to counter sea level rises, early warning systems for extreme events, better water management, improved risk management and behaviour change at personal level.

## 3 Theoretical Frameworks and Literature Review

Climate change is a complex process that is influenced by various variables. It is for this reason that Filho (2008) argues that the search for solutions to climate change related problems can never be one-dimensional. Rather, there is need to take an integrated approach to address the changes as well as communicate information relating to climate change. Communication then has to be a two-way process where society contributes to the formulation of strategies that affect their daily lives. Moser (2010) supports the foregoing assertion and argues that a communication approach that sidelines society and focus on scientific facts and technocratic solutions only without discussion with society is likely to lead to resistance and opposition. According to Filho (2008) climate information has the potential to inform the decisions related to climate change adaptation and development decisions with local ownership. Rather, at present climate change communication is seems to be stifled by the scientists and journalists who embrace a scientific approach when conveying climate change related information and projections. Ward (2008: 1) highlights the disgruntlement echoed by scientists who attended a workshop on climate change communication:

> … climate scientists were frustrated by what they saw as a failure of the general public to understand and appreciate the seriousness of the climate change issue. Many scientists said they were frustrated that the accumulated advances in understanding of climate change over more than two decades of research had not led to a better-informed public…

From the foregoing statements, it is crystal clear that climate change communication is rendered ineffective mainly due to the use of a scientific approach. It is commendable to scale down climate services and information so that individuals at the community level may easily interpret and relate with the messages and information being conveyed. This is so because community members trust their own knowledge more than they do scientific knowledge which they perceive to be too complicated (Filho 2008).

### *3.1 Agency and Action*

According to Jasper (2002) agency is best defined as anything that involves choices such that these choices are initiated or pursued by individuals or groups and result in the flow of action in a particular way or guide the actions to be made in particular ways. Agency relationship involves one or more people (principal) engaging another person (agent) to perform some services on their behalf. The most interesting thing is that although the principal and agent may be working towards the same goal, their interests may not always be the same. As a result there is need to reconcile these differences in the interests between these two groups such that there is sharing of the same interests which will ultimately result in successful execution of the project at hand. Risk sharing that may arise when the principal and agent have different attitudes toward risk usually result in the two taking diverging and contradicting actions because of the different risk preferences (Eisenhardt 1989). This argument can also be placed in the realm of climate change communication were the scientists, policy makers and those mandated to formulate climate change adaptation strategies may be perceived as the agents while the communities are the principals. Obviously the differences in the perceptions and attitudes between these two often result in two different approaches towards adapting to climate change. This is rooted in the inclination of the former on science and universalistic approaches and the latter to ITK systems, a situation which complicates the effective communication of climate change information.

## 3.2 *The 'War' Between Universal Science and ITK*

Universal science is that body of knowledge that is informed by ideas, theories and concepts that are not tied to a particular community (Kloppenburg 1991). Science establishes the rate of rising sea levels, frequency of flooding and other climate change related disasters. The knowledge is transferrable and applies in various societies across the globe. Because of this aspect of being 'immutable mobile' and being a one size fits all, universal science is not location-oriented which means that it may fail to adequately address the climate change concerns in some communities around the world due to differences in socio, economic and environmental factors. Universal science is widely used in the realm of climate change considering the international conventions and projections that are more inclined to the scientific side (The National Academy of Sciences 2010). Decisions made are heavily informed by experts who spend much of their time modelling and forecasting climate trends which results in the formulation of ideas and theories on climate change and how to adapt to its impacts. On the flip side is ITK which is a body of knowledge that belongs to communities or ethnic groups and is shaped by their culture, traditions and way of life (Mazonde and Thomas 2007). Chanza and de Wit (2015) provide a detailed definition of ITK when they define it as knowledge that is location specific, acquired in situ through gradual and progressive study of how the communities relate with the environment and orally transferred both within and between generations. ITK systems are developed from the experience which the locals gain over the centuries and adapted to the local culture and environment and are orally passed from one generation to the other. The ownership of ITK systems is usually collective and communities gain access to the practices and systems through stories, songs, folk tales, proverbs, cultural values, beliefs, rituals, community laws, local language as well as the development of plant species and animal breeds. Basically there is an array of knowledge which relates to indigenous practices that is used by communities and this includes the perceptions and understandings of climate change and related phenomenon which is held by a community such as predicting storms or floods through observing the sky; practices that involve livelihood sustainability such as diversifying livelihoods in light of climate change; survival, coping and mitigation strategies that aim at reducing damage or avoiding it which may be achieved through relocation to safer places; and cultural belief systems that include traditional rituals and ceremonies (Odero 2011; Hiwasaki et al. 2014). Therefore, ITK knowledge and practices are instrumental in guarantying successful adaptation because they are rich in cultural context; appropriate and sophisticated knowledge form, increase community buy-in, promotes equity, efficiency and environmental integrity while above all it leads to increased communication and understanding (Nyong et al 2007; Alvera 2013).

Shaw et al. (2008) provide three primary arguments that support the inclusion of ITK systems in formulation of disaster risk reduction policies (including climate change related disasters). The three arguments are:

- Community participation and empowerment with regards disaster reduction is increased when ITK systems are employed;
- ITK is instrumental in that it helps to provide invaluable information that relate to the local context such that adaptation strategies are locally embedded which will also increase their acceptance among the local communities.
- ITKs also provide non-formal means of passing on indigenous knowledge which emerges as a means of educating and informing the general public on disaster risk reduction adaptation policies.

However, despite its utility in climate change reduction indigenous knowledge remains at risk of disappearing because few people and communities use it (Alvera 2013). Science is the most common method which is used nowadays, but it is evident that science alone does not proffer effective climate change adaptation which then calls for the incorporation of ITKs (Ison and Russel 2000; Bollier 2009; Mawere 2010). These two systems have to converge rather than continue to exist in a disconnected. Table 1 presents a comparison of the two knowledge systems and how information is disseminated using each one of these.

According to Bollier (2009) modern technology (universal science) has eroded most ITKs especially among the younger generations. The result is that most of these indigenous practices exist only among the aged and elderly community members, yet it is not usually documented anywhere hence it faces extinction (see Table 1). Therefore, Ison and Russell (2000) propose for the systematic documentation, validation, standardisation and propagation of ITK among the younger generation so that the culture is passed on to future generations. With the low life expectancy in most developing countries, ITKs are threatened owing to the passing on of the elders. Furthermore, the migration of the younger generations from the

**Table 1** Comparison of ITK and universal science

| ITK | Universal science |
|---|---|
| Makes use of natural indicators such as behaviour of animals and plants | Use sophisticated weather and climate models of measuring data |
| Communication is usually oral | Communication is usually written an |
| Explanation of climate phenomenon and related issues is based on spiritual, cultural and social values | Explanation of issues is based on theoretical and scientific findings |
| Knowledge is taught and mastered through observation and experience | Knowledge is taught and mastered through lectures and reading |
| Indicators are mostly observed | Indicators are usually measurable |
| Forecast methods are rarely documented | Forecast methods are more developed and documented |
| Up-scaling and downscaling are usually complex | Up-scaling and down-scaling is usually relatively simple |

*Source* Adapted from Ziervogel and Opere (2010)

rural areas to the urban areas is also responsible for the diminishing of ITKs because once the young generations migrate to urban centres they tend to disconnect with the rural areas, their culture, beliefs and values much to the detriment of ITKs which are heavily embedded in family ties and relations (Inglis 1993; Ziervogel and Opere 2010; Mawere 2010).

Sophistication and religious influence are also threatening ITKs because ITKs are rooted in '*spiritism*' and cultural beliefs, yet contemporary religions shun these practices (see Table 1). At the end of the day, the young generation are increasingly indoctrinated and influenced by a number of socio-economic transformations such as culture dilution which means that ITKs are being endangered. Another key issue of concern in climate change communication is the role and place of women in such discourse as noted by Nellemann et al. (2011) who postulates that women are usually at the frontline in respect to climate change. Women also possess a lot of knowledge and play key roles and responsibilities in the productive processes in various sectors such as agriculture, rangelands, biodiversity and forests to households, income-generation, livelihoods and other socio-cultural and political-economic institutions and relations. Hence, it is critical to include their voices, responsibilities and knowledge in climate change adaptation strategies for developing sustainable development strategies. Odero (2011) supports the foregoing statement as he argues that ITKs must include gender defined knowledge of indigenous plant and animal species, especially drought-tolerant and pest-resistant varieties. Grantham (1996) also argues that women's' knowledge should be considered separately to that of men because men do not and possibly cannot express important aspects that involve the concerns of women, their livelihood strategies and how they cope and adapt to specific disasters.

In Asia, the Asia Indigenous Peoples' Pack (AIPP) was found in 1988 by indigenous peoples' movement with the aim of promoting, asserting and defending indigenous people's rights through focusing on critical issues affecting them.

The concerns of women and the youth at all levels of decision-making have also been considered as central in the development mandates of AIPP. Recently, climate change adaptation has become a critical area of focus for AIPP a development which shows the utility of indigenous knowledge in development initiatives in Asia (AIPP 2011). This shows that there is increasing recognition of the utility of ITK knowledge in addressing the climate change issues. Nepal presents a good case of how ITKs knowledge has been integrated into the climate change adaptation strategies for effective climate change communication. Climate change in Nepal is a reality as evident from the rapid temperature rises, extreme and irregular precipitation events and increase in the frequency of floods, landslides and droughts in the last decade (FAO 2014). This (climate change) has had negative impacts on the livelihoods of the indigenous people (especially women), property, biodiversity and economic development of the country (MoSTE 2015). Efforts aimed at dissemination information relating to climate change adaptation and mitigation has resulted in the adoption of various ITKs which have proved to be worth in promoting the climate change resilient in the country.

In Nepal there has been the establishment and strengthening of social institutions so as to systematically develop and promote the use of ITKs through maintenance of order and collective action among diverse social groups. Public gatherings such as festivals are used by the community leaders to disseminate relevant messages on climate change and emerging risks. Furthermore, various other practices are used in Nepal which includes: quick communication of distress messages within communities when climate change related disasters occur (MoSTE 2015). In the Philippines practices such as observation of animal behaviour, celestial bodies, the natural environment, traditional, faith based and cultural beliefs and practices are used to predict disasters that are related to climate change. These include use of stars and the sky to forecast the weather, relating animal behaviour to the coming of rains and storms and development of early warning systems through the use if conch shells (Hiwasaki et al. 2014).

To counter the disappearance of medicinal plants that is found in the forest areas, some communities have started ex situ conversation in their backyards or in community gardens guided by indigenous healers in countries like Nepal, Malaysia and Vietnam, etc. Some of this knowledge are documented and taught in community schools for knowledge transmission. The teaching of the ITK knowledge and practices in schools shows how ITKs are being integrated into the formal systems and eventually being passed on to the younger generations to counter their disappearance.

## 4 Methodology

The research method for this paper comprised of an intense literature review wherein climate change policies and other documents that relate to climate change have been analysed with the view of understanding how these relate to and approach the ITK and science issue in the context of communicating climate change. Considering that much of ITK remains undocumented, there was need to interact with particular key informants so as to gain a broader understanding of the complex issue of ITK that seems to be sidelined by policy makers and academics. Thus, key person interviews have been used to gather data for this study wherein various informants where interviewed and these include community leaders, environmentalists and other stakeholders who are interested in climate change issues. However, it has been difficult to interview a large number of key informants due to the limitation of funds as well as the geographical expanse of the African continent. In this regard, the intense literature review and document analysis of cases from other African countries has helped to substantiate the reliability and validity of the results.

## 5 Results: Case Studies

In Zimbabwe climate change is associated with frequent droughts and floods, heat waves and changing rainfall patterns. These new climate trends have had an effect on the livelihoods of the rural communities across the country that depends on agriculture and ecosystem services. A good example is Bikita (drought prone area in Zimbabwe) were farmers have responded to climate change by diversifying their range of crops, with a shift to drought-resistant crops like rapoko, millet and sorghum and rearing of small livestock such as goats and sheep (Gukurume 2012). Legumes are also cultivated because they mature fast and thus provide a source of food for the locals in Bikita. Alvera (2013) gives a detailed account of how ITK knowledge and practices are used in Mbire District in Zimbabwe to predict and forecast climate and weather related events. Residents in Mbire usually the elders are well versed in ITKs unlike the youth and this is attributed to the scientific observations and solutions that are being promoted as the best explanation to most phenomena in Zimbabwe. ITKs are increasingly being side-lined on the basis that they are primitive and unreliable. Yet, there were many similarities between predictions made using ITKs and universal science in the 2012/2013 with the only shortcoming of ITK being its failure to establish exact dates for the occurrence of particular climate events. In Mbire District, the community makes use of position of the moon, flowering of trees, position of bird's nests and behaviour of animals to predict amount of rainfall to be received, onset of rain season etc. (Alvera 2013). Gukurume (2012) also notes how ITK systems have been used in adaptation and mitigation strategies devised by subsistent farmers in Bikita and this has been made possible through detailed study of plant and animal behaviour to predict the likelihood of impeding climatic catastrophe.

Considering that climate change in sub-Saharan Africa has resulted in shifts in rainfall seasons, there has been need to adjust cropping seasons as well as predict and forecast the occurrence of specific climatic events and disasters such as floods and droughts. In this regard, there are various local practices that are used by communities in Africa to predict and forecast climatic events. In Zambia and

Zimbabwe weather and climate is predicted using indigenous knowledge indicators that include fruiting patterns, emergence of leaves and weeds and wind movement (Ziervogel and Opere 2010). Ajani et al (2013) also highlights how farmers in Nigeria have developed intricate systems of gathering, predicting, interpreting and making decisions based on their local experiences. However, despite this existence of diverse knowledge and experience in climate and weather patterns, there has been little incorporation of such knowledge into the national forecasts. Decisions are made on cropping patterns based on local predictions of climate and decisions on planting dates based on specific cultural models of weather that the farmers and locals are familiar with. Through these systems of predicting the future climate, farmers have been able to manage their vulnerability as well as adapt to the changing climate and the related negative problems.

In western Kenya, the Nganyi people have often relied on rainmakers who assisted local communities in deciding when to commence farming season (Kwanya 2013). The rainmakers are informed by changes in nature that include air currents, flowering and shedding of leaves of certain trees and behaviour of animals and insects (Independent 2010). However, with the realities of climate change, the rainmakers have faced some problems with regards the predictions of weather and climate events which have resulted in the traditional rainmakers collaborating with government meteorologists in forecasting weather and climate events in the region. This shows how ITK and universal science is being put to use where the two groups traditional rainmakers and government meteorologists meet and engage in discussions were together they produce a consensus forecast (Guthiga and Newsham 2011). Once they reach consensus, the Nganyi then relay the forecast to the villagers through ceremonies, public meetings and person to person communication since the majority of the villagers are illiterate and cannot read or write (Independent 2010). In this way the community have been able to make some strides towards adapting to climate change through improved climate change communication.

Climate change in Benin has compelled the local farmers and communities vulnerable to climate change to join hands with local research institutions and engage in the crafting of early warning systems. Communities have used their experiences and knowledge of the local weather and climate conditions to adapt to climate change. The success of the story in Benin is on the establishment of pre-alert committees across the country where ITK knowledge systems and practices are integrated with universal science knowledge that has been generated through research. Together the researchers, scientists and meteorologists identify areas and sectors that are more vulnerable to climate change and adaptation strategies that are more suited to local conditions so that they are developed, tested and subsequently implemented to reduce communities' vulnerability (O'Neill 2009). In the end climate policies in Benin have been informed by good quality science based work on vulnerability and adaptation and ITKs practices from the locals (Netherlands Commission for Environmental Assessment 2015).

## 6 Discussion

The occurrence of climate change in Africa is evident from the increasing rate of climate change related disasters that has compelled the formulation of climate change policies as well as climate communication. Efforts to disseminate climate change related information has been largely informed by science but it has been found that science greatly overlooks a number of local practices and experiences that have been traditionally used by indigenous people to predict and forecast weather and climate events. Disseminating information related to climate change then becomes a challenge because a large body of knowledge is sidelined which also translates in the formulation of adaptation strategies that are not recognized or easily interpreted by the local communities. The result is that these communities' vulnerabilities to climate change worsen. Various strategies such as observing the behavior of animals and plants have been used in different areas to predict and forecast climate change in Africa which concurs with the events in Nepal and other parts of the world. Moreover, there have been some great strides that have been made in Kenya and Benin with regards climate change adaptation and communication (Independent 2010; Kwanya 2013).

In Kenya, there has been the integration of ITKs and universal science knowledge a situation which is very noble considering the argument presented by Ison and Russel (2000) who advocate for the of the two knowledge systems. The case of Kenya where rainmakers and meteorologists integrate in crafting weather and climate forecasts proves to be an initiative in the right direction. Likewise, the pre-alert committee in Benin which is comprised of local communities and researchers and scientists has been a milestone with regards dissemination of climate change information due to the merging of the two knowledge systems. These cases emerge as the pacesetters in terms of how ITKs and universal science knowledge systems may be used simultaneously to enhance the adaptation strategies for climate change. The initiatives assert to the argument presented by Ajani et al (2013) who propose that modern scientific knowledge should complement ITK rather than replace it which in turn will help to achieve better results in adapting to and coping with climate change dilemma.

## 7 Conclusion

Effective climate change communication in Africa is jeopardised by an over reliance on universalistic science at the expense of ITK systems. Yet, local knowledge emerges as being vital for adapting to climate change because science alone tends to fail to comprehensively forecast future climate and weather. However, policy makers seem to be heavily inclined towards the use of universalistic science due to a number of reasons chief among this being that ITK are usually undocumented and hence inaccessible, ITK systems are perceived as being out-dated and ineffective,

ITK systems are learnt informally and also the knowledge is possessed by a few usually the elderly. As a result efforts to address the dilemma and engage in effective climate change communication are usually marred by the over reliance of the policy makers and decision makers on the universal science knowledge which in most cases is perceived as a one size fits all, yet it fails to address climate change effects holistically. In this regard there is need for future research to tap into the ITK systems used in Africa so that the knowledge system may be integrated with the universal science knowledge for effective climate change communication and adaptation in Africa. Since ITKS practices and knowledge largely remain undocumented, there is need to identify these experiences and practices so as to document the local ITKs and practices so as to safeguard them from being degraded and being lost completely especially among the youth.

## References


AIPP (2011) Indigenous peoples and climate change adaptation in Asia. AIPP, Chiang Mai

Ajani EN, Mgbenka RN, Okeke MN (2013) Use of indigenous knowledge as a strategy for climate change adaptation among farmers in sub-Saharan Africa: implications for policy. Asian J Agric Extension Econ Sociol 2(1):23–40

Alvera P (2013) The role of indigenous knowledge systems in coping with food security and climate change in Mbire District, Zimbabwe. Unpublished M.Sc. Thesis in Integrated Water Resources Management, University of Zimbabwe, Harare

BBC World Service Trust (2010) Democratic republic of Congo talks climate: the public understanding of climate change. www.africatalksclimate.com. Last accessed 23 July 2015

Bollier D (2009) Protecting indigenous knowledge. Electronic Journal of Indigenous Knowledge (Online) 12(1). http://www.organicagcentre.ca/Docs/protecting/content/vol12.org.phpid=2453. Last accessed 15 Aug 2016

Brown O, Hammill A, McLeman R (2007) Climate change as the 'new' security threat: implications for Africa International Affairs. Roy Inst Int Aff 83(6):1141–1154

Chanza N, de Wit A (2015) Enhancing climate governance through indigenous knowledge. S Afr J Sci 112(1/2):1–7

Christensen JH, Hewitson B, Busuioc A, Chen A, Gao X, Held I, Jones R, Kolli RK, Kwon W-T, Laprise R, Magaña Rueda V, Mearns L, Menéndez CG, Räisänen J, Rinke A, Sarr A, Whetton P (2007) Regional climate projections. In: Solomon S, Qin D, Manning M, Chen Z, Marquis M, Averyt KB, Tignor M, Miller HL (eds) Climate change 2007: the physical science basis. Contribution of working group I to the fourth assessment report of the intergovernmental panel on climate change. Cambridge University Press. Cambridge, United Kingdom and New York, USA

De Wit M, Jacek S (2006) Changes in surface water supply across Africa with predicted climate change—AEON—Africa Earth Observatory Network. University of Cape Town, Science Express Report, Cape Town

Eisenhardt KM (1989) Agency theory: an assessment and review. Acad Manag Rev 14(1):57–74

Eriksen S, O'Brien K, Rosentcater L (2008) Climate change in Eastern and Southern Africa impacts, vulnerability and adaptation. GECHS Report 2008:2

FAO (2014). Managing climate risks and adapting to climate change in the agriculture sector in Nepal. Rome, Italy: climate, energy and tenure division (NRC). Food and agriculture organization of the united nations (FAO), Rome

Filho WL (2008) Communicating climate change: challenges ahead and action needed. Int J Clim Change Strat Manag 1(1):6–18
Global Humanitarian forum Human impact report (2009) Climate change, the anatomy of a silent crisis. Last accessed 23 July 2016
Grantham K (1996) Gender and indigenous knowledge: the role of Nepalese women in agricultural research and development. https://assets.publishing.service.gov.uk/media/57a08db840f0b649740019e6/R6322Genderandindigenousknowledge.pdf. Last accessed 12 Nov 2016
Gukurume S (2012) Climate change, variability and sustainable agriculture in Zimbabwe's rural communities. Russ J Agric Socio-Econ Sci 2(14):89–100
Guthiga P, Newsham A (2011) Meteorologists meeting rainmakers: indigenous knowledge and climate policy processes in Kenya. IDS Bull 42:104–109
Hiwasaki L, Luna E, Syamsidik SR (2014) Local & indigenous knowledge for community resilience: hydro-meteorological disaster risk reduction and climate change adaptation in coastal and small island communities. UNESCO, Jakarta
Independent (2010) Indigenous knowledge meets science. Available at: http://www.independent.co.uk/news/world/africa/indigenous-knowledge-meets-science-1904637.html. Last accessed 24 Aug 2016
Inglis TJ (1993) Traditional ecological knowledge: concepts and cases. Internal program on Traditional Ecological Knowledge, Ottawa
IPCC (2014) Climate change 2014: impacts, adaptation, and vulnerability. Part A: global and sectoral aspects. Contribution of working group II to the fifth assessment report of the intergovernmental panel on climate change. In: Field CB, Barros VR, Dokken DJ, Mach KJ, Mastrandrea MD, Bilir TE et al (eds). Cambridge University Press, Cambridge, New York
Ison R, Russel D (2000) Agricultural extension and rural development: breaking out of traditions. Cambridge University Press, Cambridge
Jasper JM (2002) A strategic approach to collective action: looking for agency in social-movement choices. Mobil Int J 9(1):1–16
Kloppenburg J (1991) Social theory and the de/reconstruction of agricultural science: local knowledge for an alternative agriculture. Rural Sociol 56(4):519–548
Kwanya T (2013) Mainstreaming indigenous knowledge in climate change response: traditional 'Rainmaking' in Kenya. Paper presented at 2013 international knowledge management in organisations conference, Kaohsiung, Taiwan
Lim B, Spranger-Siedfried E (2004) Adaptation policy frameworks for climate change: developing strategies, policies and measures. Cambridge University Press, Cambridge
Lisk F (2009) Overview: the current climate change situation in Africa. In: Besada H, Sewankambo NK (eds) Climate change in Africa: adaptation, mitigation and governance challenges. Centre for International Governance Innovation, Onario
Mawere M (2010) Indigenous knowledge systems' (IKSs) potential for establishing a moral, virtuous society: lessons from selected IKs in Zimbabwe and Mozambique. J Sustain Dev Afr 12(7):209–221
Mazonde T, Thomas K (2007) A systems approach to comparing indigenous and scientific knowledge: consistency and discriminatory power of indigenous knowledge. J Agri Syst 4(5):291–312
Moser SC (2010) Communicating climate change: history, challenges, process and future directions. WIREs Clim Change 1:31–53
MoSTE (2015) Indigenous and local knowledge and practices for climate resilience in Nepal, Mainstreaming Climate Change Risk Management in Development, Ministry of Science, Technology and Environment (MoSTE), Kathmandu
Nellermann C, Verma R, Hislop L (eds) (2011) Women at the frontline of climate change: gender risks and hopes. A rapid response assessment. United Nations Environmental Planning, Arendal
Netherlands Commission for Environmental Assessment. (2015) Climate change profile: Benin. http://api.commissiemer.nl/docs/os/i71/i7152/climate_change_profile_benin.pdf. Last accessed 24 August 2016

Nyong A, Adesina F, Elasha BO (2007) The value of indigenous knowledge in climate change mitigation and adaptation strategies in the African Sahel. Mitig Adapt Strat Glob Change 12:787–797

Odero K (2011) The role of indigenous knowledge in responding to climate change: local-global perspectives. New voices, different perspectives. In: Proceedings of the Africa adapt climate change symposium 2011. Institute of Development Studies, Sussex

O'Neill M (2009) Inform, involve, adapt. Case study: Benin, Joto Afrika, 1: 2

Ockwell D, Whitmarsh L, O'Neill S (2009) Reorienting climate change communication for effective mitigation—forcing people to be green or fostering grass-roots engagement? Sci Commun 30(3):305–327

Roncoli C, Ingram K, Kirshen P (2002) Reading the rains: local knowledge and rainfall forecasting in Burkina Faso. Soc Nat Resour Int J 15(5):409–427

Shaw R, Uy N, Baumwoll J (eds) (2008) Indigenous knowledge for disaster risk reduction: good practices and lessons learnt from Asia-Pacific region. UNISDR Asia and the Pacific, Bangkok

Smith JB, Klein RJT, Hug S (eds) (2003) Climate change, adaptive capacity and development. Imperial College Press, London

The National Academy of Sciences (2010) America's climate choices: advancing the science of climate change. http://dels.nas.edu/resources/static-assets/materials-based-on-reports/reports-in-brief/Science-Report-Brief-final.pdf. Last accessed 23 Nov 2016

Ward B (2008) Communicating on climate change: an essential resource for journalists, scientists and educators. Metcalf Institute for Marine and Environmental Reporting, Narragansett

Weart S (2003) The discovery of global warming. Harvard University Press, Cambridge

Ziervogel G, Opere A (2010) Integrating meteorological and indigenous knowledge-based seasonal climate forecasts in the agricultural sector. International Development Research Centre, Ottawa

## Author's Biography

**Innocent Chirisa** holds a D.Phil. in Social Studies with an inclination towards regional and urban planning, an M.Sc. and B.Sc. in Rural and Urban Planning, from the University of Zimbabwe. His research interests are in the systems approach as it applies to planning, the environment and social processes. An author of over 100 refereed articles and book chapters in a space of nine years. Currently is the acting head of the Department of Rural and Urban Planning, as well as Deputy Dean of the Faculty of Social Studies at the University of Zimbabwe.

**Abraham R Matamanda** graduated from the University of Zimbabwe with a B.Sc. degree in Rural and Urban Planning in 2013 and went on to obtain a Masters in Social Ecology in 2015. Currently he is a Ph.D. student at the University of the Free State in the department of Urban and Regional Planning focusing on urban form and social challenges in The Global South. Abraham has published in international peer reviewed journals and the issues that he has addressed include poverty and housing, housing and informality, water and sanitation issues as well as on ecological planning through the lenses of ecosystem services derived from wetlands. Basically, his researches focus on urban sustainability, environmental planning and management, rural development and the planning of cities and towns.

**John Mutambwa** has research interests in development communication, academic writing, professional communication skills, climate change communication and English language teaching methodologies. He holds an M.Ed. in Language Education: English from the University of Zimbabwe. He teaches communication in the University of Zimbabwe, Department of Linguistics.

# The Impact of Climate Change on Biodiversity: The Ecological Consequences of Invasive Species in Greece

**Konstantinos Demertzis and Lazaros Iliadis**

## 1 Introduction

### *1.1 Climate and Biodiversity in Greece*

Greece is located between parallels 340 and 420 of the Northern Hemisphere. It is a country in southeastern Europe, situated on the southern tip of the Balkan Peninsula, bordering the eastern Mediterranean. The country covers an area of about 131,000 km$^2$, the coastline of which extends to 13,676 km and it corresponds to half of the total Mediterranean coastline. The Greek islands are estimated at 2500 from which only 200 are inhabited. The coastal zone includes 85% of the population, 80% of the industry, 90% of the tourism, much of the agriculture and almost all fisheries and aquaculture. A comparison between Greece and Euro-area, follows in Table 1.

The climate of Greece is Mediterranean with mild, rainy winters, relatively warm and dry summers and plenty of sunshine throughout most of the year. Various regions of Greece are characterized by a great variety of climate subtypes, always within the Mediterranean climate. This is mainly due to the topography of the country which has great elevation differences as there are high mountains along the central country. This fact causes significant climatic differences. For example, Eastern Greece has dry climate conditions whereas Northern and Western part of the country is quite humid. It is a fact that big climatic differences still occur in places that are a short distance apart. This creates a special and unique area of interest appearing rarely even in a global scale.

K. Demertzis (✉) · L. Iliadis
Lab of Forest-Environmental Informatics and Computational Intelligence,
Democritus University of Thrace, 193 Pandazidou st., 68200 Orestiada, Greece
e-mail: kdemertz@fmenr.duth.gr

L. Iliadis
e-mail: liliadis@fmenr.duth.gr



W. Leal Filho et al. (eds.), *Handbook of Climate Change Communication: Vol. 1*,
Climate Change Management, https://doi.org/10.1007/978-3-319-69838-0_2

**Table 1** Greece versus Euro-area development indicators (http://data.worldbank.org/)

| Factor | Europe | Greece |
|---|---|---|
| Cereal yield (kg per hectare) | 5898 | 4404 |
| Foreign direct investment net inflows (percent of GDP) | 2.912 | 1.528 |
| Energy use per units of GDP (kg oil eq./$1000 of 2005 PPP $) | 121.681 | 100.636 |
| Energy use per capita (kilograms of oil equivalent) | 3740 | 2707 |
| $CO_2$ emissions total (kt $CO_2$) | 2,633,269 | 97,813 |
| $CO_2$ emissions per capita (metric tons) | 7.991 | 8.704 |
| $CO_2$ emissions per units of GDP (kg/$1000 of 2005 PPP $) | 259.975 | 323.591 |
| Population in urban agglomerations >1 million (percent) | 17.81 | 28.89 |
| Nationally terrestrial protected areas (percent of total land area) | 15.451 | 13.833 |
| GDP (Euros) | 16.2 trillion | 187 billion |
| GNI per capita (Atlas $) | 38,508 | 27,810 |
| Population growth (annual percent) | 0.522439 | 0.395286 |
| Population | 329,523,510 | 11,237,094 |
| Urban population growth (annual percent) | 0.780 | 0.723 |
| Urban population | 241,234,651 | 6,854,627 |

**Table 2** Climate types of Greece based on Köppen classifications (http://climate-data.org/)

| Köppen | Climate classification | Count | Examples |
|---|---|---|---|
| Csa | Hot-summer Mediterranean | 5783 | Athens, Patras, Larissa |
| Csb | Warm-summer Mediterranean | 808 | Karpenisi, Metsovo |
| Cfa | Humid subtropical | 621 | Serres, Veria, Naoussa |
| Cfb | Oceanic | 385 | Kozani, Florina, Kastoria |
| BSk | Coldsemi-arid | 145 | Thessaloniki, Katerini |

According to the Köppen climate classification system which considers the average annual and monthly temperatures, the rainfall and the seasonal distribution of precipitation, in Greece the overall climate is Mediterranean (Csa and Csb) which is distinguished in Mediterranean coastal and in inland Mediterranean. Some typical examples illustrating the great climatic differences between Greek regions are presented in Table 2.

The climate in Athens is mild, and generally warm and temperate. The winters are rainier than the summers in Athens. According to Köppen, this climate is classified as Csa. The average annual temperature is 18.1 °C in Athens. About 397 mm of precipitation falls annually. Diagrams with annual temperature and precipitation of Athens, presents in Figs. 1, 2 and 3.

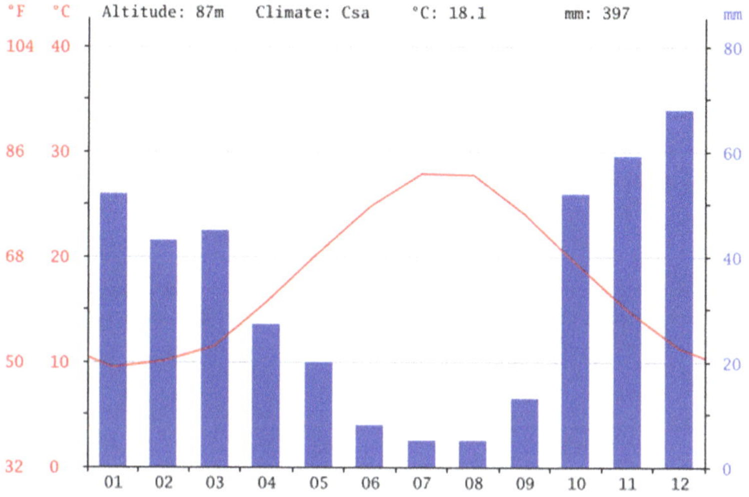


**Fig. 1** Precipitation is the lowest in July, with an average of 5 mm. The greatest amount of precipitation occurs in December, with an average of 68 mm (http://climate-data.org/)

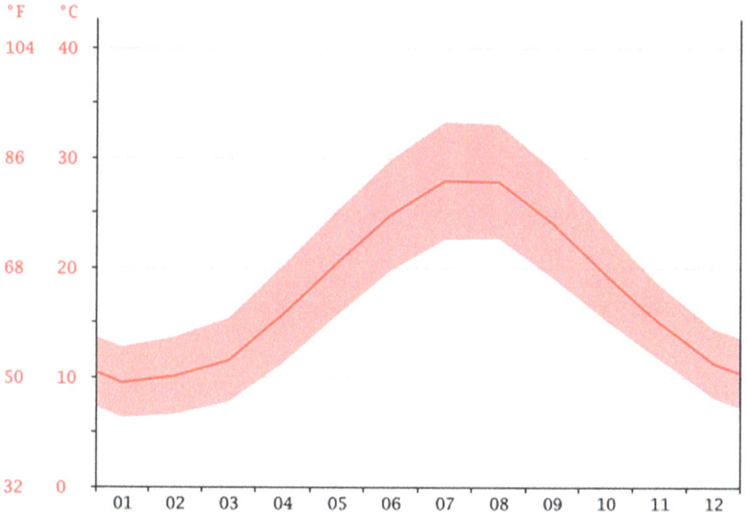


**Fig. 2** At an average temperature of 27.9 °C, July is the hottest month of the year. The lowest average temperatures in the year occur in January, when it is around 9.5 °C (http://climate-data.org/)

The climate in Karpenisi (mountainous area) is warm and temperate. The winters are rainier than the summers. The Köppen climate classification is Csb. The temperature here averages 12.0 °C. In a year, the average rainfall is 882 mm. Diagrams with annual temperature and precipitation of Karpenisi, presents in Figs. 4, 5 and 6.

| month | 1 | 2 | 3 | 4 | 5 | 6 | 7 | 8 | 9 | 10 | 11 | 12 |
|---|---|---|---|---|---|---|---|---|---|---|---|---|
| mm | 52 | 43 | 45 | 27 | 20 | 8 | 5 | 5 | 13 | 52 | 59 | 68 |
| °C | 9.5 | 10.1 | 11.5 | 15.7 | 20.4 | 24.8 | 27.9 | 27.8 | 24.1 | 19.4 | 15.0 | 11.3 |
| °C (min) | 6.3 | 6.6 | 7.7 | 11.2 | 15.6 | 19.7 | 22.5 | 22.6 | 19.1 | 15.2 | 11.6 | 8.1 |
| °C (max) | 12.8 | 13.7 | 15.4 | 20.2 | 25.2 | 29.9 | 33.3 | 33.1 | 29.1 | 23.6 | 18.5 | 14.5 |
| °F | 49.1 | 50.2 | 52.7 | 60.3 | 68.7 | 76.6 | 82.2 | 82.0 | 75.4 | 66.9 | 59.0 | 52.3 |
| °F (min) | 43.3 | 43.9 | 45.9 | 52.2 | 60.1 | 67.5 | 72.5 | 72.7 | 66.4 | 59.4 | 52.9 | 46.6 |
| °F (max) | 55.0 | 56.7 | 59.7 | 68.4 | 77.4 | 85.8 | 91.9 | 91.6 | 84.4 | 74.5 | 65.3 | 58.1 |

**Fig. 3** Between the driest and wettest months, the difference in precipitation is 63 mm. The variation in temperatures throughout the year is 18.4 °C (http://climate-data.org/)

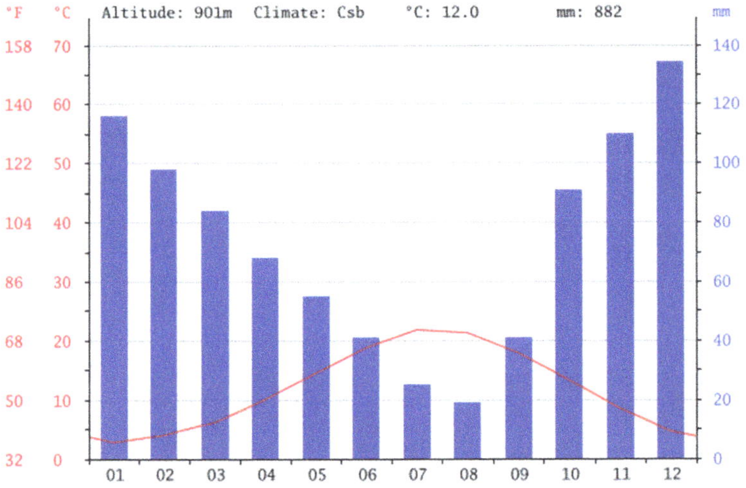


**Fig. 4** The driest month is August, with 19 mm of rain. The greatest amount of precipitation occurs in December, with an average of 134 mm (http://climate-data.org/)

In Serres (inland area) the climate is warm and temperate. The is a great deal of rainfall in Serres, even in the driest month. According to Köppen, this climate is classified as Cfa. The average temperature in Serres is 15.1 °C. Precipitation here averages 473 mm. Diagrams with annual temperature and precipitation of Serres, presents in Figs. 7, 8 and 9.

In Florina (highland area) the climate is warm and temperate. There is significant rainfall throughout the year in Florina. Even the driest month still has a lot of rainfall. The Köppen climate classification is Cfb. The average annual temperature in Florina is 11.1 °C. The average annual rainfall is 623 mm. Diagrams with annual temperature and precipitation of Florina, presents in Figs. 10, 11 and 12.

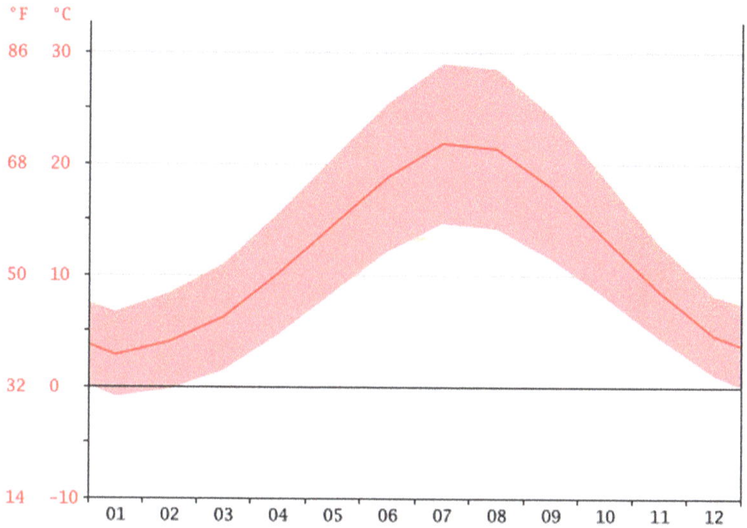


**Fig. 5** July is the warmest month of the year. The temperature in July averages 21.8 °C. The lowest average temperatures occur in January, when it is around 2.8 °C (http://climate-data.org/)

| month | 1 | 2 | 3 | 4 | 5 | 6 | 7 | 8 | 9 | 10 | 11 | 12 |
|---|---|---|---|---|---|---|---|---|---|---|---|---|
| mm | 116 | 98 | 84 | 68 | 55 | 41 | 25 | 19 | 41 | 91 | 110 | 134 |
| °C | 2.8 | 4.0 | 6.2 | 10.1 | 14.5 | 18.8 | 21.8 | 21.3 | 17.9 | 13.3 | 8.5 | 4.6 |
| °C (min) | -1.0 | -0.3 | 1.4 | 4.6 | 8.4 | 12.1 | 14.6 | 14.1 | 11.4 | 8.0 | 4.3 | 1.0 |
| °C (max) | 6.7 | 8.4 | 11.0 | 15.6 | 20.6 | 25.5 | 29.0 | 28.5 | 24.4 | 18.7 | 12.8 | 8.2 |
| °F | 37.0 | 39.2 | 43.2 | 50.2 | 58.1 | 65.8 | 71.2 | 70.3 | 64.2 | 55.9 | 47.3 | 40.3 |
| °F (min) | 30.2 | 31.5 | 34.5 | 40.3 | 47.1 | 53.8 | 58.3 | 57.4 | 52.5 | 46.4 | 39.7 | 33.8 |
| °F (max) | 44.1 | 47.1 | 51.8 | 60.1 | 69.1 | 77.9 | 84.2 | 83.3 | 75.9 | 65.7 | 55.0 | 46.8 |

**Fig. 6** There is a difference of 115 mm of precipitation between the driest and wettest months. The variation in temperatures throughout the year is 19.0 °C (http://climate-data.org/)

The climate here is considered to be a local steppe climate. In Thessaloniki, there is little rainfall throughout the year. This climate is considered to be BSk according to the Köppen climate classification. In Thessaloniki, the average annual temperature is 15.9 °C. Precipitation here averages 445 mm. Diagrams with annual temperature and precipitation of Thessaloniki, presents in Figs. 13, 14 and 15.

In general, the year can be divided in the cold and rainy season that lasts from mid-October until the end of March and the warm and dry season lasting from April to October. During the first period the coldest months are January and February, where on average the mean temperature ranges at 6.1 °C, whereas the warmest period is the last ten days of July and the first of August, when the average temperature ranges around 24.2 °C. The average monthly temperature for Greece from 1900 to 2015, presents in Fig. 16.

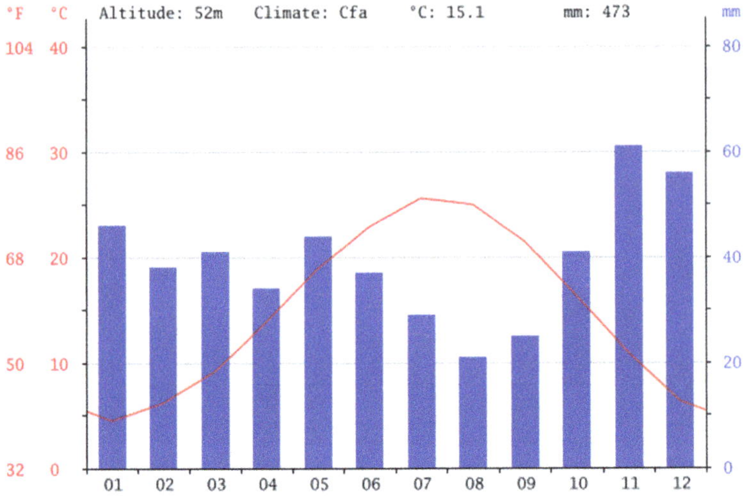


**Fig. 7** The driest month is August. There is 21 mm of precipitation in August. The greatest amount of precipitation occurs in November, with an average of 61 mm (http://climate-data.org/)

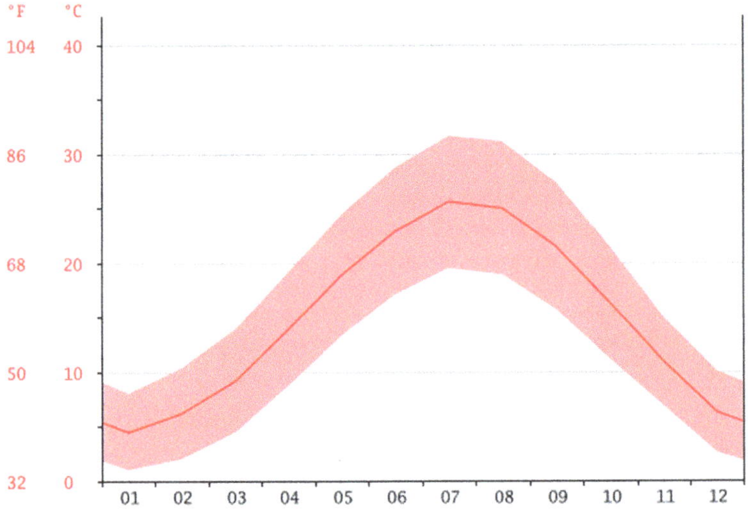


**Fig. 8** With an average of 25.6 °C, July is the warmest month. The lowest average temperatures in the year occur in January, when it is around 4.5 °C (http://climate-data.org/)

| month | 1 | 2 | 3 | 4 | 5 | 6 | 7 | 8 | 9 | 10 | 11 | 12 |
|---|---|---|---|---|---|---|---|---|---|---|---|---|
| mm | 46 | 38 | 41 | 34 | 44 | 37 | 29 | 21 | 25 | 41 | 61 | 56 |
| °C | 4.5 | 6.2 | 9.2 | 13.9 | 18.9 | 22.9 | 25.6 | 25.0 | 21.5 | 16.3 | 10.9 | 6.3 |
| °C (min) | 1.0 | 2.0 | 4.5 | 8.7 | 13.3 | 17.1 | 19.5 | 18.9 | 15.7 | 11.1 | 6.8 | 2.6 |
| °C (max) | 8.1 | 10.4 | 14.0 | 19.2 | 24.5 | 28.8 | 31.7 | 31.2 | 27.4 | 21.5 | 15.0 | 10.1 |
| °F | 40.1 | 43.2 | 48.6 | 57.0 | 66.0 | 73.2 | 78.1 | 77.0 | 70.7 | 61.3 | 51.6 | 43.3 |
| °F (min) | 33.8 | 35.6 | 40.1 | 47.7 | 55.9 | 62.8 | 67.1 | 66.0 | 60.3 | 52.0 | 44.2 | 36.7 |
| °F (max) | 46.6 | 50.7 | 57.2 | 66.6 | 76.1 | 83.8 | 89.1 | 88.2 | 81.3 | 70.7 | 59.0 | 50.2 |

**Fig. 9** The precipitation varies 40 mm between the driest month and the wettest month. The variation in temperatures throughout the year is 21.1 °C (http://climate-data.org/)

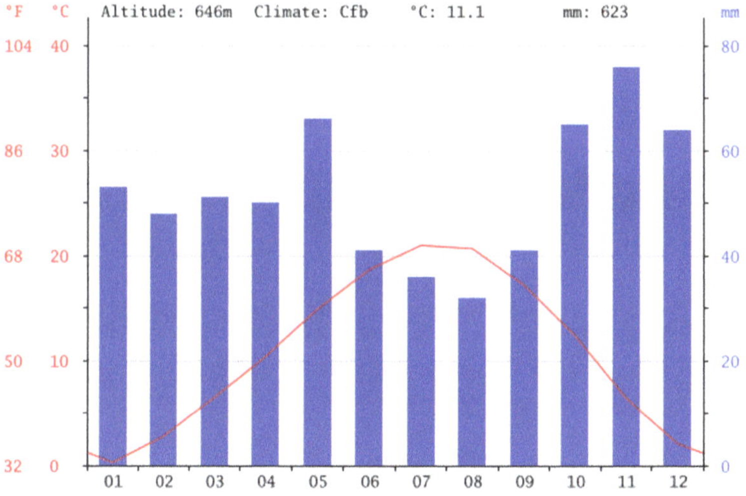


**Fig. 10** The driest month is August. There is 32 mm of precipitation in August. Most precipitation falls in November, with an average of 76 mm (http://climate-data.org/)

The rains in the country even in the winter do not last for many days and the sky does not remain cloudy for several consecutive days, as in other regions of the world. Winter storms are often interrupted during January and the first fortnight of February, with sunny days. During the warm and dry season, the weather is stable, the sky is clear, the sun is shining and rapid but short duration storms occur.

During the cold season (in December) the mean average precipitation ranges to 112.6 mm, whereas during the warmer season, namely in August the rainfall level is just 10.4 mm (http://www.worldbank.org/). The average monthly rainfall for Greece from 1900 to 2015, presents in Fig. 17.

Greece is one of the richest European and Mediterranean biodiverse countries, with many endemic species of plants and animals and important habitats. This is due to its geographical position, to its climatic and topographic characteristics, and to its diverse terrain. Another important parameter is the fact that in generally the

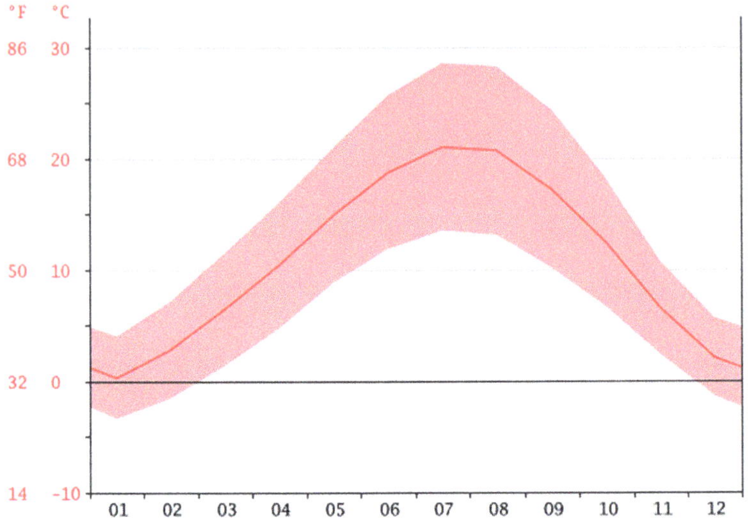


**Fig. 11** With an average of 21.0 °C, July is the warmest month. In January, the average temperature is 0.3 °C. It is the lowest average temperature of the whole year (http://climate-data.org/)

| month | 1 | 2 | 3 | 4 | 5 | 6 | 7 | 8 | 9 | 10 | 11 | 12 |
|---|---|---|---|---|---|---|---|---|---|---|---|---|
| mm | 53 | 48 | 51 | 50 | 66 | 41 | 36 | 32 | 41 | 65 | 76 | 64 |
| °C | 0.3 | 2.8 | 6.5 | 10.4 | 14.9 | 18.7 | 21.0 | 20.7 | 17.2 | 12.4 | 6.5 | 2.1 |
| °C (min) | -3.4 | -1.6 | 1.4 | 4.7 | 8.8 | 11.8 | 13.5 | 13.1 | 10.1 | 6.6 | 2.3 | -1.4 |
| °C (max) | 4.1 | 7.2 | 11.6 | 16.1 | 21.0 | 25.7 | 28.6 | 28.3 | 24.4 | 18.2 | 10.7 | 5.7 |
| °F | 32.5 | 37.0 | 43.7 | 50.7 | 58.8 | 65.7 | 69.8 | 69.3 | 63.0 | 54.3 | 43.7 | 35.8 |
| °F (min) | 25.9 | 29.1 | 34.5 | 40.5 | 47.8 | 53.2 | 56.3 | 55.6 | 50.2 | 43.9 | 36.1 | 29.5 |
| °F (max) | 39.4 | 45.0 | 52.9 | 61.0 | 69.8 | 78.3 | 83.5 | 82.9 | 75.9 | 64.8 | 51.3 | 42.3 |

**Fig. 12** The precipitation varies 44 mm between the driest month and the wettest month. The average temperatures vary during the year by 20.7 °C. (http://climate-data.org/)

traditional farming activities cause limited disturbance. More specifically, the total number of species in Greece is estimated approximately to 50,000 whereas the majority of the species are insects.

About 1500 endemic species included among the best known animal groups (Vertebrates, Echinoderms, Molluscs and Orthoptera) have been recorded in a total of 5500 (Endemism 25%) (http://biodiversity.europa.eu/). The major areas of endemism are: Crete, Cyclades, Mountains of mainland Greece and to a lesser extent the Ionian Islands and Peloponnese.

The caves are extremely important for endemism of which only a small percentage has been investigated. Moreover, Greece is home to about 6000 plant species, the largest population of endemic fish fauna species in the Mediterranean

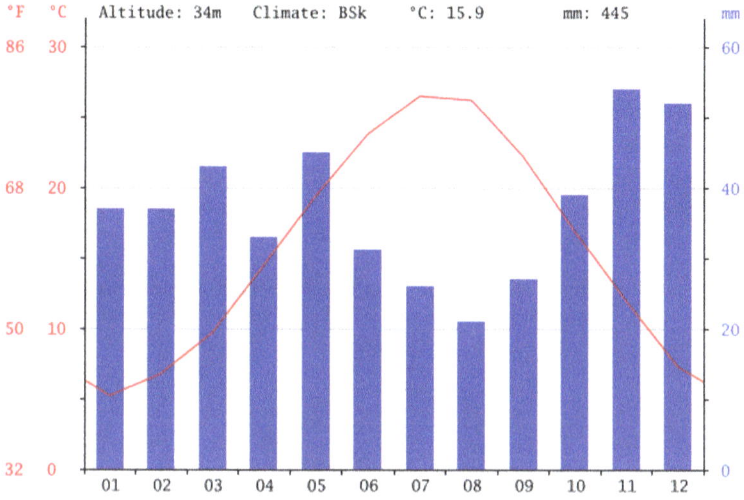


**Fig. 13** The driest month is August, with 21 mm of rain. With an average of 54 mm, the most precipitation falls in November (http://climate-data.org/)

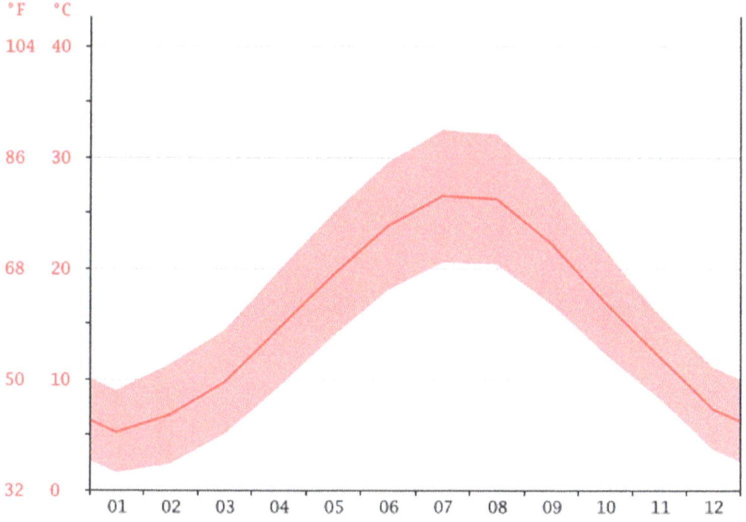


**Fig. 14** July is the warmest month of the year. The temperature in July averages 26.5 °C. January has the lowest average temperature of the year. It is 5.2 °C (http://climate-data.org/)

| month | 1 | 2 | 3 | 4 | 5 | 6 | 7 | 8 | 9 | 10 | 11 | 12 |
|---|---|---|---|---|---|---|---|---|---|---|---|---|
| mm | 37 | 37 | 43 | 33 | 45 | 31 | 26 | 21 | 27 | 39 | 54 | 52 |
| °C | 5.2 | 6.8 | 9.7 | 14.5 | 19.4 | 23.8 | 26.5 | 26.2 | 22.2 | 16.9 | 12.0 | 7.3 |
| °C (min) | 1.5 | 2.3 | 5.0 | 9.2 | 13.8 | 18.0 | 20.5 | 20.3 | 16.7 | 12.2 | 8.2 | 3.6 |
| °C (max) | 9.0 | 11.4 | 14.4 | 19.8 | 25.0 | 29.6 | 32.5 | 32.1 | 27.8 | 21.7 | 15.8 | 11.1 |
| °F | 41.4 | 44.2 | 49.5 | 58.1 | 66.9 | 74.8 | 79.7 | 79.2 | 72.0 | 62.4 | 53.6 | 45.1 |
| °F (min) | 34.7 | 36.1 | 41.0 | 48.6 | 56.8 | 64.4 | 68.9 | 68.5 | 62.1 | 54.0 | 46.8 | 38.5 |
| °F (max) | 48.2 | 52.5 | 57.9 | 67.6 | 77.0 | 85.3 | 90.5 | 89.8 | 82.0 | 71.1 | 60.4 | 52.0 |

**Fig. 15** There is a difference of 33 mm of precipitation between the driest and wettest months. During the year, the average temperatures vary by 21.3 °C (http://climate-data.org/)

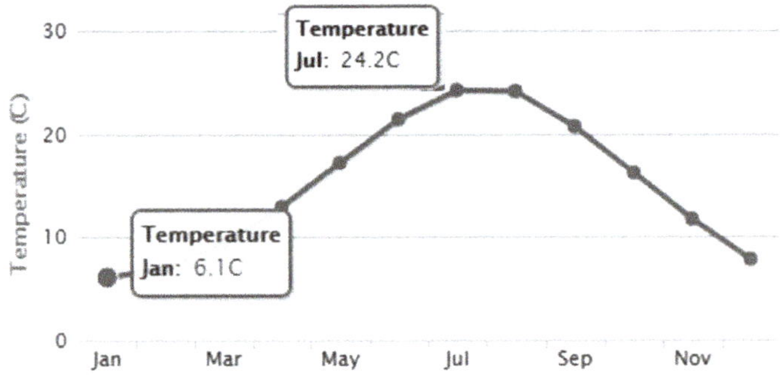


**Fig. 16** Average monthly temperature for Greece from 1900 to 2015 (http://www.worldbank.org/)

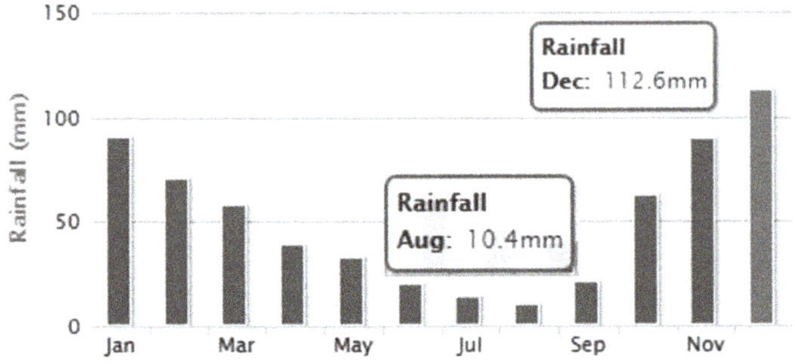


**Fig. 17** Average monthly rainfall for Greece from 1900 to 2015 (http://www.worldbank.org/)

ανδ more than 400 species of birds. Though for other organisms (microorganisms, fungi) there are no reliable data, no there are indications that there is a huge variety in the country (http://biodiversity.europa.eu/). The biodiversity of Greece, presents in the Table 3.

**Table 3** Biodiversity of Greece (http://biodiversity.europa.eu/)

| Species | TAXA | Endemicspecies | Endemic subspecies | Protected | Not protected |
|---|---|---|---|---|---|
| Mamals | 111 | 4 | 35 | 83 | 8 |
| Birds | 431 | 0 | 4 | 396 | 1 |
| Reptiles | 85 | 6 | 71 | 58 | 1 |
| Amphibia | 21 | 2 | 3 | 18 | 0 |
| Fish (fresh water) | 109 | 35 | 16 | 50 | 29 |
| Fish (marine) | 447 | – | – | 14 | – |
| Echinoderms | 107 | – | – | 1 | – |
| Orthoptera | 365 | 113 | 15 | 11 | – |
| Lepidoptera | 3197 | 142 | – | 57 | 39 |
| Hymenoptera | 2800 | 13 | – | 0 | 15 |
| Coleoptera | 6863 | 649 | – | 11 | 2 |
| Mollusks (land) | 763 | 174 | – | 24 | 5 |

Also as a result of the intense territorial dismemberment and the relatively benign human activities, Greece has a large variety of ecosystems. The most important are those of bush vegetation, the evergreen broadleaved, the phryganic, the coastal and the marine. It should be emphasized that the forest ecosystems of Greece, as in most parts of the Mediterranean region, have been exposed to human activities. This has caused degradation to the coastal forests and forests at low altitude due to urbanization and the conversion to agricultural land.

## 2 Climate Change and Its Impacts in Greece

By the term "climate change" we refer to serious global average climate alterations that include statistically significant meteorological fluctuations through a large time scale.

Greece due to its geographical position is the point where the climatic conditions favor a variety of activities and is one of the 18 most vulnerable areas of the planet due to climate change. According to climate simulations for the period up to 2100, the results suggest a continuous, gradual and relatively strong temperature increase with a steady decrease in intensity of rainfall.

In the near future (2015–2039) models forecast for Greece an increase in temperature of 1–3 °C and a decrease of in rainfall of the magnitude 5-10%. For the period (2040–2069) it is estimated that the temperature will increase from 3 to 6 °C and the rainfall will be reduced up to 20%. Finally, for the end of the century (2070–2100) the temperature will rise from 6 to 9 °C with a simultaneous reduction of precipitation up to 30%. More specifically, the Intergovernmental Panel on Climate Change (IPCC 2007) A2 scenario, estimates an increase of the greenhouse

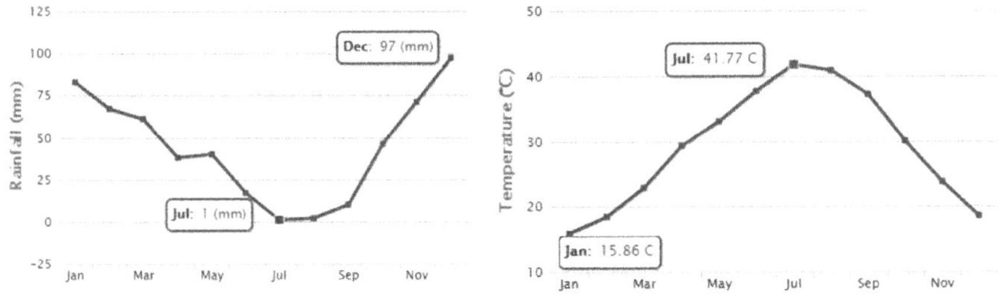


**Fig. 18** Minimum monthly rainfall and maximum temperature for Greece from 2015 to 2100 in order to Climate Change Scenario IPCC A2 (http://www.worldbank.org/)

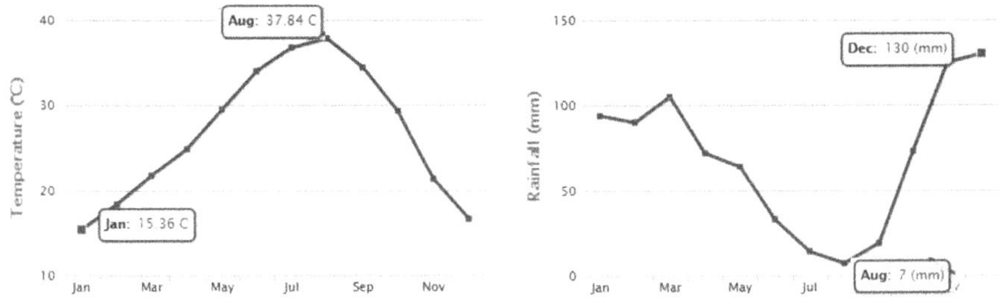


**Fig. 19** Minimum monthly rainfall and maximum temperature for Greece from 2015 to 2100 in order to Climate Change Scenario IPCC B2 (http://www.worldbank.org/)

emissions by 250% (compared to the period 1961–1990). This will cause a temperature increase of the magnitude 8–9 °C and rainfall decrease as high as 30%. (http://www.worldbank.org/). Minimum monthly rainfall and maximum temperature for Greece from 2015 to 2100 in order to Climate Change Scenario IPCC A2, presents in Fig. 18.

According to the IPCC B2 scenario, the greenhouse emissions will increase by 62% with a lower temperature of 6–7 °C and rainfall increase higher than 20%. Minimum monthly rainfall and maximum temperature for Greece from 2015 to 2100 in order to Climate Change Scenario IPCC B2, presents in Fig. 19.

Specifically, till mid-century, by warming above 2 °C, the dry areas that are mainly exposed, will particularly suffer from the reduction of water resources, while more hot and dry conditions will result in reduced forest productivity and increased forest fires. There will be changes in algal communities, large dolphin mortality and high insects spreading. The supply of groundwater will be reduced dramatically (by 70%) in southern Greece. The city residents, such as Thessaloniki, Patras, Lamia and Larissa will be subject up to 20 more days of heat per year. Meanwhile, in Lamia, Larissa, Volos, Thessaloniki and Athens, the total annual precipitation will decrease, but extreme rainfall is expected to increase by 10–20%.

The days with heat will increase by 5–15 and there will be a significant increase in the number of nights where the temperature will not fall below 20 °C mainly in islands such as Rhodes or the area of Chania in Crete. Also the ten largest rural prefectures of the country will receive heavy pressure from climate change, thereby increasing the heat days, leading in consecutive days without rain, in reduction of winter rainfall and thus causing a great increase in forest fire risk.

For example, in Evia we will expect more than 25 extra dry days compared to the period 1960 to 2000, Serres and Larissa will live 20 more days of heat, while in the areas of Heraklion and Pella rainfall in winter will decrease by 15% (http://www.worldbank.org/).

Additionally, we expect an overall decrease in the canopy of forests. Also the models forecast a spatial redistribution of species with necrosis of important Scots pine land (*Pinus sylvestris*) and fir (*Abies cephalonica*) and a corresponding invasion of Fir species of conifers (*Abies borissiregis*) and black pine (*Pinus nigra*) in broadleaf forests. It should be noted that the number of forest fires during the summer period and the total burnt area is expected to increase by 10–20% according to the same scenario. This will be due to the rise in temperature and to the corresponding reduction in rainfall, causing additional problems as quantities of timber necessarily remain in the forest after a fire, exacerbate the existing problem of bark eating insect's outbreak, that further degrade the soil quality (http://biodiversity.europa.eu/). Still, the effects of climate change will be significant at different levels of biological organizations and particularly in the variations that occur in habitats and endemic species that appear to be most vulnerable, as they are perfectly correlated with the local climate of their region.

For example, under the assumption of non-immigration the only two endemic species in the country are threatened, namely: The akanthopontikos (*Acomys minous*) and the Cretan shrew (*Crocidura zimmermanni*), as well as the Persian Squirrel (*Sciurus anomalus*) which is the only chipmunk who lives on an island and only appears in Lesvos (http://www.europe-aliens.org/).

According to the climate scenario IPCC A2 the population of the following species will be reduced significantly in Greece. More specifically, the *Quercus ithaburensis* sub macrolepis will be reduced by 56%, the *Matricaria chamomilla* by 88% and the *Sciurus anomalus* by 98%.

On the other hand, according to the scenario IPCC A2 a significant expansion of the brushwood habitat is expected (*Genista acanthoclada* by 386%) and (*Sarcopoterium spinosum* by 198%). This will be done due to the extensive soil degradation caused by climate change combined with the extreme weather phenomena and due to the forest ecosystems destruction caused by forest fires.

The areas mostly at soil degradation and desertification risk are mainly the eastern Central Greece, part of Thessaly, Evia, the Aegean islands, the eastern Peloponnese and part of Macedonia (http://www.cabi.org/cpc/).

Nevertheless, the most important consequences of climate change concerning Greece focus on the rise of sea levels and on the temperature change of surface waters afterthought the occurrence of extreme weather events in the coastal areas. The above may cause permanent flooding of coastal areas, the movement of foreshore areas and beach, coastal erosion from increased wave storms and salinization of coastal groundwater aquifers.

Specifically, according to the scenario IPCCA2, 15% of the coastal wetlands of Greece (about 1000 $km^2$) is projected to be flooded if finally, the rise level will be as high as 0.5 m by 2100. The coastal wetlands expected to be most affected are the following: The Delta of Evros, Nestos, Axios, Loudias and Aliakmonas rivers, the lagoons of Messolongi and Kyllini and the Bays of Amvrakikos and Pagasitikos.

The availability and water quality of lakes, may be affected significantly. More specifically, a reduction of water surface is forecasted for the lakes of Chimaditis 20–37%, Kerkini 5–14%, Trichonis 10%. Also the concentration of nitrogen will increase by 25%.

Regarding the Aegean Sea, it is expected a warming of the surface waters up to 3 °C, reducing runoff and changes in biochemical and physical properties of sea water, with a significant impact on marine biodiversity, productivity and food webs. This will increase the risk of spreading diseases, toxic bloom of seaweed and dissemination of thermophilic species. Specifically, coastal meadows of the endemic Mediterranean "*Posidonia oceanica*" which are place of nesting and breeding for numerous marine species and play an important ecological role, are particularly vulnerable to extreme weather events such as storms and floods but also to secondary factors such as water pollution. The climate changes of the period up to 2100 and the large amounts of solid and liquid contaminants are expected to pollute the sea, threatening with extinction the certain species.

The potential effects of the phenomena of climate change will lead to the degradation of services provided by ecosystems. As ecosystem services are defined processes and functions provided by the natural environment and benefit the people as the production of food, fuel, fiber and pharmaceuticals, water regulation, air and bioclimate, maintenance of soil fertility, cycle of nutrients (Gallardo and Aldridge 2013). Finally regarding biodiversity, the effects of climate change are multidimensional as they are related to a combination of direct effects on organisms.

The temperature affects the survival rates, the reproductive success and the dispersion and behavior patterns of the species. Also the effects are related to a combination of biotic interactions, such as the granting of competitive advantage, or they are related to abiotic factors such as bedtime with water or changes in ocean currents, changes in the level of biocoenosis, the extinction of organisms and the appearance of invasive species (Douglas and Gareth 2007).

## 3 Invasive Species in Greece

The invasive species are non-native, they have no natural predators, they are rapidly spreading and invading the endemic flora and fauna of a place, damaging the natural environment. Their social and economic impacts are considered extremely important on human health, agriculture, fisheries and food production.

Their relocation is mainly because different kinds of species such as birds, reptiles and fish, mollusks, move in search of colder climate conditions, or because their physical environment does not meet the temperature range in which they can survive. Thus they follow different kinds plants or organisms that migrate to colder climates. Excluding the human factor that may cause accidental or deliberate movement of alien species in locations that are hundreds of kilometers from their natural habitat, these non-native species adapt poorly to their new environment and most often they quickly vanish. The largest spreading factor for invasive species is the climate change, which drops the physical barriers due to which the peripheral distinct flora and fauna was developed (http://www.cabi.org/cpc/).

It is noteworthy that this migration of non-indigenous species represents one of the most serious threats to global biodiversity because they deprive from indigenous species, resources and living space. In Europe there are 10,822 non-native species under the Delivering Alien Invasive Species Inventories for Europe (DAISIE) of which 427 are invasive and pose a very serious threat to natural biodiversity as they compete aggressively with native organisms for food and habitat, they are changing the ecosystems structure by hybridization with native species, by direct toxicity and by disruption of pollination, either because they are a source of pests or pathogens and disease carrier (Keller et al. 2009). Cumulative number and groups of alien invasive species, presents in Figs. 20 and 21.

In Greece there are respectively 140 invasive species recorded, of which 48 are Alien, whereas for 9 of them the Biostatus is not specified and 91 are characterized as Native ones. A detailed presentation of the individual items by category, is given in the following diagrams (Figs. 22, 23 and 24) (http://www.cabi.org/cpc/).

The analysis of these invasive species under the habitats in which they live but also based on their taxonomic species, are presented in the following Tables 4 and 5 (http://www.cabi.org/cpc/).

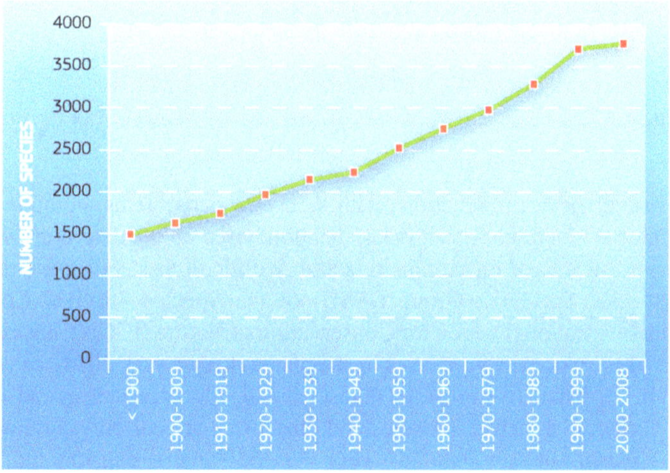


**Fig. 20** Cumulative number of alien species established in terrestrial environment in Europe (http://www.europe-aliens.org/)

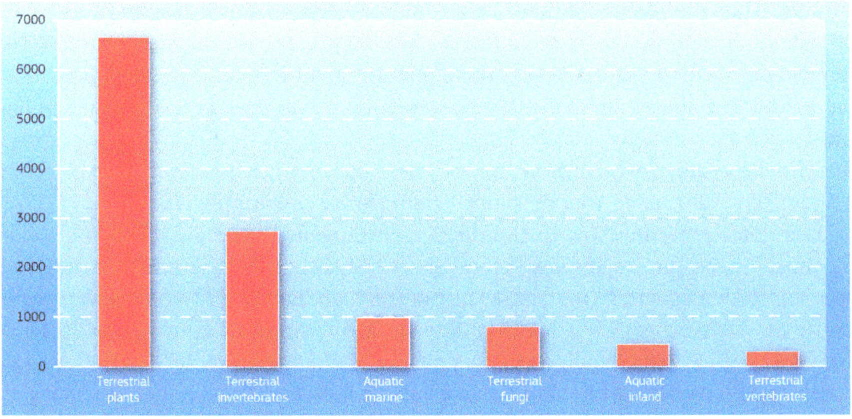


**Fig. 21** Groups of invasive alien species in Europe (http://www.europe-aliens.org/)

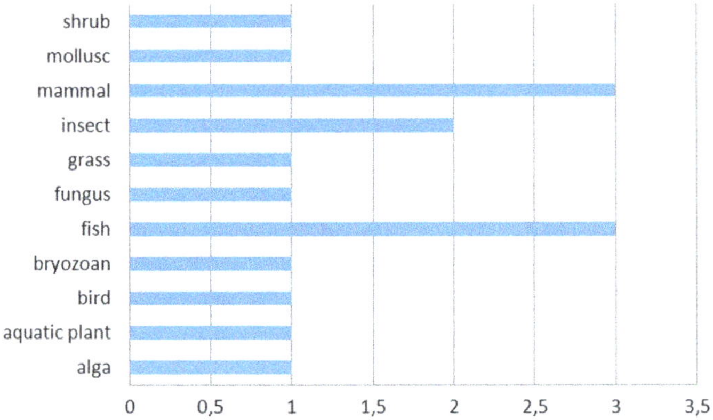


**Fig. 22** Distribution of various invasive alien species types (http://www.cabi.org/cpc/)

The invasive species interactions with their new ecosystems are quite complex. Also their global identification and classification based on the determination of their level of threat for a local environment is a very difficult task. For these reasons the Invasive Species Specialist Group (ISSG) of the Species Survival Commission (SSC) of the International Union for Conservation of Nature (IUCN), has constructed a list of the 100 worst invasive species on the planet (Luque et al. 2014), as you can see in Fig. 25. In this list there are species that are already globally widespread, others that can cause cumulative visible damages and others that are currently limited to a few parts of the world but they can have very high potential expansion and biological adaptation possibilities in new environments (Lowe et al. 2000).

It is extremely interesting the fact that 32% of the Invasive Species recorded in Greece are included in the list of the 100 worst invasive species on the planet. Of

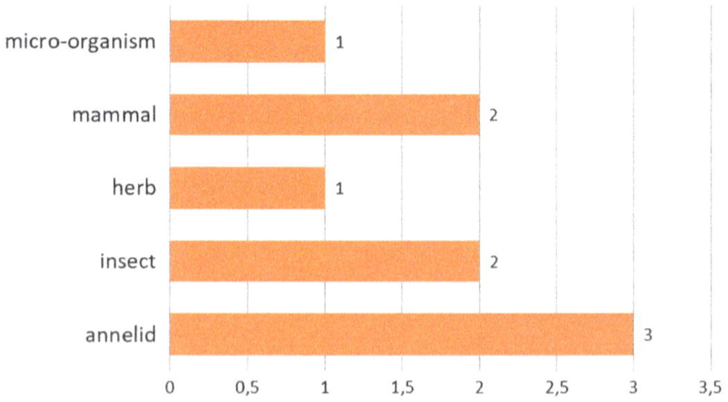


**Fig. 23** Distribution of various biostatus not specified invasive species (http://www.cabi.org/cpc/)

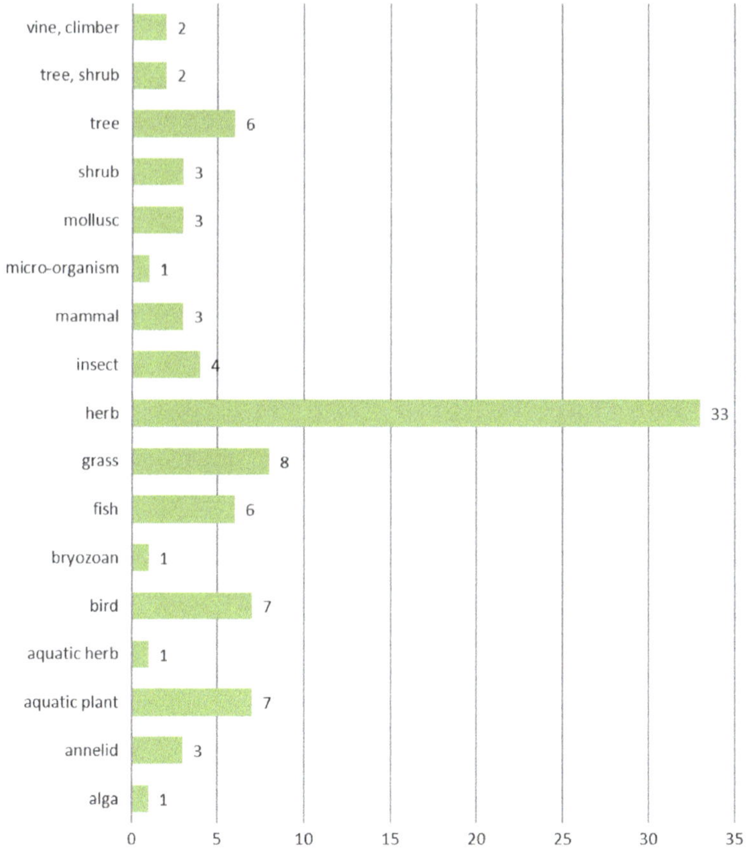


**Fig. 24** Distribution of various native invasive species (http://www.cabi.org/cpc/)

**Table 4** Invasive species in Greece by habitats (http://www.cabi.org/isc/)

| Species | Invasive alien species | Not specified | Native species |
|---|---|---|---|
| MarineHabitats | 1 | – | 1 |
| EstuarineHabitats | 12 | 1 | 12 |
| Lakes | 14 | – | 18 |
| WaterResources | 15 | – | 21 |
| Wetlands | 10 | – | 20 |
| RiparianZones | 12 | 1 | 24 |
| Coastland | 18 | 1 | 15 |
| UrbanAreas | 13 | 1 | 26 |
| Agricultural Areas | 15 | 3 | 40 |
| Ruderal/Disturbed | 17 | 2 | 45 |
| PlantedForests | 13 | 1 | 14 |
| NaturalForests | 13 | – | 23 |
| Scrub/Shrublands | 13 | 1 | 18 |
| Range/Grasslands | 13 | 1 | 34 |
| Tundra | – | 1 | – |
| Desert | 3 | – | 4 |
| Ice | – | 1 | 1 |
| Host | 3 | – | 1 |
| Vector | – | 1 | 1 |

**Table 5** Invasive species in Greece by Taxon (http://www.cabi.org/isc/)

| Species | Invasive alien species | Not specified | Native species |
|---|---|---|---|
| Micro-Organism | – | 1 | 1 |
| Fungus | 3 | – | – |
| Insect | 7 | 1 | 4 |
| Mollusc | 3 | – | 3 |
| Alga | 2 | – | 1 |
| Bryozoan | 2 | – | 1 |
| Annelid | – | 1 | 3 |
| CombJelly | 1 | – | – |
| Fish | 16 | – | 6 |
| Amphibian | 1 | – | – |
| Bird | 1 | – | 7 |
| Mammal | 5 | 1 | 3 |
| AquaticPlant | 1 | – | 3 |
| Tree | 2 | – | 8 |
| Shrub | 3 | – | 5 |
| Grass | 2 | – | 8 |

(continued)

**Table 5** (continued)

| Species | Invasive alien species | Not specified | Native species |
|---|---|---|---|
| Sedge | 1 | – | – |
| Herb | 4 | 1 | 34 |
| Vine, Climber | 1 | – | 2 |
| Succulent | 1 | – | – |
| Oomycete | 1 | – | – |

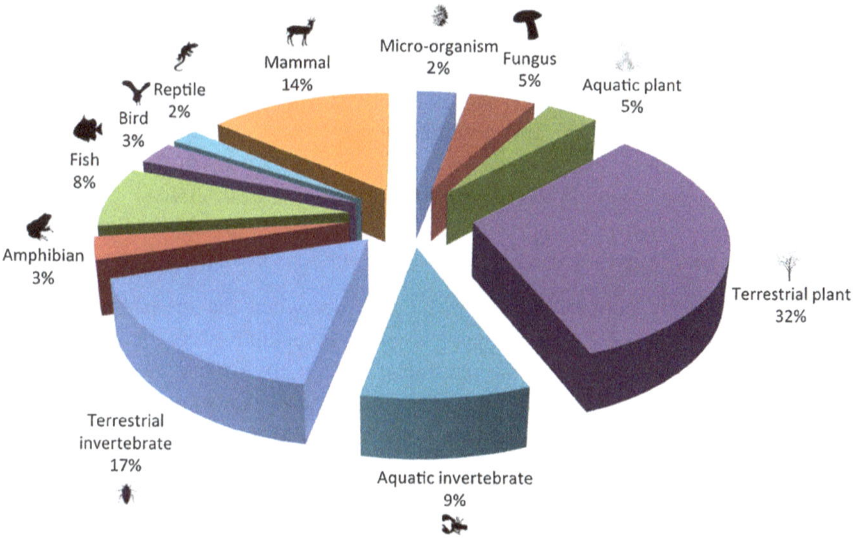


**Fig. 25** Taxonomy of 100 of the world's worst invasive alien species (http://www.cabi.org/cpc/)

these, only 5% are Native Species whereas the remaining 27% are Invasive Alien Species (http://www.cabi.org/cpc/).

## 4 Invasion of Alien Species in Greece and Climate Change Scenarios

Climate change in Greece and in particular the increase in the average temperature, the variability of rainfall (in terms of frequency and intensity) the increase of the greenhouse gases concentrations, the increased frequency and intensity of storms and the rising of sea levels, have a catalytic effect on the potential of invasive species and they create suitable conditions for this problem.

The decreasing rainfall and the increasing temperatures in southern Greece, significantly contribute to higher forest fires frequencies by 20–34% (Moriondo et al. 2006). The large number of forest fires converts the high forests in the undergrowth, which radically changes the local ecosystems, with significant consequences for biodiversity management strategies (Masters and Norgrove 2010).

The changes provided by the IPCC climate change scenario A2, will create a chain reaction in the Greek local ecosystems and especially in the environmental micro-diversity, resulting in mosaic distributions of invasive species.

It should be emphasized that it is extremely difficult to predict the size of the invasion and spread of these species, the location of the boundaries and the changing patterns of abundance in them, as well as the direct effect of micro-environmental conditions that will prevail, regarding the survival factors, the reproduction and further dispersion. What may albeit confidently predicted is that there will be variation in the abundance and distribution in relation to time, which reflects the temporal variation of parameters of the environmental niche. It is certain that in short periods of time the populations and the newly-arrived species will exhibit variations in their abundance whereas for longer periods they may appear stable or they may have a periodic growth cycle (Silva et al. 2013).

Attempting a thorough approach of possible spatial patterns related to the invasive species in Greece, the most prevalent versions are the following three:

- Invasion serially correlated with the space, i.e. when the invasions are most similar in areas nearby. This happens when an invasion reflects the suitability of the local environment and if the parameters of the environmental niche, exhibit spatial autocorrelation. Exemplifies invasive species that may use this type of invasion in the near future is the ostarkodermo *Cerco pagispengoi*. It is included in the list of the 100 worst invasive species on the planet (Zenetos et al. 2006). Is a competitor to other plankton fed invertebrates and smaller fish. It breeds in large numbers against the zooplankton it blocks fishing nets and trawls and is generally associated with serious economic consequences. It is spot in Turley, under a BS—Steppe climate. The favorable annual precipitation is between 430 and 860 mm. Corresponding climate in the immediate area if found in east Macedonia and Thrace, which creates spatial invasion conditions in these Greek regions.
- Invasion ranging systematically, near the borders of a species expansion geographical area. In other words, limited invasion appears to the boundaries of the invasion area, reflecting the fact that in this place there are usually unfavorable parameters. Example of such a case is the species *Trogoderma granarium*, which is considered one of the most serious pests of grain, which also can affect oilseeds and oilcake. It is currently located on the northern border of Greece and in particular in Albania and Fyrom, while the basis of climate change is likely to invade the regions of western Macedonia and Epirus in Greece.
- Invasion above the limits of the geographical range of a species (if the spread limits are determined by a single parameter of the environmental niche that suddenly becomes unfavorable). An example of such an invasion is the mollusk *Musculista senhousia*. It will happen if the surface water temperature in the

Aegean rises by 2 °C, which is certain based on forecasting for the period 2040–2050 (Weber 2003).

In summary, it should be noted that taking into account only environmental and climatic factors for the period up to 2100, over 25 invasive species are expected to invade Greek territory, creating serious problems to the native ecosystem with incalculable socio-economic consequences.

## 5 Discussion

Climate change is recognized as an additional threat to biodiversity, both in terms of the habitats and the ability of species to survive. It is obvious that due to the special landscape of Greece and because of its geographical location, ecosystems of the country would be substantially affected by the constant increase in the average temperature, the more frequent occurrence of extreme weather events, by the changes in precipitation and by the potential reduction of the quantities of available water.

Preventive measures should be initiated in order to investigate and identify those that exist already in the country (Demertzis and Iliadis 2015). This might help towards spread avoidance (bio-security). Therefore, citizens and authorities' awareness in matters related to alien and invasive species is absolutely necessary. Additionally, addressing impacts in biodiversity requires monitoring, early detection, long-term control, consequences containment measures and effective application of the international conventions.

In situations where biodiversity has been affected by invasive species, measures and actions should be imposed. The aim will be the preservation and restoration of healthy ecosystems in order to enhance their capability to withstand the pressure of climate change and thus to halt the loss of biodiversity due to this cause.

Generally, as a basic precaution, measures should be promoted, capable of addressing climate change. The healthy ecosystems can perform functions related to the regulation of climate. For example, forests, seas and wetlands can contribute to carbon sequestration and they can help towards the reduction of carbon dioxide concentrations and other greenhouse gases in the atmosphere.

## 6 Conclusions

Climate change will have significant negative impacts on several sectors in Greece, with agriculture, forestry, fisheries, tourism, transport, coastal activities and the urban built environmental expected to be affected by the rise in temperature, drought, extreme weather events and sea level rise. These impacts will lead to reduced productivity, loss of capital and additional expenditure for damage repair. Negative impacts will also affect biodiversity, ecosystems and health.

Also, climate change will disrupt ecosystems due to the invasion of new species. This problem has already been detected especially in marine and terrestrial ecosystems. This paper deals with the climate change impacts on the biodiversity of ecosystems and it presents a spatio-temporal analysis and recording of the invasion of alien invasive species in the flora and fauna of the Greek territory. It also discusses future invasions based on likely climate change scenarios. The study produced projections for Greece, in a detailed geographic breakdown up to the year 2100.

Species abundance is expected to decrease in Southern Europe, in parts of the Iberian Peninsula, Italy and Greece, and species distribution will depend on habitat suitability. It has been recorded that 60 of the 127 native freshwater fish (about 47%) found in Greece are threatened by climate change. With regard to Greece's forest ecosystems, climate change is expected to lead to a contraction in distribution of cold temperate conifers (spruce, forest pine, etc.) and to warm temperate conifers invading into deciduous oak forests. Turning to wetlands, several ephemeral ecosystems are expected to disappear, while other permanent ones will shrink (http://www.bankofgreece.gr/).

The success for biodiversity conservation and adaptation to climate change will require good cooperation between the government and various stakeholders at local and national level, as well as providing guidelines for those who make the decisions.

It is important, in further examining the impact of anthropogenic climate change, to take the sectoral studies deeper, by disaggregating their analysis by geographical area and focusing on the more vulnerable areas and the more vulnerable social groups. Pursuing this research in greater depth, enlarging its scope and disseminating its findings could all contribute decisively to creating a "critical mass" in society, which would, whenever necessary, push for proper policy decisions on climate change on the basis of long-term planning, and not in a myopic manner dictated by short-term political cost-benefit considerations, which would only multiply the burden on future generations.

The present study does not examine issues related to the distributional effects of climate change and of the mitigation and adaptation policies. The impact of climate change has social dimensions which deserve to be explored, especially as regards possible increases in poverty and migration, given that the effects of climate change and of policy responses to it will be most strongly felt by the lower-income population groups, which lack the necessary resources to address the problems caused by climate change and to finance measures for emission abatement and adaptation. Specific policies will be needed to avert the exclusion of any social groups from access to clean energy and technologies and to ensure adequate protection against damage from climate change.

Future studies will have to look deeper into the fundamental strategic question of how dealing with climate change and reducing invasions of alien species can help boost growth in all sectors of the economy. This opportunity, if properly seized, could help reduce the costs of the mitigation and adaptation policies. New technologies, new activities, new standards, need to become the focus of the new growth effort aiming at a high-biosecurity economy and protection against possible climate change-induced damage.

Finally, the continuation of this project will provide both a challenge and an opportunity to improve the mathematical or fuzzy models of integrated assessment and to incorporate re-distributional effects, uncertainty and non-linear modes of the ecosystems' response to climate change.

## References

Demertzis K, Iliadis L (2015) Intelligent bio-inspired detection of food borne pathogen by DNA barcodes: the case of invasive fish species *Lagocephalus sceleratus*. In: Engineering applications of neural networks, vol 517 of the series communications in computer and information science 16th international conference, EANN 2015, pp 89–99. doi:10.1007/978-3-319-23983-5_9, Springer International Publishing

Douglas K, Gareth EJ (2007) Invasive species policy and climate change: social perceptions of environmental change in the Mediterranean. Environ Sci Policy 10(3):230–242

Gallardo B, Aldridge DC (2013) Evaluating the combined threat of climate change and biological invasions on endangered species. Biol Conserv 160:225–233

Intergovernmental Panel on Climate Change (IPCC) (2007) Fourth assessment report: climate change 2007 (AR4). IPCC, Geneva (https://www.ipcc.ch)

Keller RP, Lodge DM, Lewis MA, Shogren JF (2009) Bioeconomics of invasive species: integrating ecology, economics, policy, and management. Oxford University Press ISBN: 9780195367980

Lowe S, Browne M, Boudjelas S, De Poorter M (2000) 100 of the world's worst invasive alien species a selection from the global invasive species database. Published by The Invasive Species Specialist Group (ISSG) a specialist group of the Species Survival Commission (SSC) of the World Conservation Union (IUCN), Indicator codes: SEBI 010, 12 p (http://www.eea.europa.eu/)

Luque GM, Bellard C, Bertelsmeier C, Bonnaud E, Genovesi P, Simberloff D, Courchamp F (2014) The 100th of the world's worst invasive alien species. Biol Invasions 16:981–985. doi:10.1007/s10530-013-0561-5

Masters G, Norgrove L (2010) Climate change and invasive alien species. CABI Working Paper 1, 30 p (http://www.cabi.org/Uploads/CABI/expertise/invasive-alien-species-working-paper.pdf)

Moriondo M, Good P, Durao R, Bindi R, Giannakopoulos C, Corte-Real J (2006) Potential impact of climate change on fire risk in the Mediterranean area. Published in CR vol 31, No. 1. Online publication date: June 26, 2006, Print ISSN: 0936-577X; Online ISSN: 1616-1572

Silva P, Sopeña A, Sliva J, Toland J, Nottingham S, Jones W, Eldridge J, Thorpe E, Thévignot C (2013) LIFE and invasive alien species, European commission and coordinated by the environment directorate-general, LIFE Units - E.3. and E.4. (http://ec.europa.eu/)

Weber E (2003) Invasive plant species of the world: a reference guide to environmental weeds 2003, pp viii + 548 pp, ISBN 0-85199-695-7

Zenetos A, Cinar ME, Pancucci-Papadopoulou MA, Harmelin JG, Furnari G, Andaloro F, Bellou N, Streftaris N, Zibrowius H (2006) Annotated list of marine alien species in the Mediterranean with records of the worst invasive species. Mediterr Marine Sci 6(2):63–118

http://biodiversity.europa.eu/
http://www.cabi.org/cpc/
http://www.europe-aliens.org/
http://www.iucngisd.org/gisd/
http://www.worldbank.org/
http://www.bankofgreece.gr/

# Evaluating the Suitability of Community-Based Adaptation: A Case Study of Bangladesh

**Ranjan Roy**

## 1 Introduction

Despite adaptation is often the 'poor cousin' of the climate change challenge (Pielke et al. 2007), recently it has received more emphasis in the international policy agenda as a potential solution to curb climate impacts, particularly under the pessimistic circumstances around international forums increasingly fail to deliver against mitigation targets. Currently, adaptation studies are fragmented due to disparate studies both in disciplinary terms and geography, and inconsistent use of terminologies, diverge methods, and tools (Ash and Stafford Smith 2013). However, there is a growing convergence that adaptation literature—in both developed and developing countries—is locally focused (Lavell et al. 2012). The UN Climate Change Summit 2014 has 'scaled up adaptation' by drawing lessons on 'locally-led adaptation solutions'. Although the concepts of 'local' and 'community' are subjected to criticism (due to, for example, localized politics of resource allocation), a raft of studies (Baas and Ramasamy 2008; Ayers 2011; UNDP 2011a, b; Ayers and Huq 2013) explores the reasons for why locally-led or community-centric adaptation is more effective. Key reasons are, inter alia, this adaptation planning and implementation process: addresses 'social vulnerability perspective' that focuses climate change is a global risk, yet vulnerability is locally experienced; promotes 'social learning', which is central to climate change adaptation; and hinges on 'locally embedded understandings of perceptions and experiences of risk' as well as 'social processes' and 'development pathways' that shape no regrets adaptation actions.

Unlike conventional approaches to adaptation planning (predominantly rely on expert advice and credible science), 'CBA is a community-led process, based on

R. Roy (✉)
Department of Agricultural Extension & Information System, Sher-e-Bangla Agricultural University, Sher-e-Bangla Nagar, Dhaka 1207, Bangladesh
e-mail: ranjansau@yahoo.com



W. Leal Filho et al. (eds.), *Handbook of Climate Change Communication: Vol. 1*, Climate Change Management, https://doi.org/10.1007/978-3-319-69838-0_3

communities' priorities, needs, knowledge, and capacities, which empowers people to plan for and cope with the impacts of climate change' (Reid et al. 2009). CBA recognises that reducing vulnerability and exposure, building adaptive capacity, and increasing climate resilience are socially and culturally embedded. The bedrock of CBA is, therefore, to focus on empowering communities and improving sustainable livelihoods in order to assess, take action, and manage based on their own decision making process, inculcating a guiding principle 'helping people help themselves'. Characteristically, it strengthens communities' adaptive capacity—a system's ability to take action to reduce vulnerability to climate change—that is dependent on awareness and information, economic resources, local institutions, and infrastructure. CBA is an emerging approach, which likely requires to conducting more trials, eliciting errors, replicating the successes and deriving good lessons as well as influencing decisions from the ground up to 'scaling up' (vertical scale-up) and 'scaling out' (horizontal scale-up) CBA. Still, it's planning and action remits largely remains at the community scale, but can be scaled up with the regional and national levels by mainstreaming into government processes (Reid and Huq 2014). Effective CBA requires planning and strategy formulation at the national level is rooted in evidence-based needs of communities in response to local livelihood and environmental contexts (Ensor 2010).

Tailored to local cultures and conditions (see King 2014), CBA is a promising way to address the challenges of top-down adaptation planning that particularly fails to engage with the needs of the world's most vulnerable communities (Boyd and Ibarraran 2009). CBA has the potential to transform power relations between individual, communities and institutions, leading to an effective and holistic strategy for responding to climate change (Dodman and Mitlin 2011), and offer synergies with poverty and sustainable development (SD) objectives (Heltberg et al. 2009). Considering social causation—social vulnerability causes more adaptation research, a growing number of CBA studies have been conducted in developed countries (e.g., UNISDR 2009) and developing countries (e.g., Baas and Ramasamy 2008).

CBA has brought a notable change in coping with numerous and varied effects of climate change in many LDCs. CBA projects of the Global Environment Facility's Small Grants Programme (GEF SGP) have a very high rate of 'sustainability' compared to other 'environmental grant projects' (UNDP 2011b). Despite project-based CBA is effective in reducing the climate vulnerability, diversifying livelihoods, and building resilience of communities, it is an immature approach to policy makers. The conceptual limitations and procedural shortcomings of CBA are documented in the next section. These shortcomings present the rationale of conducting this study. Moreover, still donors' and stakeholders' understandings and perceptions of CBA application are largely restricted to the local or community levels, which proliferate this approach as a piecemeal approach that fails to attract more strategic and wider-ranging climate policies, measures, and climate-resilient development investment. Therefore, determination and assessment of key factors that influenced multi-stakeholders perceptions about the suitability of CBA for improving adaptive capacity of communities' are important to draw broader set policy implications for adaptation, mitigation, and SD.

Keeping the above perspectives in mind, this study perceptually evaluates the suitability of CBA to determine the positive and negative factors of it so that the application of CBA can contribute more to improve communities' adaptive capacity in Bangladesh—one of the world's most vulnerable countries to climate impacts. Floods, droughts and sea level rise are the main climate hazards in Bangladesh and these hazards have been negatively impacting socioeconomic condition of this country (MoEF 2009). Bangladesh plays a pioneering role in developing CBA approach, and this country is regarded as a 'field laboratory' and 'adaptation knowledge hub' for climate practitioners (Rashid and Khan 2013). A SWOT (strengths, weaknesses, opportunities and threats) technique in combination with the analytical hierarchy process (AHP) was applied to achieve the aforementioned objective. The next sections present an overview of CBA challenges and subsequent section consecutively describes methodology, results and discussion. The final section draws conclusions and policy implications.

## 2 An Overview of CBA Challenges

### *2.1 Conceptual Challenges of CBA*

Firstly, the term 'community' is insignificantly conceptualised in the CBA literature. Yet, the concept 'community' is in itself multi-functional and has context-specific meanings and interpretations. Many studies (e.g., Cooke and Kothari 2001) have mentioned that the uncritical use of the term 'community' is questionable in participatory projects that aim to empower vulnerable community members. Mansuri and Rao (2004) stated the notion of community is problematic in two ways (i) drawing the geographic boundaries of a community is not straightforward, i.e., administrative boundaries can be meaningless when increasing mobility have transformed community boundaries, and (ii) adopting the concept 'community' without understanding the local structures of social and economical power. Literature on CBA broadly conceives this term as politically and culturally homogenous social systems. Without a clear understanding of local power structures the effectiveness of participatory studies is 'good-for-nothing', for example, the unequal distribution of power helps elites to maintain control over resources and reduces the potential productivity of the poor (Mansuri and Rao 2013). Yet, Dasgupta and Beard (2007) showed where power was the most evenly distributed; resource allocation to the poor was restricted. 'A note of community'—an explicit understanding of community's values, exposures, networks, conflicts, and power structures—is therefore required to present before conducting CBA.

Secondly, there is no established theory of CBA, which is founded on both conceptual underpinnings and best-practice principles. As proponents' state universities or training workshops are not suitable for learning CBA theory, rather it comes from the practice itself (Huq and Reid 2007). The building blocks, namely, 'empowerment' is the core of CBA theory. Besides, how much community-based development differs from CBA is unclear. This lack of distinction has posed problems for CBA practitioners, policy makers, and funders who require signposts of successful CBA projects to distinguish good practice (Jones and Rahman 2007). Based on the past studies (e.g., Rashid and Khan 2013), three salient features of CBA project are: (i) community-driven decision making blended with scientific predictions on the impacts of climate change, (ii) community's broader understanding of adaptation (and mitigation) and its implications on socio-economic activities, and (iii) strengthening community's adaptive capacity and fostering climate resilient development.

Thirdly, the impacts of climate change are local; therefore, planning for climate change remains rooted in local priorities and understanding. CBA offers the ability to tailor interventions to the local context. However, it overemphasises around problems framing (or defining problems) on local/community level raises criticisms. CBA advocates report that CBA portrays adaptation to climate change as a fundamentally community issue, raising further questions about the adequacy of the conceptual framework (Dodman and Mitlin 2011) and privileges the local scale for its planning at the expense of the national and global nature of adaptation (Huq and Reid 2007). The national level plays a key role in adaptation planning and implementation, whereas adaptation responses have diverse processes and outcomes at the subnational and local levels (Mimura et al. 2014). In fact, without coordination and cooperation of administrative levels, adaptation is unlikely to be successful (Adger et al. 2005). The interface and interaction between different levels of actors, roles, and functions with community outcomes are a sine qua non of successful CBA.

To provide more autonomy and control over planning decisions to community members to improve accountability and minimise corruption, graft, and elite capture, CBA should be renamed as 'community-driven adaptation' (CDA). Revisiting the evolution of community-driven development (CDD), several studies (Dasgupta and Beard 2007; Mansuri and Rao 2013) indicate that CDD is an efficient and effective mechanism for making development more inclusive, building social capital, empowering poor people, and strengthening local governance. Recently, a flagship report of the World Bank concludes that "there is great potential for a CDD approach to serve as a powerful tool for building disaster and climate resilience at the local level" (Arnold et al. 2014). Particularly, CDD has objective of reversing power relations through creating an agency and voice for marginalized people and allowing them to have more control over development assistance (Mansuri and Rao 2004). Shedding light on the foundations and attributes of CDD, CDA can achieve its objective by reducing information problems, expanding the resources available to the poor, and strengthening the civic capacities of communities.

Moreover, renaming of CBA can be justified by the following two reasons. First, unlike CBA, most community-based programme is designed to use community resources to design, implement, monitor, and evaluate the programme (Rashid and

Khan 2013). However, in the case of adaptation, the fund is not necessarily to be local and from community, which mainly provides by the industrialised country to the climate vulnerable countries (accepting the 'polluter pay' principle). Based on this, it is more rational to rename CBA as CDA, in which community members drive the whole adaptation activities. Second, several studies (for example, Reid et al. 2009) indicate community members are in the driving seat in the CBA process, where they apply their collective perceptions, knowledge and capacities, and complementarily, CBA practitioners incorporate climate information. The term 'community-driven' is thus more consistent with its fundamentals.

## 2.2 *Procedural Challenges of CBA*

CBA can be short-, medium- and long-term project-based and its planning and implementation is largely reactive, i.e., adaptation that takes place after impacts of climate change have been observed, and incremental, which is opposite of transformational adaptation. Several reports unequivocally indicate that future climate change impacts are expected to have a lot of damages and destructions that exacerbate the existing vulnerability, poverty, and other development challenges, where incremental adaptation options are not enough, which requires longer-term strategies and transformative adaptation solutions. However, it does not mean incremental adaptation is not important—proactive and well planned incremental adaptation options are essential in dealing with many climate impacts (Palutikof et al. 2013). Considering the current and future consequences of global change and vulnerability, CBA practitioners must explore more strategic and wider ranging climate policies and measures to maximise CBA's influence on adaptation and mitigation.

Adaptation (including CBA) to climate impacts has been punctuated across the LDCs owing to shortage of funding. Appraising the agreed fund of the United Nations Framework Convention on Climate Change (UNFCCC), Parry et al. (2009) states the financial assistance needed by developing nations may be two to three times higher overall and many more times higher for certain sectors. Yet, the UNFCCC calculated the adaptation costs to address the impacts resulting from 1.5 °C of warming, despite the slow progresses of mitigation approaches 2 °C of warming above preindustrial levels by 2050 (World Bank 2010). Literatures (e.g., World Bank 2010) related to climate finance recommend for addressing three specific issues of adaptation: exploring innovative financial mechanisms, involving private sectors in adaptation, and utilizing adaptation fund effectively. In the wake of shortage of climate funds, the exclusion of mitigation from CBA is a notable obstacle for receiving donors' funding. Hence, policy oriented and action research is warranted to enhance CBA's (procedural) theory and practice, in which mitigation is strongly considered.

CBA inherently employs participatory research approaches. However, application of participatory approaches generates criticisms, and CBA literature has paid very little attention on this issue. Critics mention participatory tools are increasingly being used in: (i) 'research as usual'—research that remains 'top-down' and Eurocentric (Kindon et al. 2009), (ii) cases where 'community' is considered a homogeneous entity, without understandings of 'power structures' of community members (Mercer et al. 2008), and (iii) overly 'romantic terms', ignoring social divisions and without acknowledging vulnerable people might be transient in the vulnerable areas (Forsyth 2013). Observations experience indicates CBA process entirely led by experts, who not only designed projects but also managed in a purely 'outsider-down'—it arrives with outside agencies that promise to achieve something 'with' the local people—manner, although outsider-insider dynamic is rarely explored. Community members' participation can be defined as 'incentive-driven'—participation happens for the sake of incentives, and 'person-driven'—participation occurs due to the pressures from persons such as local leaders and officers.

Effective CBA is required a broad participation of stakeholders and institutions to minimise the gaps between indigenous and scientific knowledge, and bottom-up and top-down actions. However, experiences from the studies on 'Community-Driven Development'—a popular and useful development intervention of the World Bank (Arnold et al. 2014)—suggests that community members are in a better position to drive the CBA process, particularly to ensure community control over adaptation planning and implementation, while CBA practitioners facilitate the process by incorporating scientific predictions of climate change and putting a balanced emphasis on social, economical, and biophysical dimensions to improve foresight and develop adaptation options. Effective local participation in CBA through adopting the tenets of CDD requires consideration of several aspects (i) a responsive state that does not necessarily has to be democratic but has to be responsive to socio-economic demands of communities (e.g., World Bank, 2010); (ii) tailored to local context such as the nature of social interactions, political systems, history, geography, and networks (e.g., King 2014); and (iii) effective civic engagement through adjustments of the nature of social interactions and modifications of the norms and local cultures, promotions of monitoring and evaluation, and infusion of the spirit of learning by doing (Mansuri and Rao 2013).

# 3 Methodology

## *3.1 Case Study Area*

This study was conducted in Bangladesh—one of the most vulnerable countries in the world to climate risks. CBA has become a much talked issue in climate discourse in this country. This is because several national and international organizations, inter alia, ActionAid, FAO, and UNDP have been involving with many CBA projects in this country and propagating adaptation lessons across the world.

Moreover, Bangladesh is very active in organising the International Conference on CBA to Climate Change and hosted conferences six times out of the total ten times from 2005 to 2016. Furthermore, NGOs and vibrant civil societies have been organising national and divisional CBA conferences in Bangladesh. Adaptation experts' reports Bangladesh is 'one of the most adaptive countries' in the world and an 'adaptation capital of the world' (www.iied.org/climate-change-expertshead-adaptation-capital-world). Besides the support the international community; over the last three decades, the Government has invested over $10 billion (at constant 2007 prices) to make the country more adaptive and less vulnerable to natural disasters (MoEF 2009). It is, therefore wise choice to select this country as a case study area for assessing the suitability of CBA.

## 3.2 Data Collection and Analysis

This section determines SWOT factors, conducts pair-wise comparisons, and analyses data, which are consecutively presented below. A review of the literature (Rashid and Khan 2013; Reid and Huq 2014; Reid et al. 2014) was conducted to understand and assess the different issues of CBA such as adaptation options, constraints, and draw an initial list of strengths, weakness, opportunities and threats (SWOT), which was complemented to the factors that derived by employing other tools, like field observation, community meeting and discussion with informants. Strengths here refer to characteristics of CBA that give it an advantage over others, and elements of CBA that could exploit to its advantage are meant opportunities. To understand community perceptions of CBA, six communities meeting were organised in six villages of three sub-districts: Fulchhari (Gozria and Chandia), Gaibandha; Porsha (Shavapur and Saharandha), Naogaon; and Dumuria (Kholshi and Sholua), Khulna that represents flood-affected, drought-affected and coastal area, respectively in Bangladesh (Fig. 1). Coastal area is also flood-affected, but mainly salinity intrusion and more disaster-prone characteristics are considered in selecting a coastal area as a part of the case study area. A number of 10–12 interested people participated at each meeting, followed by discussion and factor prioritisation lasted for 110–120 min. Local CBA practitioners and Sub-Assistant Agricultural Officers (SAAO) played a key role in selecting respondents that constituted male farmers, female farmers (25%), sharecroppers, day labourers, young growers, small businessmen and members of ethnic community. Community members were selected by fulfilling two criteria (i) previous involvement in CBA and (ii) having an understanding of the method, if got it correctly. Meetings were conducted in a way that ensured equal spaces for each respondent to give his/her opinion and grievances in factor determination were resolved democratically.

Each meeting was assisted by SAAOs and CBA practitioners, besides the author. Meetings were divided into two parts: initial discussion on CBA and identification of helpful and harmful factors of adapting climate impacts. Factors derivation was categorised into internal (strength and weakness) and external (opportunity and threat)

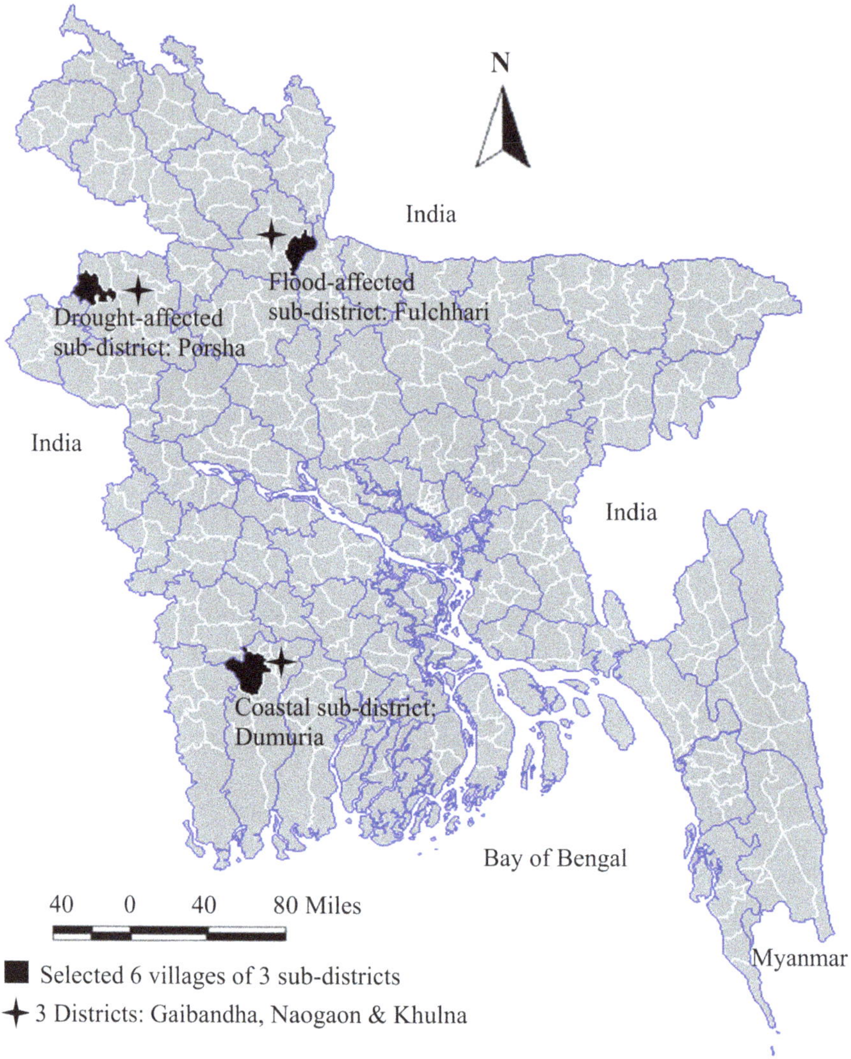


**Fig. 1** Map of Bangladesh showing the case study areas. *Source* Author

origins. Broadly, two questions were asked: what are the strengths and weaknesses of your community, and what are the opportunities and threats facing your communities in adapting climate change impacts? Besides these, to deduce SWOT factors of CBA, informal interviews were conducted with 60 CBA informants including NGO workers, input dealers, local leaders, businessmen, civil society members, CBA practitioners, SAAOs, women entrepreneurs, school teachers, senior community members, and policy makers. These informants had a direct and indirect relation with the CBA projects as local advisers, observers, implementers, decision makers, organizers, watchers and supporters. Around 18% CBA informants were female. A number of open-ended questions related to CBA's methodological rigour, shortcomings, barriers, outputs, implementation and outcomes were used in informal interview. Diversified CBA informants and community members were selected in order to ensure that the evaluation is independent, e.g., gathering data also from the people who have not a stake in the success of CBA. Moreover, author gathered good insights through project areas observation to derive an inclusive and refined list of SWOT factors (Table 1).

SWOT is a structured planning method for conducting situation analysis of internal and external factors, influencing particular decisions that help to inform the later steps in planning to achieve the predetermined objective, e.g., evaluating CBA. However, application of the conventional SWOT analysis cannot quantify the magnitude of factors to determine the effects of each factor on a plan or strategy (Masozera et al. 2006). The employment of AHP, a basic approach for decision making, is useful to overcome this shortcoming by assigning relative priority to each factor through pair-wise comparisons (Saaty and Vargas 2012). In determining important factors of something (e.g., organisation) and measuring quantified information of these factors, the combined application of SWOT with AHP is the most suitable approach (for details, see Masozera et al. 2006; Kurtilla et al. 2000). SWOT analysis was combined with the AHP by following three steps (i) identifying possible SWOT factors, (ii) conducting pair-wise comparison of factors within each SWOT group and (iii) carrying out pair-wise comparison of four SWOT groups. Only the most relevant SWOT factors were identified and retained in Table 1 (Step 1), since the number of factors within each SWOT group should desirably not exceed ten to smoothly manage pair-wise comparisons (Kurtilla et al. 2000).

Pair-wise comparisons (Eq. 1) were made with 40 community members (including 14 female members and members of ethnic community) and 36 CBA informants (including 8 female informants), who were known about the activities of CBA activities and projects. A total of 76 community members and CBA informants participated in pair-wise comparisons who previously attended in community meetings and interviews, respectively. At first, pair-wise comparisons of factors within each SWOT group were conducted (Step 2). Figure 2 presents a sample of questionnaire used for pair-wise comparisons. The factor with the highest priority value under each SWOT group was brought forward for further comparisons. In the second step, the four groups were brought forward for pair-wise comparisons (Step 3). After all pair-wise comparisons, priority values of the group, priority values of the factors within the group and consistency ratios (Eq. 2) were calculated using the Eigen value method. This method is applied to find the principal Eigen vector of a pair-wise comparison matrix. Principal Eigen value is obtained from the summation of products between each element of Eigen vector (also called priority vector) and the sum of columns of the reciprocal matrix (see Saaty and Vargas 2012 for details). The consistency ratio (CR) was kept below 10% to keep the inconsistency within acceptable levels. If the CR is greater than 10%, a revision of the subjective judgment is required. Detailed explanations on AHP and pair-wise comparisons are

**Table 1** The consistency ratio (CR), and factor priority score (FPS) and overall priority score (OPS) of SWOT factors*

| SWOT group and factor | Community Member | | | CBA Informant | | |
|---|---|---|---|---|---|---|
| | CR (%) | FPS | OPS | CR (%) | FPS | OPS |
| **Strengths** | | | | | | |
| S1: CBA has become a forefront of the climate change agenda | 7.2 | 0.12 | 0.03 | 4.67 | 0.18 | 0.06 |
| S2: CBA expands livelihood strategies | | 0.31 | 0.08 | | 0.28 | 0.09 |
| S3: CBA strengthens climate resilience | | 0.48 | 0.13 | | 0.37 | 0.11 |
| S4: CBA can be integrated with DRR tools | | 0.09 | 0.02 | | 0.17 | 0.05 |
| **Weaknesses** | | | | | | |
| W1: CBA does not address mitigation | 5.89 | 0.13 | 0.02 | 7.31 | 0.16 | 0.04 |
| W2: Weak monitoring and evaluation systems | | 0.17 | 0.03 | | 0.24 | 0.05 |
| W3: CBA is largely unable to alleviate poverty | | 0.38 | 0.07 | | 0.32 | 0.07 |
| W4: Lack of coordination and integration amongst the sectors and institutions | | 0.32 | 0.06 | | 0.28 | 0.06 |
| **Opportunities** | | | | | | |
| 01: Strengthening local institutions | 7.82 | 0.29 | 0.10 | 8.26 | 0.26 | 0.07 |
| 02: Strengthening national and international finance | | 0.29 | 0.10 | | 0.28 | 0.08 |
| 03: Private sector involvement in CBA | | 0.29 | 0.10 | | 0.27 | 0.08 |
| 04: Approaching incremental to transformational adaptation | | 0.13 | 0.04 | | 0.19 | 0.06 |
| **Threats** | | | | | | |
| T1: Country increasingly faces adverse impacts of climate change | 4.98 | 0.42 | 0.08 | 7.59 | 0.24 | 0.04 |
| T2: High dependency on foreign aids/ donors' support | | 0.27 | 0.06 | | 0.25 | 0.05 |
| T3: Limited infrastructure | | 0.22 | 0.05 | | 0.23 | 0.04 |
| T4: Potential collapses of CBA | | 0.09 | 0.02 | | 0.28 | 0.05 |

(*Source:* Author)

* Community meetings and discussion with CBA informants generated about 35% SWOT factors

Note: (i) The consistency ratio (see Equation 2) of the comparisons between four SWOT groups was 7.89%, (ii) According to the community members' score, the priority values of strengths, weaknesses, opportunities and threats (SWOT) were 0.26, 0.19, 0.34 and 0.21, respectively; whereas based on CBA informants' score the priority values of SWOT were 0.31, 0.22, 0.29, and 0.18. (iii) Factor priority score is the normalized Eigen vector of a comparison matrix, (iv) The overall priorities of the factors were computed by multiplying the priority of the factor within the group by the priority of the group, and (v) DRR means disaster risk reduction

provided by Saaty and Vargas (2012) and Kurtilla et al. (2000). Masozera et al. (2006) has described how AHP is linked with SWOT analysis. The matrix (A) of pair-wise comparison:

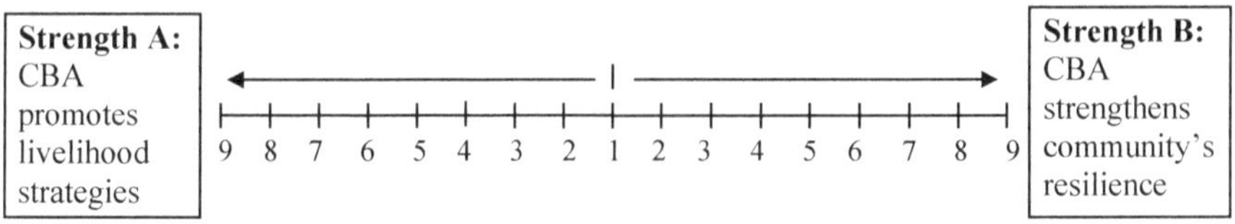


**Fig. 2** A sample questionnaire of a pair wise comparison between strength factors. Respondents were requested to compare the importance of the two factors and based on his/her preferences of the particular factor he/she was asked to assign a weight from 1 to 9 for the relatively important factor to reflect the magnitude of the importance. The same process was followed for all pair wise comparisons

$$A = (a_{ij}) \begin{bmatrix} 1 & w_2/w_1 & \dots & w_1/w_n \\ w_2/w_1 & 1 & \dots & w_2/w_n \\ \vdots & \vdots & \dots & \vdots \\ w_n/w_1 & w_n/w_2 & \dots & 1 \end{bmatrix} \tag{1}$$

Information derived from pair-wise comparisons can be presented as a reciprocal matrix of weights (w), where the assigned relative weight enters into the matrix as an element αij and reciprocal of the entry 1/αij goes to the opposite side of the main diagonal. In the matrix, when i = j, αij = 1 (Masozera et al. 2006).

$$\text{Consistency Ratio(CR)} = \text{Consistency Index(CI)}/\text{Random Consistency Index(RI)} \tag{2}$$

CI = ($\lambda_{max}$ − n)/(n − 1), Where $\lambda_{max}$ means the largest Eigen value and n is the size comparison matrix. RI is derived from a sample of randomly generated reciprocal matrix using the scale 1/9, 1/8, …., 1, …. 8, 9 (Saaty and Vargas 2012).

## 4 Results

Table 1 presents the overall priority scores and the factors priority scores. The scores for weaknesses and threats indicate the negative, whereas the scores for strengths and opportunities indicate the positive of CBA regarding its suitability as an approach for climate change adaptation. The overall priority scores for all factors sum up to 1, which means it can be interpreted as the weight of the general perceived suitability of CBA. For example, the overall priority score of the factor named ‘CBA expands livelihood strategies’ under the ‘strength’ group was 0.08, which indicates that as a strength of CBA this factor account for about 8%, i.e., community members perceive as an approach CBA has notable roles in expanding livelihood strategies that helps them to adapt to climate impacts. Moreover, the factor priority score of a specific factor within each SWOT group can be interpreted as its relative contribution to the group concerned, e.g., CBA informants perceive that 'CBA strengthens climate resilience'—by diversifying non-farm income sources, improving access to markets and building social assets within and between stakeholder groups—alone accounts about 37% of the total strength of CBA, which indicates that informants perceive CBA as a suitable strategy for enhancing climate resilience and adaptation; the more resilient a community is, the greater its capacity to adapt to climate impacts.

The positive factors of CBA outweighed the negative factors. Community members perceive 'CBA strengthens climate resilience' is the dominant strength of CBA as well as 'strengthening local institutions' and 'strengthening finance' are key opportunities around CBA for broader adaptation planning and implementation. According to them, five out of seven of the biggest factors represent strengths and opportunities, and other two factors belong to threats. The perceptions of CBA informants were almost similar with the perception of the community members on prioritising SWOT factors. In determining CBA's threats, CBA informants somewhat differed from the perceptions of community members. Overall, results overwhelmingly explain the presence of strengths and opportunities of CBA to climate risks, and the low importance of weaknesses and threats around CBA (Table 1). It is better to mention here that the results of 'factors determination' and 'pair-wise comparison' of the female community members/informants were virtually the same as the male members/informants that declined the opportunities of gendered analysis. Accordingly, results were presented under the heading of 'community member' and 'CBA informants' (see Table 1).

To provide more empirical explanations, both points of view are demonstrated in Figs. 3 and 4. The lines in the top two quadrants indicate the relative values of the factors of strength and opportunity, whereas the lines in the bottom two quadrants represent the relative values of the factors of weakness and threat. Additionally, the line in each quadrant indicates the total value of the factors of each SWOT group

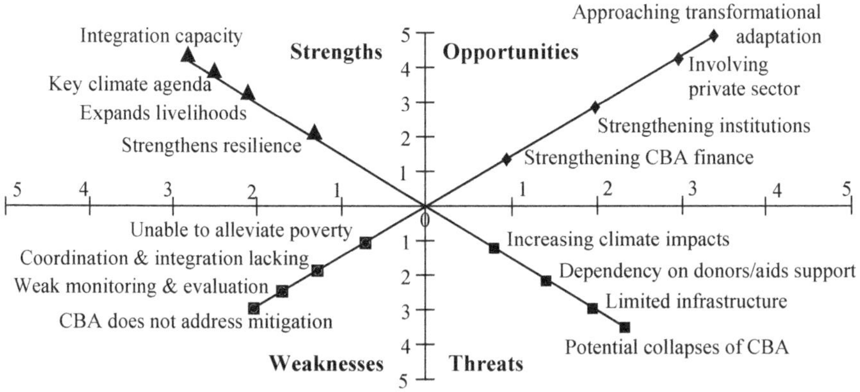


**Fig. 3** Graphical interpretations of the results of pair wise comparisons of SWOT groups and factors by the community members. *Source* Author

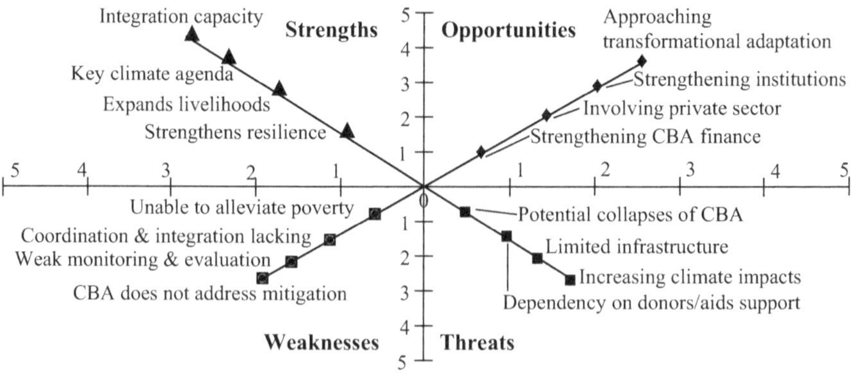


**Fig. 4** Graphical interpretations of the results of pair wise comparisons of SWOT groups and factors by CBA informants. *Source* Author

and the points on each line show the values of each factor within the group. Figures are self-explanatory and factors having more values can be picked out for formulating strategy to improve CBA.

## 5 Discussion

CBA strengthens community resilience is consistent with the findings of other studies, e.g., Bryan and Behrman (2013). The underlying reasons are that the fundamental structure of CBA is based on the premises that generate local knowledge, incorporate climate information, promote collective actions, and emphasis local decision making processes, which strengthen the capacity (i.e., social and human capitals) of community people to adapt to climate impacts. The government of Bangladesh has invested about US$10 billion dollars over the last four decades to make the country less vulnerable to climatic disasters (MoEF 2009). Recently, the government has set up multi-million dollars two pioneering climate change funds, namely, the Bangladesh Climate Change Trust Fund and Bangladesh Climate Resilience Fund. However, advocates raises allegations in misusing funds from the country's adaptation trust fund as well as using political influence, nepotism and corruption in the selection of NGOs to conduct projects on the ground. Most of the NGOs (e.g., ActionAid) who receive fund from the government predominantly adopt CBA approach for uplifting socio-economic status of vulnerable people to climate change. Transparency International Bangladesh proposed 6 measures to improve climate finance governance including improving transparency in decision making processes, strengthening civic engagement, and establishing institutional instruments, e.g., ombudsman systems (Khan et al. 2013).

To support particularly the vulnerable communities, both of the funds have allocated 10% towards community level adaptation activities through NGOs. The rest of the 90% is delivered through the government agencies and ministries to invest on country's overall development. It can be assumed that very little of that is targeted for the development of local communities. To use climate fund properly and to save the most vulnerable people, government needs careful thinking about how to better target the climate change funds. Having a close connection with the local level, the Ministry of Local Government and Rural Development & Co-operatives; Road Transport and Bridge; Disaster Management and Relief; and Posts, Telecommunications & Information Technology can play a significant role in developing local infrastructures of the affected areas that curb the consequences of climate impacts.

CBA plays an important role in diversifying income sources of communities and 'good' lessons are emerging. To uplift socioeconomic condition these lessons have to be communicated with wider communities by providing information, motivating, engaging and organizing dialogues with the relevant stakeholders. King (2014) draws a notable example on how CARE (Cooperative for Assistance and Relief Everywhere)—a humanitarian agency—builds climate resilience of about 65,000 households in Ethiopia through CBA (i.e., adopting promotion of climate-resilient livelihoods strategies, social mobilization, and fostering local entrepreneurship). CBA project enhances adaptive capacity of communities and promote innovations in the design and implementation of adaptation on the one hand and promote numerous livelihood strategies on the other (UNDP 2011b). Organic farming to protect biodiversity in Jamaica and resistant crop varieties to increase productivity in Vietnam are some 'promising' livelihood strategies of CBA. In Bangladesh, the 'Forest, Fish and Fruit' (FFF) project conducted by the UNDP in the four coastal sub-districts is a good example of CBA that substantially expands local livelihood strategies (UNDP 2011a). The FFF model not only provides additional source of income, but also established a 'green shield' surrounding some most vulnerable communities by planting protective and productive vegetation, with an elevated mound and ditch structure interspersed with fish nursery ponds.

The critical roles of institutions are widely discussed in the literatures around adaptation and building adaptive capacity. Similarly, discussion with respondents illustrates local institution's key role in promoting the transition from planning to implementation of CBA, albeit these institutions are often limited by lack of skills and capacities. Reportedly, officials at the local institutions are bemoaning their lack of funds (Roy and Chan 2014). Strengthening local institutions is inevitable to improve: institution's 'coordination' between different administrative and political levels; 'leadership' for initiating, planning, mainstreaming and implementing CBA; and 'flexibility' mainly for rational decision making (Roy et al. 2015).

Financing adaptation has become one of the most crucial issues of international climate policy agenda. A number of limitations are there around financing adaptation. First, a huge gap exists between the amount what are available (and pledged) and required for adapting the impacts of climate change. Second, the concept 'adaptation' itself has operational ambiguities such as the lack of an agreed operational definition of adaptation (Fankhauser and Burton 2011), and the inextricable links of adaptation with SD and poverty reduction (McGray et al. 2007). To allocate more national and international finance for adaptation, effective approaches (e.g., showing a short documentary) for climate change communication can play a constructive role in upholding dire consequences of local people due to climatic hazards, raising political leaders' concerns, and reminding and keeping world leaders' pledges for providing official development assistance as well as climate funds to the most vulnerable people. Strengthening CBA finance is imperative to help the poorest and most vulnerable people and communities to be able to adapt to climate risks. Financing CBA is a useful means to mobilise adaptation funds that direct them to the benefit of the most vulnerable people to climate change.

Strengthening international finance for local adaptation requires proactive action, cooperation and bold pledges from the developed countries. Yet, the fulfillment of their current pledges to the UNFCCC and its related protocols are crucial. According to the Katmandu Declaration (2014) on financing local adaptation, international financing bodies need to ensure at least 50% of funding for tackling climate change for local adaptation, including CBA. Besides, developing suitable climate policy framework, creating flexible funding mechanisms, local government's capacity building, ensuring transparency, and mobilising political will are essential for strengthening national finance for adaptation. More importantly, national governments should be allocated at least 50% of funding for locally-led adaption realizing the depth and breadth of the consequences of climate hazards (Katmandu Declaration 2014). For example, the Government of Nepal has taken an effort to direct 80% of adaptation funding towards local communities (Reid et al. 2014). A key message of this declaration is at least 50% of climate finance for adaptation must go to local level actions and local communities.

Introducing micro-finance and insurance are likely to be useful option for financing adaptation at the regional and local levels (Agrawala and Carraro 2010), where private sectors would be a prominent stakeholder in climate change actions; albeit, it needs more explorations and understandings about the products, services and other conditions that are suitable for adaptation (Ball et al. 2013). Micro-finance in fostering adaptation requires monitoring and evaluation (M&E) following best practice in terms of accountability and transparency. In particular, careful observations are needed on interest rates as well as adaptation products and services that need to be conducive for the vulnerable people. Developing a set of complementary measures (e.g., appointing an expert for overseeing) is hence required, and the Micro-credit Regulatory Authority can be assigned for overall M&E. Importantly, climate change requires adjustment of existing micro-finance programs and practices as well as harnessing of new opportunities. Financing CBA through micro-finance is particularly important for Bangladesh, where there is a well established network for providing micro-credit across the country developed by a number of NGOs, like Grammen Bank.

Private sector, including insurance companies and commercial bank's involvement in local adaptation is now gaining ground in many regions. This issue received a strong emphasis in the latest International Conference of CBA to Climate Change and the UN Climate Summit 2014, and these fora derives private sectors are potential and important stakeholders for supporting climate actions, and their involvements have multiplier effects that can spur climate financing for achieving climate-resilient development. Proponents stated communication on climate change can strengthen private sector involvement in CBA through generating up-to-date information on the current and imminent climate impacts and vulnerability, which eliminate skepticisms and builds confidence for private investment. Private sector involvement is often seen as a way to boost the efficiency of climate adaptation (Tompkins and Eakin 2012) and their initial outputs of using market-based approaches in resilience building are evolving (IPCC 2014). For instance, Kenya has developed tools to leverage private sector investments in climate-resilient development (Reid et al. 2014). In the same way, collaborating with private companies' Gambian government is investing in the carbon market in tree plantations.

Literatures show private sector is a potential partner that can significantly contribute to adaptation (IPCC 2014), but without attention to major design challenges, they can also provide disincentives, cause market failure, and decrease equity. Cities are attractive to private enterprises because so much business activity, private investment, and demand are concentrated there. However, private investment or insurance markets will not protect low income urban dwellers because of the low ability to pay, high risks, and the high transaction costs for companies of administering many small policies (Hallegatte et al. 2010). Experts determine barriers to private sector investment in the developing countries, inter alia, lack of relevant risks information, lack of capacity and skills required, poor regulatory environment, and infrastructure limitations (Reid et al. 2014). Private sector becomes more prominent when local governments and civil society actions are limited, but this raises the issue of government incentives in regard to the most vulnerable communities (Chu and Schroeder 2010). Public sector can significantly contribute to provide educational and skill development programmes with the latest decision tools, assessment methods, and practices to curb climate risks. In the case of private sector investment, government needs to play key roles as regulators, providers, or insurers of last resort.

In the context of widespread poverty, respondents reveal CBA is largely unable to deal with social problems, e.g., alleviating poverty (Figs. 3 and 4). Vulnerability and poverty are closely interconnected in the case study areas, which is situated in geographically vulnerable regions that expose to flooding, river bank erosion, salinity intrusion, and sea level rise. Yet, these problems have been compounded by over population, very few income generating activities, and persistent inequalities, which in turn arise poverty traps. Most of the short and medium term CBA projects have lost their development continuity after projects duration completed, despite community members marked projects were initially 'beneficial' in terms of their inclusion in decision making, creating awareness and learning, and addressing measures that perpetuate poverty and vulnerability. Experience indicates CBA is likely to be the most effective, if CBA puts emphasis more on the wellbeing of people, synergies between adaptation and local development, and social transformation. Similarly, respondents pointed that CBA has potential for approaching incremental to transformational adaptation (Table 1 and Figs. 3, 4). A notable example of these types of adaptation is 'floating gardens', which are prepared by using freshwater weeds, water hyacinth, soil, and cow dung. Floating gardening is increasingly becoming a useful way to grow food on waterlogged or flooded land in Bangladesh. To seize the opportunity, calibration (in the theory and practices) of CBA is necessitated. However, without addressing the pressing needs of the poorest people and structural inequalities, CBA will have partial effects on climate change adaptation.

Although government is trying to shore up coordination and integration amongst the sectors and institutions, the preponderance of evidence suggests these limitations are existed across administrative levels (Roy et al. 2013a), and that badly affect local adaptations. Overly centralised environmental governance is one of the main reasons for that. Moreover, marginalisation of the local governments in receiving funds and decision making has discouraged such coordination and integration at the local level (Roy et al. 2013b), whereas at the national level, this coordination and integration has not brought desired outcomes in adaptation planning and implementation owing to either lack of leadership or wrong leadership.

Community members strongly seem that this country increasingly encounters the negative impacts of climate change due to recent experience of recurrent floods and cyclones. Informants opine 'potential collapse of CBA' as a threat, particularly in the face of increasing global temperature up to a 2 °C and beyond, expecting more intense cyclones, unprecedented heat-waves and widespread food shortages in the LDCs. Both groups assume that dependency on foreign aids and donors' support is a big threat for local adaptation, and emphasis on national level solutions by transforming social, economical and political systems that require investment on creating mass awareness, social learning, local networks, researches, innovations, and extension, as well as developing effective institutional scaffoldings and policies. As a regional CBA informant said "it is uncertain how long climate funds will smoothly come to the LDCs, but it is certain that this fund will substantially be reduced when industrialised countries will severely encounter the adverse impacts of climate change like the LDCs face today……perhaps will get either development assistance or climate funds, but not both (Anil Kumar, a High School Teacher and local leader).

## 6 Conclusion and Policy Implications

Despite CBA approach is still evolving, it has produced several successful episodes of adaptation and development. Conceptual and procedural challenges of CBA are documented earlier (see 'An Overview of CBA Challenges') and in the context of increasing concerns of imminent climate impacts, the applicability of CBA has invoked the improvement of CBA's theory and practice. This study evaluates the suitability of CBA to determine the positive and negative factors of it so that the application of CBA can contribute more to improve communities' adaptive capacity, drawing case study lessons from flood affected, drought affected, and coastal areas of Bangladesh. An important conclusion is that the CBA has ample strengths (e.g., strengthening climate resilience) as well as opportunities (e.g., approaching transformational adaptation) for strengthening communities' adaptive capacity in order to adapt current and future climate risks, although with some shortcomings, specifically threats such as potential collapses of CBA in the face of extreme climatic events has posed limitations to its wider application. Moreover, specific attention of practitioners and policy makers is required in exploring the opportunities for CBA as well as overcoming its weaknesses. To overcome the weaknesses, CBA planning primarily needs to: recognise diverse social-cultural contexts, interests, and expectations, and consider the dynamics of vulnerability and exposure and their linkages with other dimensions, namely socioeconomic processes. Yet, planning and implementation of CBA is necessitated to be improved through complementary actions across levels, from individuals to governments.

Climate change adaptation is place- and context-specific. This case study presents one example of the suitability of CBA that deduces straightforward policy implications to improve the CBA approach. Based on the findings, policy should mainly emphasis: (i) 'strengthening CBA finance' through proper utilisation of existing and emerging economic instruments, making adaptation finance effective for the most vulnerable, and fostering collective commitment by developed countries to provide new and additional funds, (ii) 'involving private sectors in CBA' by facilitating public-private coalitions and partnerships to develop adaptation products and services, co-financing of infrastructure development, and developing a set of complementary measures, like appointing experts who can work with the funding recipients and co-financiers in order to broaden private sector engagement, (iii) 'strengthening local institutions' through upgrading capacity, providing resources, and specifying roles for scaling up and scaling out CBA. However, climate change communication is an important factor for policy information implementation, since it demonstrates impacts of extreme weather events, fosters a mutual understanding of development between scientists, civil society, political leaders and businessmen, and facilitates effective and meaningful public engagement. The principles of 'development communication' can be incorporated into CBA to enhance its effective and efficient communication (i.e., implementation), which can be an area of future research.

The study findings are based on a case study research conducted in flood-affected, drought-affected and coastal areas. Case situations are seldom comparable and findings from these case studies cannot be generalized to evaluate the procedure of CBA and to fit a whole country population. Moreover, as it is an evaluative study based on perceptions of community members and CBA informants, so there is a possibility of expressing self-interest. Findings of this study are applicable considering research limitations.

In sum, to enhance CBA's suitability as a more promising adaptation (and mitigation) tool for current and future climate impacts, practitioners must find ways to meet the pressing needs of the most vulnerable people to climate change, and simultaneously, to foster iterative learning, deliberative processes, technological innovations, and social transformation at the local level.

## References

Adger N, Arnell NW, Tompkins E (2005) Successful adaptation across scales. Global Environ Change 15:77–86

Agrawala S, Carraro M (2010) Assessing the role of microfinance in fostering adaptation to climate change, OECD Environmental Working Paper No. 15. OECD

Ash AJ, Stafford Smith M (2013) Adaptation research: community, science or discipline? In: Palutikof et al (eds) Climate adaptation futures. Wiley, West Sussex, pp 49–55

Arnold M, Mearns R, Oshima K, Prasad V (2014) Climate and disaster resilience: the role for community-driven development. Social Development Department, World Bank, Washington

Ayers J (2011) Resolving the adaptation paradox: exploring the potential for deliberative adaptation policymaking in Bangladesh. Global Environ Polit 11:62–88

Ayers J, Huq S (2013) Adaptation, development and the community. In: Palutikof et al (eds) Climate adaptation futures. Wiley, West Sussex, pp 203–214

Ball T, Werritty A, Geddes A (2013) Insurance and sustainability in flood-risk management: the UK in a transitional state. Area 45:266–277

Baas S, Ramasamy S (2008) Community based adaptation in action: a case study from Bangladesh. Environment and Natural Resources Service Series, No. 14, FAO, Rome

Boyd R, Ibarraran ME (2009) Extreme climate events and adaptation: an exploratory analysis of drought in Mexico. Environ Dev Econ 14:371–396

Bryan E, Behrman J (2013) Community-based adaptation to climate change: a theoretical framework, overview of key issues and discussion of gender differentiated priorities and participation, CAPRi Working Paper No. 109. IFPRI, Washington, D.C

Chu SY, Schroeder H (2010) Private governance of climate change in Hong Kong: an analysis of drivers and barriers to corporate action. Asian Stud Rev 34:287–308

Cooke B, Kothari U (2001) Participation: the new tyranny?. Zed Books, London, p 432

Dasgupta A, Beard VA (2007) Community driven development, collective action and elite capture in Indonesia. Dev Change 38:229–249

Dodman D, Mitlin D (2011) Challenges for community-based adaptation: discovering the potential for transformation. J Int Dev 2011. https://doi:10.1002/jid.1772

Ensor J (2010) Governance for community based adaptation. Discussion paper. Practical Action. UK

Fankhauser S, Burton I (2011) Spending adaptation money wisely. Clim Policy 1:1037–1049

Forsyth T (2013) Community-based adaptation: a review of past and future challenges. WIREs Clim Change. https://doi:10.1002/wcc.231

Heltberg R, Siegel PB, Jorgensen SL (2009) Addressing human vulnerability to climate change: toward a 'no regrets', approach. Global Environ Change 19:89–99

Hallegatte S, Henriet F, Patwardhan A, Narayanan K, Ghosh S, Karmakar S, Patnaik U, Abhayankar A, Pohit S, Corfee-Morlot J (2010) Flood risks, climate change impacts and adaptation benefits in Mumbai: an initial assessment of socio-economic consequences of present and climate change induced flood risks and of possible adaptation options. OECD Environment Working Papers No. 27, OECD

Huq S, Reid H (2007) Community based adaptation: A vital approach to the threat climate change poses to the poor. IIED Briefing, London

IPCC (2014) Climate change 2014: impacts, adaptation, and vulnerability. Part A: Global and sectoral aspects. Contribution of Working Group II to the AR5 of the IPCC [Field et al (eds)]. Cambridge University Press, UK

Jones R, Rahman A (2007) Community-based adaptation. Tiempo 64:17–19
Katmandu Declaration (2014) www.KathmanduDeclaration.org
Khan MZH, Haque M, Rouf M (2013) An assessment of climate finance governance Bangladesh. Transparency International Bangladesh, Dhaka
King S (2014) Community-based adaptation in practice: a global overview of CARE international's practice of CBA to climate change. Poverty Environment and Climate Change Network for CARE International
Kindon S, Pain R, Kesby M (2009) Participatory action research. In: Thrift N, Kitchin R (eds) International encyclopedia of human geography. Elsevier, Oxford, pp 90–95
Kurtilla M, Pesonen M, Kangas J, Kajanus M (2000) Utilizing the analytic hierarchy process (AHP) in SWOT Analysis—a hybrid method and its application to a forest-certification case. Forest Policy Econ 1:41–52
Lavell A, Oppenheimer M, Diop C, Hess J, Lempert R, Li J, Muir-Wood R, Myeong S (2012) Climate change: new dimensions in disaster risk, exposure, vulnerability, and resilience. In: Field et al (eds). Managing the risks of extreme events and disasters to advance climate change adaptation. A special report of working groups I and II of the IPCC. Cambridge University Press, UK, pp 25–64
Mansuri G, Rao V (2004) Community-based and -driven development: a critical review. World Bank Res Obs 19:1–39
Mansuri G, Rao V (2013) Localizing development: does participation work?. World Bank, Washington, D.C., p 435
Masozera MK, Alavalapati JRR, Jacobson SK, Shrestha RK (2006) Assessing the suitability of community-based management for the Nyungwe Forest Reserve Rwanda. Forest Policy Econ 8:206–216
McGray H, Bradley R, Hammill A (2007) Weathering the storm: options for framing adaptation and development. World Resources Institute, Washington, D.C.
Mercer J, Kelman I, Lloyd K, Suchet-Pearson S (2008) Reflections on the use of participatory research for disaster risk reduction. Area 40:172–183
Mimura N et al (2014) Adaptation planning and implementation. In: Field CB et al (eds) Climate change 2014: impacts, adaptation, and vulnerability. IPCC WGII AR5. Cambridge University Press, UK, pp 869–898
MoEF [Ministry of Environment and Forest] (2009) Bangladesh climate change strategy and action plan. MoEF, Government of the People's Republic of Bangladesh, Dhaka
Palutikof J, Parry M, Stafford Smith M, Ash AJ, Boulter SL, Waschka M (2013) The past, present and future of adaptation: setting the context and naming the challenges. In: Palutikof J, Boulter SL, Ash AJ (eds) Climate adaptation futures. Wiley, West Sussex, pp 3–30
Parry ML, Arnell N, Berry P (2009) Assessing the costs of adaptation to climate change: a review of the UNFCCC and other recent estimates. IIED and Grantham Institute for Climate Change, London
Pielke R, Prins G, Rayner S, Sarewitz D (2007) Lifting the taboo on adaptation. Nature 445:597–598
Rashid AKM, Khan MR (2013) Community based adaptation: theory and practice. In: Shaw R, Mallick F, Islam A (eds) Climate change adaptation actions in Bangladesh. Springer, London, pp 341–364
Reid H, Alam M, Berger R, Cannon T, Huq S, Milligan A (2009) Community-based adaptation to climate change: an overview. In: Participatory learning and action. IIED, London, pp 11–38
Reid H, Huq S (2014) Mainstreaming community-based adaptation into national and local planning. Clim Dev 6:291–292
Reid H, Fenton A, Kirkby P, Parry E, Roberts E, Smith B, Stott C (eds) (2014) Community based adaptation: financing local adaptation. In: 8th international conference on CBA proceedings, 24–30 Apr, Nepal
Roy R, Chan NW, Uemur T, Imura H (2013a) The vision of agri-environmental sustainability in Bangladesh: how the policies, strategies and institutions delivered? J Environ Prot 4:40–51

Roy R, Chan NW, Rainis R (2013b) Development of an empirical model of sustainable rice farming: a case study from three rice-growing ecosystems in Bangladesh. Amer-Eur J Agric Environ Sci 13:449–460

Roy R, Chan NW (2014) A multi-level evaluation of policy integration of human resource development in agriculture sector. Nat Resour 5:119–129

Roy R, Chan NW, Xenarios S (2015) Sustainability of rice production systems: an empirical evaluation to improve policy. Environ Dev Sustain 18:257–278

Saaty TL, Vargas LG (2012) Models, methods, concepts & applications of the analytic hierarchy process, 2nd edn. Springer, New York

Tompkins E, Eakin H (2012) Managing private and public adaptation to climate change. Global Environ Change 22:3–11

UNISDR [United Nations International Strategy for Disaster Reduction] (2009) After action review: second session global platform for disaster risk reduction. Switzerland

UNDP (2011a) CBA to climate change through coastal afforestation in Bangladesh. UNDP-ALM Project. UNDP Regional Centre, Bangkok

UNDP (2011b) Empowering vulnerable communities to address climate change adaptation. GEF SGP delivery of CBA Projects. The GEF SGP, New York

World Bank (2010) Economics of adaptation to climate change. Synthesis Report. The World Bank, Washington

# Getting Buy-In for Climate Change Adaptation Through Urban Planning: Climate Change Communication as a Multi-way Process

Leslie Mabon and Wan-Yu Shih

## 1 Introduction

The purpose of this paper is to assess the role of communication in attaining climate change adaptation through urban spatial planning. The city scale is crucial for action on climate change. This is not only due to the rapidly urbanising nature of the world's population, but also because it is municipal and/or local governments who have the precision to put national- or international-level climate change policies into practice through processes such as building regulations and land use planning (Puppim de Oliveira et al. 2011). Nonetheless, cities and their planning is an intrinsically complex issue (Crawford 2016) which has given rise to calls for more communication between people and sectors (van Dijk and Ubels 2016) to synthesise the range of knowledges and considerations at play. When one adds to this the complexities and uncertainties inherent in climate change, spatial planning becomes a good test case for assessing the value and limitations of multi-way climate change communication.

In this paper, we therefore evaluate how communication between sectors has helped to bring to fruition municipal climate change adaptation initiatives in Durban, South Africa. What makes Durban an even more valuable case study is the ability of eThekwini Municipality (the metropolitan municipality within which Durban is located) to instigate a suite of programmes for adapting to climate change through spatial planning in spite of a challenging socio-economic context. Through

L. Mabon (✉)
School of Applied Social Studies, Robert Gordon University, Aberdeen
AB10 7QG, Scotland, UK
e-mail: l.j.mabon@rgu.ac.uk

W. -Y. Shih
Department of Urban Planning and Disaster Management,
Ming-Chuan University, Gui Shan District, Taoyuan City 333, Taiwan
e-mail: shih@mail.mcu.edu.tw

 
W. Leal Filho et al. (eds.), *Handbook of Climate Change Communication: Vol. 1*,
Climate Change Management, https://doi.org/10.1007/978-3-319-69838-0_4

a narrative literature review of planning documentation and peer-reviewed literature relating to climate change adaptation in Durban, we aim to identify lessons which other cities may draw on for how communication can help to get support for their climate change initiatives. By extending this narrative review to include social science critiques of urban planning in Durban and of some concepts drawn on by eThekwini Municipality, however, our intention is to go beyond producing a 'best practice' guide and also flag up limitations or cautions for the role of climate change communication in the urban spatial planning context.

We hope this will be of value to decision-makers (municipal governments, local planners, community groups, practice-focused academics) tasked with actually having to implement climate change adaptation 'on the ground', and to academics wishing to develop a critical yet practically-focused approach to means of attaining cross-sector support for climate action. Through the use of a middle-income country case study, we hope to illustrate how momentum for climate change adaptation through urban planning can be sustained even in challenging socio-political contexts. However, we also encourage scholars and practitioners to continue to reflect on inequalities and power imbalances within which climate change communication may take place.

## 2 Literature Synthesis: Communication as a Multi-way Process

We understand climate change 'communication' as a multi-way exchange of ideas and opinions on the most appropriate way for society to respond to the problems raised by climate change, rather than a one-way transfer of information on the scientific basis for action. Bearing in mind that our interest is in how support may be built for evidence-based strategies for assessing risks and adapting to climate change, it is useful to look to science and technology studies literature to understand why 'communication' ought to be understood in this way. Wynne (1992) argues the 'information deficit' model of passing scientific information from experts out to society at large is an inadequate means of governing science in society. There are at least three reasons for this. One is that contemporary technical and scientific issues are very complex, meaning even 'experts' may disagree among themselves (Pellizzoni 2003). A second is the decline in societal trust in the scientists and decision-makers previously entrusted to make decisions on society's behalf (Kasperson 2014). A third is that new technologies and the knowledge on which they are based (including climate change mitigating technologies and policies) can have profound, far-reaching and potentially irreversible effects on how people can live their lives. This may include the cost and availability of low-carbon energy, or the extent to which people have to live with environmental change because of the success or otherwise of climate policies (Mabon et al. 2015).

In sum, science has profound effects on—and may be received in a number of different ways by—society. There may therefore be a range of perspectives at play on what is considered an appropriate course of action in response to scientific knowledge. In response, social scientists have imagined a range of processes like upstream engagement (Rogers-Hayden et al. 2007), co-production (Jasanoff 2004), 'opening up' (Stirling 2008) and 'heating up' (Sundqvist 2014). All of these processes are about fostering dialogue between science and society, in order to take societal concerns into account alongside scientific knowledge in decision-making. This is reflected in social science literature specific to climate change, where there is a growing interest in techniques that encourage discussion on how actions taken on the basis of scientific knowledge about climate change will influence the kind of society we live in (e.g. Einsiedel et al. 2013; Roberts and Escobar 2015; Vercelli et al. 2014).

This does not mean that 'anything goes' with regard to what constitutes valid knowledge, or that scientific knowledge is somehow less valuable than the other belief systems that may be at play (Wynne 1996). However, there is a need to pay attention to the grounds on which different sections of society are actually concerned about an issue like climate change, and to acknowledge that concerns such as social justice (Shrader-Frechette 2002), trust in decision-makers (Kasperson 2014), connection to place (O'Neill 2007) and morality or spirituality (Northcott 2015) can influence actors' standpoints on climate change in addition to their technical and scientific understanding. Discussions on climate change are therefore likely to involve a range of viewpoints and value systems, not all of which may buy into a 'pure science' rationale for taking action. The challenge is thus to ground communication in broad-based rationales which may facilitate agreement on practical action, even if more fundamental value disagreements remain (Mabon and Shackley 2015). More practically, communication which gives a number of reasons for climate action with the potential to appeal to a range of constituencies can help to sustain policymakers' attention. This is crucial as the attention-cycle issue (Downs 1972) has been argued to be equally applicable to climate change (UNEP 2009). Finding framings and arguments that widen out the number of reasons for taking actions that have climate benefits may thus help to make rationales more durable and appealing when climate change is not a top political issue.

## 3 Methodology

The rationale behind this paper is to assess the role of communication in attaining climate change adaptation through urban spatial planning. To do this, the case study of Durban, South Africa is selected for two reasons. Firstly, it is widely held up as an exemplar of climate change adaptation through spatial planning (e.g. ASSAf 2011). Secondly, the challenging social context of high unemployment, significant poverty and low life expectancy (eThekwini Municipality 2010) means Durban may yield lessons for how to build and sustain momentum for urban climate change planning in the face of other pressing societal issues. In keeping with the argument in Sect. 2 that there may be a range of views on what constitutes an appropriate course of action on climate change, however, this paper also aims to scrutinise the grounds on which Durban's climate change adaptation initiatives may actually be considered examples of 'best practice'. Doing so entails assessing potential limitations or challenges for climate change communication raised by the Durban case study, through reference to social research carried out in Durban and also through engagement with underpinning social theory.

A narrative review was undertaken of both peer-reviewed and policy literature pertaining to climate change adaptation in Durban. Jones (2004) argues that narrative reviews of this nature are more appropriate than systematic literature reviews when the material being reviewed is itself qualitative in nature—as much of the material reporting the planning process for climate adaptation in Durban is. The reason for this, Jones argues, is that a narrative review allows the researcher to engage more flexibly with the stories contained in the qualitative material itself, and also to extend their survey to include 'grey' literature such as policy documents which lie outside the peer-reviewed literature but are nonetheless significant and influential. Moreover, given the overarching aim of engaging with the broader critical social science literature around planning and governance in Durban as well as the material specific to climate adaptation in the city, a narrative literature review was selected on account of the interpretative flexibility it gave to explore a new and potentially broad-reaching topic.

Selection and subsequent analytical guidelines were developed to address Green et al. (2006) critique that narrative literature reviews can tend towards haphazard sampling from the literature. Relevant peer-reviewed and policy documents were identified by searching in Google Scholar (scholar.google.com) with the terms 'Durban' and 'climate change adaptation' in the range from 1 January 2000 to 31 August 2016, excluding results discussing the Durban Adaptation Charter (which refers to local climate action globally) rather than climate adaptation in Durban specifically. To get a sense of the wider societal context into which Durban's planning for climate change adaptation occurred, Google Scholar was again searched over the same time periods for the terms 'Durban', 'society', and 'environment' and/or 'climate change', again excluding results referring to the Durban Adaptation Charter. This material was then further supplemented with policy documents produced by eThekwini Municipality referred to in the peer-reviewed articles, such as the Municipal Climate Protection Programme (eThekwini Municipality 2010), the latest eThekwini Municipality Integrated Development Plan (eThekwini Municipality 2015) and documentation on specific projects. For the sampled texts, thematic analysis was then undertaken to identify relevant aspects of 'communication' as a dialogic process for decision-making about climate change. Four aspects in particular were drawn out of the reading: (1) Key messages and frames used to justify climate adaptation actions; (2) Internal communication within eThekwini Municipality; (3) External communication between the Municipality and other actors such as research institutes and private sector; and (4) Civil society participation in both adaptation initiatives and also decision-making processes.

Following this model of a narrative review with thematic analysis therefore allowed for flexibility to draw in relevant material from outside a narrow 'climate change adaptation' sample, but at the same time retained a certain degree of rigour in the analysis and interpretation process. The aspects of the narrative review identified above—messaging, internal communication, external communication, and civil society participation—are now discussed in turn.

## 4 Durban Case Study

### *4.1 Overview*

South Africa ranked 116th in the 2014 Human Development Index, which measures length and healthiness of life; knowledgeability; and quality of living standard (UNDP 2015). This is reflected in the societal context of Durban which, despite making a significant contribution to South Africa's economy (ASSAf 2011), still faces high unemployment, housing shortages and low life expectancy (eThekwini Municipality 2010). For Durban, climate change therefore has the potential to intensify existing societal issues and can thus be considered a serious risk (Roberts and O'Donoghue 2013).

Durban proposed a Municipal Climate Protection Programme (MCPP) in 2004 to respond to both adaptation and mitigation. Within this programme, ecosystem-based adaptation came to have an important role (Shih 2017). Ecosystem-based adaptation is the use of biodiversity and ecosystem services as part of an overall adaptation strategy to help people adapt to the negative effects of climate change (SCBD 2009). The value of such an approach is argued to lie in its potential to produce multiple benefits such as carbon sequestration, temperature regulation and flood control (Munang et al. 2013) and frame the climate challenge at a municipal or local scale where fine-scale recommendations can be made (Roberts et al. 2012). Within Durban's MCPP, the Durban Metropolitan Open Space System (D'MOSS) came to be viewed as an available, cost-effective and sustainable means of coping with climate change, and indeed to act as an ecosystem-based adaptation strategy. D'MOSS is an interconnected green space system comprising ecologically valuable areas in both private- and public-owned lands. It was first developed in 1979 to protect important natural areas from urban development, but has evolved from these conservation-oriented roots to serve multiple functions (Shih 2017). It is argued to be able to deliver several benefits at once: biodiversity conservation, environmental hazard reduction, carbon sequestration, and poverty elimination (Roberts et al. 2012). Given the potential for such benefits, we focus on the communication aspects of this more adaptation-focused iteration of the open space system in this section as well as providing analysis of communication in Durban's overall climate adaptation discussions.

## 4.2 Framing and Messaging

The importance of climate change adaptation in Durban is framed within a much wider context of societal and environmental challenges in the city. The MCPP, for instance, argues that high unemployment, limited access to education, housing and infrastructure shortages and low life expectancy "increase the vulnerability of many of Durban's residents, and this situation is likely to be exacerbated as the impacts of climate change place further strain and demands on limited resources and infrastructure" (eThekwini Municipality 2010: 3). Climate change is positioned not as a distant and unrelated phenomenon, but rather as something with the potential to have implications for poverty intensification, urbanisation and environmental deterioration in Durban (Roberts and O'Donoghue 2013). Specific to ecosystem-based adaptation, justifications from eThekwini Municipality centre on 'win-win-win' outcomes with potential to achieve dual adaptation and mitigation; prevent biodiversity loss; and improve human wellbeing (Roberts et al. 2012). The value of broad-based rationales for climate change adaptation initiatives becomes even more important in the South African context, where concerns with social justice post-apartheid continue (Smith 2008). Understanding the value of climate change adaptation in terms of not undermining or losing development gains post-apartheid can thus help to gain political traction (Roberts 2010) in a way purely science-based arguments may not.

At the same time, however, decision-making for climate change adaptation through spatial planning remains firmly grounded in 'expert' scientific evidence. Freund (2001) notes the economic values assigned to D'MOSS are derived from environmental functions like temperature control and water cleansing. Indeed, eThekwini Municipality (2011a) explain that D'MOSS is "mapped by the Biodiversity Planning Branch of the Environmental Planning and Climate Protection Department (EPCPD) in consultation with relevant experts." Whilst the messaging around climate change adaptation and indeed open space planning in Durban has shifted from science-driven to communicating societal benefit, the range of solutions and outcomes proposed are therefore in the first instance informed by a scientific evidence base (Shih and Mabon 2017). We return to this balancing of the scientifically appropriate with the socially acceptable in Sect. 5.

## 4.3 Internal Communication

Communication within eThekwini Municipality has been a significant driver in determining the success or otherwise of Durban's climate change adaptation initiatives. There is emerging understanding that spatial planning for climate change adaptation requires cross-sectoral cooperation (Crawford 2016). In the case of Durban, however, there is an ongoing need to build skills and capacity within government—especially outside the environmental protection sector—to allow engagement with the complexity of data and issues involved (Roberts et al. 2012). Linking back to Sect. 4.2, this reinforces the point that a dialogue between different standpoints as to the most appropriate course of action for climate change adaptation may still need to be informed by a basic common understanding of the scientific basis for action. Moreover, there are questions over how much the rhetoric of sustainability seen in Durban is backed up by financial commitment (Aylett 2010) or supported in other government sectors (Bond and Dada 2007), demonstrating the challenge of aligning goals within the organisation.

In a context of differing levels of knowledge and varying extents of support for climate change adaptation initiatives, the role of the individual in communication has been key in Durban. Freund (2001) believes individual 'champions' within the Municipality have been vital in affecting public consciousness and building support among decision-makers. Leck and Roberts (2015) see the informal interactions at the personal level between departments—and indeed between sectors—as a powerful influence on the success of climate change adaptation projects across Durban alongside formal platforms for communication. Whilst Leck and Roberts (2015) admit such processes run the risk of being exclusionary, the fact adaptation projects have come to fruition through such processes illustrates Hajer and Kesselring's (1999) observation that under the right conditions, informal 'power brokering' can produce outcomes amenable to all.

Communication within eThekwini Municipality around spatial planning for climate change adaptation demonstrates that at different points in the planning process, different types of 'communication' may be appropriate. Sometimes this will be 'one-way' transfer of knowledge to build the basic capacity required for informed debate, sometimes it may be an individual or group leading based on political nous, and sometimes it may be less formal 'power brokering.' Just as Roberts et al. (2012) note in the context of ecosystem-based adaptation that changes in the roles, responsibilities and functions of existing local government institutions may be required, so too may be different understandings of how internal communication ought to be undertaken depending on context.

## *4.4 External Communication*

Given our interest in the implementation of evidence-based strategies for adapting to climate change, we focus here on the role of knowledge in communication between institutions. The complexity of climate change and its interaction with environmental and social systems means data must be drawn in from a range of sources, necessitating targeted partnerships with research institutions (Roberts et al. 2012). This can be seen for water sector adaptation with the involvement of the University of Kwa-Zulu Natal in research into infrastructure and security (eThekwini Municipality 2010), the involvement of 'relevant experts' in mapping D'MOSS (eThekwini Municipality and ICLEI Africa Secretariat 2007), private sector organisations Green Roof Designs and Ecoman in preparing Durban's

Guidelines for Green Roof Habitats (eThekwini Municipality 2011b), and the engagement of a local non-profit organisation, Wildlands Conservation Trust, in training people for the Buffelsdraai Landfill Site Community Reforestation (Douwes et al. 2015). The knock-on effect of this, however, is a need for local government to increase capacity in order to be able to use information of an academic and technical nature (Roberts et al. 2012) and for even more work to be done before civil society more widely can engage with sophisticated data (Aylett 2010). In short, governing climate change adaptation processes inevitably entails the bringing together of different knowledge systems, hence communication between different actors will also entail the translation and interpretation of these different knowledges.

External communication around knowledge also entails managing relations with those whose immediate priorities may not directly relate to climate change. Roberts and O'Donoghue (2013) remind us that climate change is just one of many political issues in Durban, competing for attention with economic pressures and societal challenges, hence there may be value in linking climate protection to sustainable livelihoods, a 'green' economy, job creation and improvements in living standards. This serves as a reminder of the value of broad-based rationales for particular courses of spatial planning (e.g. poverty alleviation, biodiversity protection) which do not require one to immediately and directly connect with the 'pure science' knowledge basis for climate adaptation. Nevertheless, 'win win' situations may not always be possible. Roberts et al. (2012) concede that there may have to be 'losers' at the individual level in order to attain outcomes which benefit Durban more broadly. It is the limitations of attempts to reach consensus, and the possibilities for civil society to intervene in this science-driven process, which we now address.

## 4.5 *Civil Society Participation*

A key concern in Durban, intensified by high levels of poverty in the municipality, is civil society participation via building the ability of local communities themselves to adapt (Roberts 2010). This has more recently developed into community-based ecosystem adaptation, which emphasises local benefits from climate actions (Douwes et al. 2015). An example of this is the *Tree-Preneur* programme associated with the Buffelsdraai Landfill Site Community Reforestation Project, where working with the Wildlands Conservation Trust NGO, unemployed community members are engaged as 'Tree-Preneurs' to grow trees which will be used for the reforestation project. The seedlings can then be exchanged for credit notes for food, basic goods and school fees (eThekwini Municipality 2011c). Whilst the underlying aim of this project is carbon sequestration, it offers a number of other rationales which can be drawn on to get buy-in for the initiative—ecological restoration more generally, capacity building, engagement between government and civil society. All of these have the potential to provide pathways to societal support for a spatial planning initiative grounded in a 'scientific' rationale.

Whilst such initiatives are of course welcome in terms of alleviating poverty and developing local capacity, it is important to note their limitations. Some communication to civil society on climate change in Durban seems to focus on "message transfer" (Roberts 2010: 409) and "effective communication" (Roberts et al. 2012), albeit more partnership-based actions are emerging (Cockburn et al. 2016). This in part may be explained by the need to provide basic knowledge to allow citizens to make informed decisions of their own about climate change action. However, alongside this there is a need to ensure communities have a means of participating in knowledge-making (not just the subsequent planning and decision-making) processes, as communities may in cases have their own understandings of ecosystems and conservation (De Georges and Reilly 2009) which can complement more 'formal' scientific knowledge. Moreover, in the context of Durban, Aylett (2010) criticises a constant drive for consensus in civil society engagement, arguing that conflict can drive innovation, identify weaknesses and find better solutions. Aylett illustrates this point with reference to implementation of improved air quality monitoring systems in Durban, which came about due to NGO efforts to draw attention to the air quality situation through protest and citizen monitoring rather than consensus.

As well as reinforcing the value of imagining climate adaptation actions which could get societal buy-in on a number of grounds, then, the nature of civil society engagement on spatial planning for climate change adaptation in Durban also raises dilemmas and challenges for climate change communication. One of these is the extent to which effective 'communication' needs to provide citizens with the knowledge to engage with climate change debates, versus designing processes that allow scientists and planners to tap into the environmental understandings local communities may already hold. The other is to question whether consensus always ought to be the aim of communication, or whether provoking debate—like Sundqvist's (2014) 'heating up'—might lead to better solutions. We now consider these more critical dimensions of climate change communication raised by the Durban case.

# 5 Discussion

## *5.1 Framings*

Our first discussion point concerns what the social implications may be of the way in which a climate change debate is framed. In Durban, for instance, ecosystem services thinking has come to be central to understanding and demonstrating the value of the open space system to the city (eThekwini Municipality and ICLEI 2007; Roberts et al. 2012). However, the ecosystem services concept more generally has faced criticism for thinking of ecosystems and the environment in the same monetary and capitalist terms which arguably contribute to societal inequalities in the first instance, and for ignoring justice concerns (Kosoy and Corbera 2010; Matulis 2014). Although the ecosystem services framing is clearly important for drawing attention to the value of Durban's open space, especially given the issues of poverty within the city, caution must be exercised not to deflect attention away from the social processes which may render some communities vulnerable in the first place. Communicating the value of climate change adaptation initiatives in terms of ecosystem services may engage a wider range of actors who may be more persuaded by economic arguments than environmental ones, but care has to be taken to ensure this does not inadvertently 'close down' the terms of the debate and exclude concerns over justice which may not so easily be quantified.

## 5.2 Balancing the Scientifically Appropriate with the Socially Desirable

In the case of Durban and the implementation of D'MOSS, the effects of a country-wide shift towards more democratic governance on environmental planning are noticeable. eThekwini Municipality's guidance on D'MOSS makes explicit mention that the council will not take ownership of land in the D'MOSS layer, and that the South African Constitution's statement that no one will be deprived of property rights is respected (eThekwini Municipality 2011d). The expectation of 'democratic' decision-making hence restricts municipal government ability to zone or re-claim land for environmental protection purposes on the basis of scientific evidence alone. This should never be interpreted as a sign that democracy has negative effects. Rather, it demonstrates that in a democratic society, there is perhaps an expectation that communication on climate change decisions will be a multi-way process. In cases, this expectation that communication channels are opened to allow a range of perspectives to enter the decision-making process might rule out courses of action—like the automatic control of certain pieces of land within Durban—even though they appear technically superior. The ways in which D'MOSS in Durban has had to be adapted to take into account societal benefit and ecosystem services as well as biodiversity protection hence serves as a reminder that the spatial planning 'solutions' reached for climate change adaptation need to be socially acceptable as well as technically viable. This extends too to planning policy and processes, which attempt to respond to land ownership concerns through processes such as environmental servitudes (eThekwini Municipality 2015). The key lesson for climate change communication arising from this is to pay attention to whether the format and nature of communication fits with wider societal expectations on how the environment ought to be governed, and what may happen if it is not.

### 5.3 Limitations and Constraints

We finish by acknowledging limitations and constraints of our work. One is the limited data available, especially on the societal dimensions of climate adaptation in Durban. Freund (2001: 736) notes "(a)ny evaluation of environmental policy, let alone environmental awareness or developments on the ground, for post-apartheid Durban must inevitably be highly circumspect due to lack of data […] and one's fear of being overly influenced by one influential opinion-maker or another." One must therefore exercise caution when using case studies like Durban as 'best practices' in climate change adaptation, and pay attention to the possibility of 'epistemic communities' (Haas 1992; Toke 1999) setting the terms of discussion on what constitutes 'successful' climate adaptation based on what they themselves consider to be valid knowledge and an appropriate range of outcomes. In other words, if places like Durban and projects like the Buffelsdraai Landfill Site Community Reforestation Project come to be described through academic and policy literature as exemplars of best practice, there remains a need for the research community to critically scrutinise *who* it is that is making such claims, what kind of evidence they use to ground their assertions, and what they stand to gain from communicating about the success of initiatives. This is especially true when claims to societal benefits such as 'poverty eradication' or 'improvement in quality of life' are made in reports and peer-reviewed literature. A literature-based review of the kind we have undertaken is constrained by the data available online. There is hence a need for ongoing critical social science research—ideally involving empirical or field components—that can help to question *who* it is that makes a judgement as to whether an intervention like stakeholder engagement has been successful. Academic research may also evaluate the extent to which it an initiative may truly be considered a 'successful' example for others to follow, by undertaking independent follow-up assessment on environmental and social impact.

Another constraint may be the nature of the narrative literature review itself. As Green et al. (2006) point out, reviewing literature with the aim of interpreting and evaluating from it inevitably runs the risk of subjectivity. In addition to our suggestion of empirical social science research outlined above, subsequent assessments of the social dimensions of climate adaptation in Durban may wish to undertake more structured systematic reviews, for instance setting specific research questions to be addressed through the review. Due to spatial constraints, we have also not fully explored the significant literature around climate change mitigation in Durban (e.g. Bond and Dada 2007) which may offer further analogous examples on communication in municipal climate governance. Nonetheless, as a wide-ranging critical evaluation of the existing literature in order to draw out lessons and pitfalls for climate change communication in urban planning and environmental management, we believe a narrative literature review is an appropriate starting point which may provide context and a basis for subsequent research.

# 6 Conclusion

In this chapter, we have assessed the role of communication in attaining climate change adaptation outcomes through urban spatial planning and environmental management, focusing on the case of Durban in South Africa. Our key lessons are as follows:

- **Climate change 'communication' should often be more than a one-way transfer of information**. In Durban, spatial planning for climate adaptation requires communication *between* actors. This entails those responsible for putting spatial planning into action (municipal governments and urban planners) taking into consideration the concerns, priorities and capabilities of other government departments, communities, developers, non-governmental organisations, citizens and others with a stake in planning actions undertaken;
- **Framings which can build broad-based rationales for climate change actions may stand a greater chance of success**. The messaging used in Durban around climate change adaptation and ecosystem-based adaptation in particular demonstrates how benefits of planning decisions such as poverty alleviation, disaster mitigation more generally or sustainable economic development can foster consensus for specific spatial decisions. Whilst there is of course a robust scientific basis for the actions which may be taken to mitigate or adapt to climate change, decision-makers need support from a range of institutions who may not buy into 'pure science' rationales for climate change adaptation, or may have concerns grounded in economic, cultural or social values;
- Alongside this drive for consensus and buy-in, **there is a need for scholars of climate change communication to keep up their critical guard**. This entails reflecting on which institutions are setting the terms of the discussion and to what effect; and where our attention may be diverted from the bigger structural issues which make communities vulnerable to climate change in the first place.

Above all else, there is a need in climate change communication to strike a balance between what is scientifically appropriate and what is socially acceptable. For both the 'science' and 'society' aspects, it is crucial to question who is involved in the communication process, who may be excluded, and to question who may benefit as a result of decisions taken. Retaining this healthy scepticism will help to ensure that an understanding of climate change communication as a dialogue between different value positions can be a force for good in getting buy-in and galvanising support for climate change adaptation through spatial planning.

**Acknowledgements** The ideas and arguments in this paper were developed through a grant from the RGU Foundation received by Leslie Mabon, and a Japan Society for Promotion of Science (JSPS) Postdoctoral Fellowship (Grant No. 23.01774) held by Wan-Yu Shih.

## References

Academy of Science of South Africa (ASSAf) (2011) Towards a low carbon city: focus on Durban. Report of the ASSAf Consensus Study on Low Carbon Cities. ASSAf, Pretoria

Aylett A (2010) Participatory planning, justice, and climate change in Durban, South Africa. Environ Plan A 42:99–115

Bond P, Dada R (2007) A death in Durban: capitalist patriarchy, global warming gimmickry and our responsibility for rubbish. Agenda 73:46–56

Crawford R (2016) New Zealand Productivity Commission Research Note 2016/2: what can complexity theory tell us about urban planning?. New Zealand Productivity Commission, Wellington

Cockburn J, Rouget M, Slotow R, Roberts D, Boon R, Douwes E, O'Donoghue S, Downs CT, Mukherjee S, Musakwa W, Mutanga O, Mwabvu T, Odindi J, Odindo A, Procheş S, Ramdhani S, Ray-Mukherjee J, Sershen, Schoeman MC, Smit AJ, Wale E, Willows-Munro S (2016) How to build science-action partnerships for local land-use planning and management: lessons from Durban, South Africa. Ecology and Society 21(1). doi: 10.5751/ES-08109-210128

De Georges PA, Reilly BK (2009) The realities of community based natural resource management and biodiversity conservation in Sub-Saharan Africa. Sustainability 1(3):734–788

Douwes E, Rouget M, Diederichs N, O'Donoghue S, Roy K, Roberts D (2015) Buffelsdraai Landfill Site Community Reforestation Project. XIV World Forestry Congress, 7 Sept 2015, Durban. doi:10.13140/RG.2.1.3577.9286

Downs A (1972) Up and down with ecology—the issue-attention-cycle. The Public Interest 28:38–50

Einsiedel EF, Boyd AD, Medlock JE, Ashworth P (2013) Assessing socio-technical mindsets: public deliberations on carbon capture and storage in the context of energy sources and climate change. Energy Policy 53:149–158

eThekwini Municipality (2010) Durban's municipal climate protection programme: climate change adaptation planning for a resilient city. eThekwini Municipality, Durban. http://www.durban.gov.za/City_Services/development_planning_management/environmental_planning_climate_protection/Publications/Documents/Durbans%20Municipal%20Climate%20Protection%20Programme_Climate%20Chan.pdf. Last accessed 08 Nov 2016

eThekwini Municipality (2011a) What is the Durban Metropolitan open space system? http://www.durban.gov.za/City_Services/development_planning_management/environmental_planning_climate_protection/Durban_Open_Space/Pages/-What-is-the-Durban-Metropolitan-Open-Space-System.aspx. Last accessed 08 Nov 2016

eThekwini Municipality (2011b) Creating space for biodiversity in Durban: guideline for designing green roof habitats. Durban, South Africa. http://www.durban.gov.za/City_Services/development_planning_management/environmental_planning_climate_protection/Publications/Documents/Guideline%20for%20Designing%20Green%20Roof%20Habitats1.pdf. Last accessed 08 Nov 2016

eThekwini Municipality (2011c) Buffelsdraai Community Reforestation Project. Durban, South Africa. http://www.durban.gov.za/City_Services/development_planning_management/environmental_planning_climate_protection/Projects/Pages/Buffelsdraai-Community-Reforestation-Project.aspx. Last accessed 08 Nov 2016

eThekwini Municipality (2011d) Durban Metropolitan open space system FAQ. Durban, South Africa. http://www.durban.gov.za/City_Services/development_planning_management/environmental_planning_climate_protection/Durban_Open_Space/Pages/MOSS_FAQ.aspx. Last accessed 08 Nov 2016

eThekwini Municipality (2015) 2015/2016 Final integrated development plan. Durban, South Africa. http://www.durban.gov.za/City_Government/City_Vision/IDP/Documents/Final%20Adopted%20IDP%2027May2015.pdf. Last accessed 08 Nov 2016 eThekwini Municipality, ICLEI Africa Secretariat (2007) Durban Biodiversity Report 2007. Local Action for Biodiversity (LAB) Report. eThekwini Municipality, Durban. http://www.durban.gov.za/City_Services/development_planning_management/environmental_planning_climate_protection/Publications/Documents/Durbans%20Biodiversity%20Report%202007.pdf. Last accessed 08 Nov 2016

Freund R (2001) Brown and green in Durban: the evolution of environmental policy in a post-apartheid city. Int J Urban Reg Res 25(4):717–739

Google (n.d.) Google Scholar, scholar.google.com. Last accessed 08 Nov 2016

Green BN, Johnson CD, Adams A (2006) Writing narrative literature reviews for peer-reviewed journals: secrets of the trade. J Sports Chiropr Rehabil 15:5–19

Haas PM (1992) Introduction: epistemic communities and international policy coordination. Int Org 46(1):1–35

Hajer M, Kesselring S (1999) Democracy in the risk society? Learning from the new politics of mobility in Munich. Environ Polit 8(3):1–23

Jasanoff S (2004) Science and citizenship: a new synergy. Sci Public Policy 31:90–94

Jones K (2004) Mission drift in qualitative research, or moving toward a systematic review of qualitative studies, moving back to a more systematic narrative review. Qual Rep 9(1):95–112

Kasperson R (2014) Four questions for risk communication. J Risk Res 17(10):1233–1239

Kosoy N, Corbera E (2010) Payments for ecosystem services as commodity fetishism. Ecol Econ 69(6):1228–1246

Leck H, Roberts D (2015) What lies beneath: understanding the invisible aspects of municipal climate change governance. Current Opin Environ Sustain 13:61–67

Mabon L, Shackley S (2015) More than meeting the targets? The ethical dimensions of carbon dioxide capture and storage. Environ Values 24:465–482

Mabon L, Shackley S, Vercelli S, Anderlucci J, Boot K (2015) Deliberative decarbonisation? Exploring a framework of ethical governance for low-carbon energy. Environ Plan C Gov Policy 33(2):256–271

Matulis BS (2014) The economic valuation of nature: a question of justice? Ecol Econ 104:155–157

Munang R, Thiaw I, Alverson K, Mumba M, Liu J, Rivington M (2013) Climate change and ecosystem-based adaptation: a new pragmatic approach to buffering climate change impacts. Current Opin Environ Sustain 5(1):67–71

Northcott M (2015) Place, ecology and the sacred: the moral geography of sustainable communities. Bloomsbury, London

O'Neill J (2007) Markets, deliberation and environment. Routledge, London

Pellizzoni L (2003) Uncertainty and participatory democracy. Environ Values 12(2):195–224

Puppim de Oliveira JA, Balaban O, Doll CNH, Moreno-Peñaranda R, Gasparatos A, Iossifova D, Suwa A (2011) Cities and biodiversity: Perspectives and governance challenges for implementing the convention on biological diversity (CBD) at the city level. Biol Cons 144(5):1302–1313

Roberts D (2010) Prioritizing climate change adaptation and local level resilience in Durban, South Africa. Environ Urban 22(2):397–413

Roberts JJ, Escobar O (2015) Citizens' juries on wind farm development in Scotland. ClimateXChange, Edinburgh. http://www.climatexchange.org.uk/files/5614/3213/1663/Citizens_Juries_-_Full_Report.pdf. Last accessed 08 Nov 2016

Roberts D, O'Donoghue S (2013) Urban environmental challenges and climate change action in Durban, South Africa. Environ Urban 25(2):299–319

Roberts D, Boon R, Diederichs N, Douwes E, Govender N, McInnes A, McLean C, O'Donoghue S, Spires M (2012) Exploring ecosystem-based adaptation in Durban, South Africa: "learning-by-doing" at the local government coal face. Environ Urban 24(1):167–195

Rogers-Hayden T, Mohr A, Pidgeon N (2007) Introduction. Engaging with nanotechnologies: engaging differently. NanoEthics 1:123–130

Secretariat of the Convention on Biological Diversity (SCBD) (2009) Connecting biodiversity and climate change: report of the second ad hoc technical expert group on biodiversity and climate change. UNEP, Montreal

Shih W-Y (2017) Greenspace planning for climate co-benefits in Durban, South Africa. In: Doll CNH, Puppim de Oliviera JA (eds) Urbanization and climate co-benefits: implementation of win-win interventions in cities. Routledge, London pp. 88–95

Shih W-Y, Mabon L (2017) Land use planning as a tool for balancing the scientific and the social in biodiversity and ecosystem services mainstreaming? The case of Durban, South Africa. J Environ Planning Manage. doi: 10.1080/09640568.2017.1394277

Shrader-Frechette K (2002) Environmental justice: creating equality, reclaiming democracy. Oxford University Press, New York

Smith L (2008) South African social work education: critical imperatives for social change in the post-apartheid and post-colonial context. Int Soc Work 51(3):371–383

Stirling A (2008) "Opening Up" and "Closing Down": power, participation, and pluralism in the social appraisal of technology. Sci Technol Human Values 33(2):262–294

Sundqvist G (2014) 'Heating up' or 'Cooling Down'? analysing and performing broadened participation in technoscientific conflicts. Environ Plan A 46(9):2065–2079

Toke D (1999) Epistemic communities and environmental groups. Politics 19(2):97–102

United Nations Development Programme (UNDP) (2015) Human development report 2015: work for human development. UNDP, New York

United Nations Environment Programme (UNEP) (2009) IEA training manual: a training manual on integrated environmental assessment and reporting: training module 3: Developing an impact strategy for your IEA. UNEP, New York http://www.unep.org/ieacp/_res/site/File/iea-training-manual/module-3.pdf. Last accessed 08 Nov 2016

van Dijk T, Ubels H (2016) How Dutch professionals conduct interactive design sessions to foster 'shared understanding'. Environ Plan 43(3):464–479

Vercelli S, Battisti N, Dolcetti F, Ciesielska J, Barkved L, van der Panne G, Engen S, Provoost M (2014) Dialogue and mutual learning towards a low carbon society—experiences from 10 countries across Europe. Energy Procedia 58:30–35

Wynne B (1992) Misunderstood misunderstandings: social identities and public uptake of science. Public Underst Sci 1:281–304

Wynne B (1996) May the sheep safely graze? A reflexive view of the lay-expert knowledge divide. In: Lash S, Szerszynski B, Wynne B (eds) Risk, environment and modernity: towards a new ecology. SAGE, London, pp 44–83

# How Aesthetic Style Can Influence Reception of Visual Communications of Climate Change

**Rebecca Green**

## 1 Introduction

Subscription to the belief that humans are contributing to climate change in Australia is far lower than expected, providing impetus for the importance of effective communication of the phenomenon. In Australia, 81% of Australians surveyed agreed that climate change was happening, but more than half weren't concerned about the implications, and only 47.3% attributed these changes to anthropogenic influence (Leviston et al. 2014). With similar results returned in studies in the USA (Leiserowitz et al. 2015) and UK (Capstick et al. 2015), the importance of effective communication of this complex, long-term issue cannot be overstated. At the same time, the rapid advancement of digital media technology has led to the dominance of an even more image-based style of visual communication, yet as Domke et al. (2002, in Jqffe 2008) identified, we know little about how these visuals affect public reception of messages.

As an experienced graphic designer, one of my research goals is to design a set of visual artefacts using different aesthetic styles for investigations into their influence on engagement in climate change mitigation and adaptation. However, a review of literature exposed a paucity of studies into graphic design and visual communication, let alone visual communication of the issue of climate change. New theory must be built in order to progress to any type of future design-led inquiry.

Much of graphic design literature focuses on a more technical perspective of artefact creation, and little theory exists outside the scope of advertising and consumer-based visual communication. Knowledge that does exist seems to shun the key language of graphic design—aesthetic style—focusing more on borrowed

R. Green (✉)
UNSW Art & Design, F217G, Paddington CampusCnr Oxford St & Greens Rd, Paddington, Sydney, NSW 2021, Australia
e-mail: bgreen@unsw.edu.au

 
W. Leal Filho et al. (eds.), *Handbook of Climate Change Communication: Vol. 1*, Climate Change Management, https://doi.org/10.1007/978-3-319-69838-0_5

academic domains such as rhetoric and semantics from a message-only or frame perspective. This preference is reflected in wider studies of social communication strategy, whereby empirical studies focus primarily on message frame and content, divided between two broad types: strong emotion, and values. The first, a fear-based frame, or threat appeal, attempts to evoke emotions such as fear and/or disgust in order to influence viewers (Foust and Murphy 2009; Brulle 2010). The second frame is one that attempts to use a more rational argument, or coordinated appeals to viewer values to effect change (Tonkinwise 2011; Lakoff 2010; Hulme 2008, 2009 O'Neill and Nicholson-Cole 2009; de Hoog et al. 2005). These studies focus on the message, frame and the strategists creating them, but not the visual language used to promote these messages. When images are considered in extant studies, for example using Q method (Lorenzoni et al. 2007; O'Neill and Nicholson-Cole 2009), the research centres on the object that is being represented, rather than the aesthetic style used to represent these objects or concepts.

Approaching research into the communication of messages relating to climate change and environmental sustainability from a graphic design perspective allows for new ways of understanding the influence of visual communication on the two key types message, or frame strategies found in literature. By granting equal weight to aesthetics as well as to the message, or frame, new theory can be built that spans sociology and graphic design fields (Bartmanski 2015). In particular, a graphic design perspective is well-suited to this challenge, not only as the field that produces much of the world's visual communication artefacts, but as one that regularly deals with both aesthetic and message types of rhetorical language. This paper will discuss the exposition of the two types of frame found in literature to phenomenological inquiry, using visual artefacts as representative of these types.

## 2 Study Design

This research formed part of a larger cultural study involving three reconstructed instances of the case of climate change communication. Visual artefacts were theoretically selected according to their adherence to two different types of framing and visual communication, as discussed in the previous section (Figs. 1, 2 and 3). Frames using emotional claims involving fear and disgust were sought, as well as more rational, values-based approaches to message framing strategies. Visual artefacts were also required to represent aesthetic styles ranging from photographic to more abstract illustrated imagery. The designers of selected visual artefacts were then approached both for consent and copyright to use the artefacts within the study, as well as in subsequent dissemination. These visual artefacts were then shown to selected viewers within individual semi-structured interviews using a phenomenological approach, leading to analysis of stated individual experiences of real world artefacts. This simulated reality does not replicate the truth of a real-world encounter with these artefacts, however the method does allow for the elicitation of intensive responses for interpretive analysis.

**Fig. 1** Visual artefact 1: *Keep Buying Shit*. *Source* Reproduced with permission from Diego Gutierrez, 2008

**Fig. 2** Visual artefact 2: *Apology From the Future*. *Source* Reproduced with permission from Toby Cotton, Arc Communications, 2009

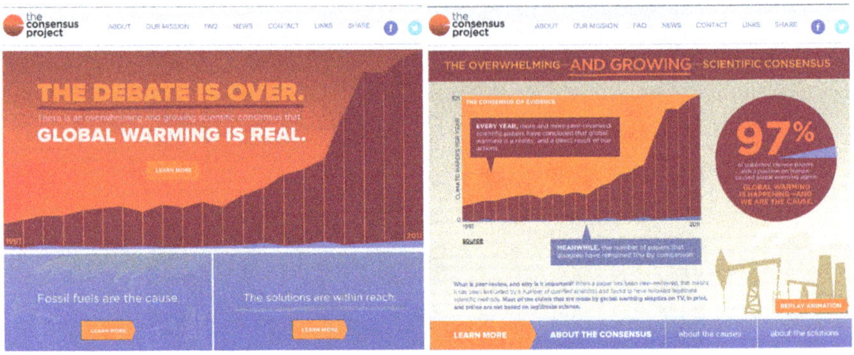


**Fig. 3** Visual artefact 3: *The Consensus Project* (with alternative screen). *Source* Reproduced with permission from Matt Birdoff, SJI Associates, 2013

The research presented in this paper focuses on the way the selected participants made meaning by decoding the visual artefacts to which they were exposed, appropriating Bourdieu's logic of practice (Bourdieu 1992; Schatzki 2003). The paper is limited to analysis of how the viewers discussed the artefacts to which they were exposed, in particular their perceptions of the messages and authors. It does not approach other parts of the cultural study, which included other analyses of the interview data, analysis of the interviews with the designers, or visual analysis of the artefacts themselves.

## 3 Selection of Viewers

Viewers were selected using the snowball method (Thoyre 2011). A request was sent to friends and colleagues to recommend potential respondents based on several qualifying criteria:

1. Middle class, English-speaking professionals
2. Aged between 18 and 65
3. Not holding extreme views on ACC at either end of the spectrum
4. No experience in graphic design, marketing or allied media professions.

The first criterion was based on the idea that middle class professional are better suited to questions that relate to aesthetic style (Snelders et al. 2014; Nicholson-Cole 2005; Kenney and Scott 2003; Bourdieu 1984). In particular, Kenney and Scott proposed that viewers at a higher level of intelligence placed visuals into types, or codes. The Nicholson-Cole study found that this group had both a good exposure to media and broad knowledge about climate change. It also found that respondents under the age of 18 were more influenced by their education environment, making them a less ideal group for this study. This resulted in a secondary criterion that delimited participation to between the ages of 18–65.

The third criterion was that the respondent hold some views on anthropogenic climate change, but not what might be considered an extreme position. This excluded both ends of the spectrum: environmental activists as well as climate change deniers, as it was perceived that respondents from these extreme positions may not be open to influence, visual or otherwise. The nine selected respondents positioned themselves at varying points on a spectrum of subscription to anthropogenic climate change at the start of the interview. The fourth criterion was also revealed during the interviews, where any respondents who had experience or education with the graphic design or allied industries were excluded due to the likelihood of bias in their experience of the visual artefacts.

Recommended respondents were contacted via email and those who agreed were interviewed between June and August 2014. During the interview period, an opportunity arose to attend a conference near London, which led to the interview of several respondents there. These London interviews were teamed with Brisbane interviews to obtain nine constructed experiences with the visual artefacts for analysis. Informed consent was obtained from each respondent, followed by a one-on-one semi-structured interview taking between 26 and 48 min. Each interview was conducted in person between the researcher and the respondent, and was recorded. The viewer interviews were comprised of five sets of interview questions: the first set discussed the respondent, the next three sets addressed each of the three visual artefacts in turn, followed by a final set of more general questions. The first question was a framing question, which asked where the respondent sat on the spectrum of anthropogenic climate change subscription. This response helped the interviewer to frame the questions in closer alignment to the respondent's position. The final two questions asked questions around the viewer's background, such as education, age and professional experience.

After the first set of questions, each respondent was then exposed to the three visual artefacts (Figs. 1, 2 and 3), which were mounted to A4 black cardboard. Viewers selected, in turn, which of the artefacts they would like to discuss first, resulting in a secondary exposure. This secondary exposure involved only the selected visual artefact, plus any alternatives displayed on a computer screen (Fig. 3).

Viewer utterances were transcribed and subjected to two types of coding: provisional coding (Miles et al. 2014), where themes were isolated into groups informed by existing theory such as decoding (Williamson 1978); followed by pattern and thematic analysis. Narrative analysis of these patterns and themes followed, using the technique of "describe, understand, explain" (Ragin 2006; Hamel et al. 1993). This analysis was directed at exposing the logics that these viewers used in order to create meaning from the visual artefacts. Bourdieu, through his logic of practice (Bourdieu 1992; Schatzki 2003) proposes "systems of principles" that users develop and deploy in experience of phenomena, and which were borrowed for this study as a means of understanding the experiences of the viewer agents.

## 4 How Viewers Make Meaning by Decoding Signs in Visual Artefacts

This section focuses on how viewers make meanings from their decodings of the three climate change visual artefacts. A Bourdieusian approach allows for investigation of the systems of principles, or common group of conditions that viewers apply when decoding visual artefacts, and how these systems contribute to heuristic perception of symbolic representations of the dominant hegemony, as proposed by Antonio Gramsci (Kurtz 1996).

The interplay between corporate brands and grassroots organisations in the communication of anthropogenic climate change is a much-discussed topic in the literature surrounding communication of climate change (Regniez and Custead 2011; Nerlich et al. 2010; McKie and Galloway 2007). These authors indicate that relationships between these groups is key, with clear boundaries between the types of organisations required to engender trust. Viewer "decoding" of visual artefacts can impact the perception of what type of principal authority is behind a communication. This perception has the potential to alter the perceived message (Kahneman et al. 1982) potentially rendering any high-level communication strategy ineffective in practice. How these conditions relate to the meaning made by viewers of the visual artefacts follows this discussion.

Viewers decoded the three visual artefacts into two broad types aligning with theory that discusses the professional coding of visual communications (Berger 2010; Williamson 1978). The first of the two types is the hegemonic code (Kurtz 1996), which represented for the viewers the dominant hegemony in the context of anthropogenic climate change: corporations, governments, news media outlets, fashion and retail sectors (Table 1), in many cases using Scott's assessment that styles of visuals link them to other visuals (Scott 1994). The second type is the counter-hegemonic code (Rose 2011), which represented community and grassroots organisations, charities, universities and individuals (Table 2). Four causal conditions contributed to viewer discussion of visual artefacts as belonging to one or

**Table 1** Types of hegemonic code entities stated by viewers

| Types of hegemonic code entities | |
|---|---|
| News and information | Newspapers and news sites |
| | Advertising |
| | Billboards |
| Government | Regulatory and legislative bodies |
| Political | Political parties independent of government duty |
| Corporate | Food corporation |
| | Oil company |
| | Corporate general: material generally used in corporate environment |
| Fashion and retail | Fashion magazine |
| | Shopping centre/shopping mall |

**Table 2** Types of counter-hegemonic code entities stated by viewers

| Types of counter-hegemonic code entities |
|---|
| Community groups |
| Charities |
| Environmental groups |
| Individuals |
| Universities |

other of these types, and those conditions will each be addressed in the following paragraphs:

1. Emotional response and perceived emotional claims
2. Hidden agenda 1: rhetorical representations
3. Hidden agenda 2: representation of principal
4. Colour signifiers.

## 4.1 *Emotional Response and Perceived Emotional Claims*

Emotional responses to the visual artefacts played a key part in determining professional code types by viewers. Much of the literature that guided theoretical selection of the visual artefacts pointed to emotions such as fear and disgust as barriers to communication, resulting in avoidance of the issue or "turning away" from disgust-evoking images (O'Neill and Nicholson-Cole 2009; Joffe 2008; de Hoog et al. 2005). For some viewers, the presence of emotional claims was actually a signifier of the counter-hegemonic code, which led to assignation of professional code type as part of the decoding of the visual artefact:

> It doesn't look like it's a group of Green activists, there's no sort of emotion, and 'We're killing everybody', you know.
>
> —viewer B7 referring to visual artefact 3: *The Consensus Project*
>
> Because they're passionate about it. They just come at it from a different angle than your more regulatory or authority stance.
>
> —viewer L4 referring to visual artefact 1: *Keep Buying Shit*

Viewers had the strongest emotional response to visual artefact 1 with an array of negative emotions described. Viewers had a lesser emotional response to visual artefact, and the least emotional response to visual artefact 3. With a small sample size, little can be drawn from a comparison of numbers other than a general pattern of how strongly the viewers felt. What can be argued is that this emotion did not deter any of the viewers from understanding and interpreting the visual artefact with the preferred reading intended by the designer.

One viewer responded with a lack of trust of the message (indicated by "No"), but defined this as a result of there not being any statement of the principal authority behind the message:

> No. And that's because who is this?
>
> —B8 referring to visual artefact 1: *Keep Buying Shit*

In contrast, six of the nine viewers reported no emotional response at all to visual artefact 3. One viewer, however, did record the sensation of hope, the only instance of this reported in response to any of the visual artefacts. This lack of perceived emotional response contributed to the perception of hegemonic code for this artefact in several respondents:

> I mean Climate Change to me brings vision of pollution in third world countries, and weather changes, so it doesn't really indicate anything like that. It doesn't indicate anything emotive.
>
> —B7 referring to visual artefact 3: *The Consensus Project*

This finding also brings to light the proposal that a lack of emotional claim has a more negative effect in viewers than negative claims. As the visual artefacts did not provoke high levels of reported fear, this proposition would bear further testing in future studies. The causal condition for the viewers was therefore that the presence of an emotional claim, either perceived as present by the viewers, or validated by the presence of an emotional response, is a signifier for the counter-hegemonic coding of visual artefacts.

## *4.2 Hidden Agenda 1: Rhetorical Representations*

The second causal condition relates to the use of different types of images. More representational imagery, in particular, the photograph seen in visual artefact 1: *Keep Buying Shit* (Fig. 1), was identified as a truthful representation of the referent. This truth perceived through photographic representations of object appears to be a marker for transparency, and—dependent on the referent object or objects—helps define the visual artefact as belonging to one or other of the two types of professional code.

> It's just kind of like, this is the actual reality of what you can do. You can make a big pile of shit and seagulls eat it. —L2 viewer referring to visual artefact 1: *Keep Buying Shit*

The contribution of this perception of truth in photographic imagery is much-discussed (Morris and Sayler 2014; Baudrillard in Harden 2011; Cole 2010; Rose 2011; Tversky and Kahneman 1974), and makes a contribution on its own to decoding the visual artefacts. However, it exists here as a key part of a set of causal conditions, or system of principles that determine how visual artefacts are coded between the two contrasting authority types. Within this system, photographic representation of object appeared to work in tandem with a second 'hidden agenda' causal condition, which is the visual representation of the principal authority behind the message.

## 4.3 Hidden Agenda 2: Representation of Principal

Visual artefacts 2 and 3 make a clear statement of principal using a visual symbol (or a logo using graphic design terminology) while visual artefact 1 makes no such visual statement. The difference between these two signifiers is that the Greenpeace logo in visual artefact 2 is known to belong in the counter-hegemonic category of viewer coding, whereas the logo in visual artefact 3 is not well-known. As only visual artefact 2 can assume coding based on this signifier, visual artefacts 1 and 3 rely more heavily on types of imagery to create the "hidden agenda" condition.

As London viewer 1 (L1) describes below, without a clear representation of the principal authority, the type of rhetorical images displayed plays a key factor in determining from which end of the professional code spectrum the ad is to be experienced. Abstract representations, viewed without a clear statement of principal combined to affect trust, while the photographic representation caused L1 to code visual artefact 1 as counter-hegemonic.

> I wouldn't necessarily trust it because there's a lot of statistics in there, there's a lot of information in there. The previous one [visual artefact 1] is very simple. There's no hidden agenda. Whereas this one would appear to possibly have a hidden agenda with regards to who is driving it. Whereas the other one [visual artefact 1] just looked like it was an interested party trying to help.
>
> —L1 viewer referring to visual artefact 3: *The Consensus Project*

Most viewers of visual artefact 1 made their own determination of the principal authority behind the message. Without a logo signifier, they relied on a combination of other principles of the professional code, such as emotional claims or reactions, and the use of photographic representations of object as a rhetorical signifier. The imagery provided viewers with what they perceived as a truth or reality that they could actively relate to, negating the "hidden agenda" perception that could have clouded decoding of this visual artefact.

Four of the viewers of visual artefact 3 indicated a lack of trust for the message, with confusion surrounding the principal as a key cause. Speculation as to the principal authority ranged from political parties and corporations within the hegemonic code, to universities or groups of scientists within the counter-hegemonic code. Photographic representation was not a key causal condition for coding, but the lack of the perception of truth evidenced by photographic representation, combined with a lack of indication of principal, did contribute to a negative comparison to the counter-hegemonic code by the "hidden agenda" condition.

## 5 Colour

The final causal condition that viewers used to make decisions on professional coding was colour. Viewers reported strong correlations between colours and corporations, colours and signified concepts, as well as colour that did not fit their expectations of professional design application. While these findings cannot contribute a generalised system or typology of colour for environmental messages (indeed, such a typology would be difficult to produce with any type of academic rigour), they do indicate that colour selection can have a significant impact on viewer perception in terms of the professional code.

Intentions of the designers when deciding colours as part of their process were varied. Diego, producer of visual artefact 1: *Keep Buying Shit* used a referential technique during his process, where a mental connection was made between visual and written referents in the artefact: seagulls and the word "shit", resulting in a particular yellow that for him referenced "bird poop". The colour also represented both the "raw", urgent style of the artefact, and a rebellious positioning towards his perception of the hegemony within the design field:

> It's also a no-no for type usually, against a background like that. I always got in trouble with the purist Swiss teachers. They hated me. You can't do that with type. I'm like, well I just did and you're dealing with it.
>
> —Diego, Producer of visual artefact 1: *Keep Buying Shit*

In contrast, Matt (producer of visual artefact 3: *The Consensus Project*) used a conscious application of his understanding of colour meanings:

> Psychologically, red was kind of the colour to go to, just because it's the alarming colour. Blue and green are the happy earth colours, but that's really not the message that we're trying to convey here. We wanted something that would convey some sense of alarm. —Matt, Producer of visual artefact 3: *The Consensus Project*

The decodings of the viewers contrasted to both of these producer assumptions. Five of the nine viewers made a point of discussing that the yellow made the text hard to read, according to their own experience of the artefact, or to their understanding of colour perception, such as issues with dyslexia. This would support the design theories against which Diego was rebelling, but as noted by London viewer 3, that difficulty and disturbing sensation was likely the intention.

The heavy use of red colours used in visual artefact 3 were interpreted as relating to global warming, heat and doom rather than warning. Viewer B9, who subscribed to anthropogenic climate change noted that the message did not use a more counter-hegemonic colour:

> It's interesting there's no green or anything like that
>
> —B9 referring to visual artefact 3: *The Consensus Project*

The colour scheme for visual artefact 3 was also related to specific companies that sit within the dominant code, such as McDonald's and Sainsbury's (a UK supermarket chain), or financial news outlets such as The Economist or The Financial Times. These findings suggest that colours can be as polysemic and emotion-provoking as images when used for a rhetorical purpose.

The causal conditions for assignation of either the hegemonic or counter-hegemonic code to visual artefacts are by no means exhaustive, rather they constitute the conditions that were apparent in the reconstructed instances of the visual communication of three visual artefacts. These conditions contributed to the creation of a code, which in turn guided the position of viewers during their exposures to the visual artefacts. Discussion now turns to how this code influenced the viewer perception of the three visual artefacts.

## 6 How Perception of Hegemonic or Counter-Hegemonic Code Influences Viewer Perception

### *6.1 Contribution of Professional Code to Perception of Principal or Authority*

As Bourdieu argues, language does not exist merely as a system of signs to be decoded by viewers, it also relies on signified "authority and wealth" (Loesberg 1993) for communication. This signification of authority relates to Tversky and Kahneman's (1974) concept of "representativeness" as a heuristic, where an actor is judged to be part of one group or another, and is therefore cast in the same light as that group. Again, viewer B7's response to visual artefact 3: *The Consensus Project* highlights this issue, where the decoding as hegemonic code becomes a signifier for the perception of the authority issuing the message contained in the artefact. Viewer B7, like most of her fellow Brisbane viewers, was also influenced by reading rhetorical purpose into supporting imagery. She likened the aesthetic style of the visual artefact to those used by hegemonic entities:

> Well, it's very simplistic looking, so it looks similar to maybe a corporate type of website, a corporate design. —viewer B7 referring to visual artefact 3: *The Consensus Project*

As indicated in the previous section, her position in relation to visual artefact 3 became oppositional, despite her aesthetic preference for this artefact in the initial exposure. As the interview progressed, her assessment of the signified facts and statistics were influenced by this position:

> I'd have to really read it and make up my own mind, but when they throw things like ninety-seven percent makes me a little bit sceptical, because it says of published climate papers. Well, that's just a throwaway line.
>
> —viewer B7 referring to visual artefact 3: *The Consensus Project*
>
> The other one was very vague, you know, ninety-seven percent of all Climate Change papers, well how many of those have been written? That's just a made up statement really, isn't it? —viewer B7 referring to visual artefact 3: *The Consensus Project*

What was considered compelling evidence—clear communication of the scientific consensus on climate change (Lewandowsky et al. 2013)—displayed in a simple, understandable aesthetic style by the designer, and received well aesthetically by a viewer already aligned to the message, still resulted in an oppositional stance and a lack of trust. The hegemonic coding, in particular the "corporate style" and the support imagery that was decoded as rhetorical and a synecdoche of the principal authority appear to have influenced this viewer's positioning in relation to the artefact, and accordingly her subscription to the message itself.

This influence of professional code on Brisbane viewer 7 (B7) is further illustrated by responses to visual artefact 2: *Apology from the Future*. Although visual artefact 2 stated a well-known environmental group as the principal authority —"Greenpeace", itself a descriptor for this viewer's conditions for counter-hegemonic code—viewer B7 indicated confusion due to her perception of the aesthetic style as representing the hegemonic code:

> Well, they'd like you to think it's Greenpeace... but whether that really is, it is Greenpeace or not, I'm not sure.
>
> Well, I guess you think Greenpeace are sort of activists and this looks quite corporate as well and they don't give the impression of being in a corporate environment but maybe they are now, that's who they're targeting. Maybe changed from being wild activists...
>
> —B7 viewer referring to visual artefact 2: *Apology from the Future*

While visual artefact 2 was the first to be selected for discussion, this suspicion could also be attributed to bias created by the interview and reconstructed experience of the artefacts. Discussion of visual artefact 3 was the last of the three to occur, and the respondent had already been asked questions regarding trust of message twice, as well as one question referring to a general trust in science by this stage of the interview. This may have contributed to an increasingly oppositional stance. However, this context does not necessarily discredit the viewer's perspective, as this type of suspicion could be generated by any number of variables within a natural environment. It does raise questions about the assumption that a simple representation of principal such as a logo is enough to guarantee perception of the correct principal authority, without consideration of the professional codes that aesthetic styles can represent.

While B7 was a good example of this theory of influence of professional code, it was important to examine other responses with reference to the causal conditions to detect other patterns or occurrences. B7 and B8 preferred visual artefact 3: *The Consensus Project* on first exposure, the only respondents out of the nine who did. B6 and L3 elected to discuss visual artefact 3 last, and therefore act as non-ideal cases for comparison using the causal conditions. B6 was the most sceptical of anthropogenic climate change and stated trust of the message of visual artefact 3. L3 also selected to discuss visual artefact 3 last, but subscribed to both the visual artefact's message and to anthropogenic climate change. The dispositions and statements of these four respondents are summarised in Table 3.

B8 stated no emotional response to visual artefact 3, referencing the first of the four causal conditions found in the previous section. This was a possible condition

**Table 3** Influence of professional code on viewers of different dispositions

| Viewer | Agreement with ACC ←→ | | | | | Personal alignment to either hegemony or counter-hegemony | Stated trust |
|---|---|---|---|---|---|---|---|
| B7 | + | | | | | Counter-hegemony (coded science as hegemonic, in relation to corporate influence) | Trusted only counter-hegemonic, emotionally-charged artefacts |
| L3 | + | | | | | Counter-hegemony | Trusted all 3 visual artefacts, decoded as counter-hegemonic |
| B8 | | + | | | | Dominant hegemony | Trusted only hegemonic code artefacts |
| B6 | | | | - | | Dominant hegemony | Trusted only hegemonic code artefact |

for perceiving it as belonging to the hegemonic code. The second condition—colour—did not appear to have a bearing, as the red was decoded only as "doom and gloom", and the hidden agenda condition showed that absence of a recognisable logo was a positive for visual artefact 3, as it did not represent a government-type authority. Research-based universities were assumed as the principal, with a lean towards these institutions belonging in the hegemonic code evident. Interestingly, B8 highlighted the lack of recognisable principal in visual artefact 1 as the key negative influence on trust, and perceived visual artefact 1 as belonging to the counter-hegemonic code. Visual artefact 2 was perceived as falling within the hegemonic code, and B8 stated trust for both hegemonic code visual artefacts 2 and 3.

> I think that's more because I see this as being corporate.
>
> —B8 referring to visual artefact 2: *Apology from the Future*

The first of the non-ideal viewers—L3—aligned with anthropogenic climate change, and a counter-hegemonic disposition. L3 had a non-emotional response to visual artefact 3, but perceived the principal as a counter-hegemonic environmental organisation. There was mention of hegemonic coding prompted by colour: specifically, what financial magazine the aesthetic style might suit, but this was not as explicit as the perception of principal. L3 did not consider there to be a hidden agenda, and assigned the counter-hegemonic code to all three visual artefacts, stating trust for all three. He had a slight mistrust of visual artefact 2, as some doubt was cast on whether it was actually Greenpeace as principal, but discussion moved quickly to trust assuming this was not misrepresentation.

The second non-ideal viewer—B6—had the most sceptical stance on anthropogenic climate change and described a non-emotional response to visual artefact 3, likening the artefact to a "powerpoint presentation". His discussion suggested the artefact's alignment with the dominant hegemony, which correlated with the trust stated in the visual artefact 3 message.

> It seems to be more for a formal presentation as opposed to an advert or something to get your attention.—B6 referring to visual artefact 3: *The Consensus Project*

Visual artefacts 1 and 2 were coded counter-hegemonic, and B6 did not state trust in either of the decoded messages or perceived principals. The table below summarises these dispositions, the decoding and the trust the viewer stated for each visual artefact.

The comparative findings suggest that a combination of disposition and professional code perceived through aesthetic style directly affected viewer uptake of message. This parallels Bourdieu's conclusions that rhetoric must fit the doxa of the viewer to be successful (Schirato 1998, 1997). Understanding doxa through the hegemonic and counter-hegemonic code contributes to understandings about producing rhetorical visual artefacts, in particular, challenging the assumption that for climate change communication, viewer doxa may not be assumed to align with the dominant hegemony of economic rationality and consumerism (Scheufele and Tewksbury 2007; Tonkinwise 2011; Agid 2011; Loesberg 1993; Buchanan 1992).

Joffe (2008: 660) states that the information found in visual artefacts is "likely to play a powerful role in 'positioning' public conceptions of climate change." She also proposes in a later paper that visuals add validity to messages on climate change (Smith and Joffe 2009). While this study neither sought nor achieved refutation of these claims, it demonstrates that pre-existing beliefs and dispositions, in concert with coding influenced by aesthetic style, may in turn 'position' the information translated by the visual artefacts. This finding supports assessments that many studies assume a level of ignorance in viewer audiences (Nerlich et al. 2010), rather than approaching with a respect for the complexity of human capabilities (Scott and Batra 2003).

Viewer B6 was the only viewer who sat within the stated target audience of visual artefact 3: *The Consensus Project*. The positive reception according to the stated position of trust suggests that the hegemonic coding may be successful as long as the viewer's disposition is to trust the dominant hegemony. For those who align with a more counter-hegemonic position, opposing effects may be experienced with a hegemonically-coded visual artefact. While the target audience viewer is addressed and reception is positive, the designer of visual artefact 3 noted that the actual target audience was far wider and included scientists and those who agree and align with anthropogenic climate change. The designer also noted that photographic representation imagery had more of an impact in subsequent social media material produced within the campaign. This invites further discussion on influences of different image types affecting viewer decodings.

## 7 Conclusion

This study has shown that aesthetic style directly influenced the reception of visual communications for the nine viewers. These findings challenge assumptions surrounding message framing, although the presence of a more emotional claim contributed to one key condition in this influence. As a phenomenological investigation, this study did not seek to present population-wide evidence one way or another on how viewers read visuals surrounding these issues. Indeed, each individual viewer's personal experiences, learning and understanding will have a major impact on the reception of any type of communication. What is illustrated is that each respondent aligned to the visual artefacts that they perceived as matching their stated disposition: not always in terms of their stated alignment with the concept of anthropogenic climate change, but in terms of their alignment with the dominant hegemony or counter-hegemony. Analysis of how the viewers read the visual artefacts using four main causal conditions led to a perception of either a hegemonic or counter-hegemonic principal authority behind the communication artefact, with whom the viewer would then align. This heuristic was a major influence on both the viewer interpretation and trust of the message contained within. This finding opens a path for future investigations, such as design-led testing, using causal conditions to create hegemonic and counter-hegemonic-styled visual artefacts. Messages that are positioned differently using this aesthetic language could then add variables such as message content, representation of principal and frame to investigate viewer response. Given the increasing importance of visual media as a result of the rapid developments in technology and global communication, a better understanding of how visual communication can influence attitudes towards anthropogenic climate change is crucial. Informed use of the visual language of graphic design may make a major contribution to the effective communication of climate change.

## References


Agid S (2011) How do we transition people from a system that doesn't want to let them go?: social design and its political contexts. Des Philos Pap 9:187–206

Bartmanski D (2015) Refashioning sociological imagination: linguality, visuality and the iconic turn in cultural sociology. Chin J Sociol 1:136–161

Berger AA (2010) The objects of affection: semiotics and consumer culture. Palgrave Macmillan, USA

Bourdieu P (1984) Distinction: a social critique of the judgement of taste. Harvard University Press, USA

Bourdieu P (1992) The logic of practice. Stanford University Press, USA

Brulle RJ (2010) From environmental campaigns to advancing the public dialog: environmental communication for civic engagement. Environ Commun J Nat Cult 4:82–98

Buchanan R (1992) Wicked problems in design thinking. Des Issues 8:5–21

Capstick SB, Demski CC, Sposato RG, et al (2015) Public perceptions of climate change in Britain following the winter 2013/2014 flooding. Understanding Risk Research Group Working Paper 15-01: Cardiff University, Cardiff, UK

Cole SJ (2010) Baudrillard's ontology: empirical research and the denial of the real. Int J Baudrillard Stud 7:2

de Hoog N, Stroebe W, de Wit JBF (2005) The impact of fear appeals on processing and acceptance of action recommendations. Pers Soc Psychol Bull 31:24–33

Foust CR, Murphy WOS (2009) Revealing and reframing apocalyptic tragedy in global warming discourse. Environ Commun J Nat Cult 3:151–167

Hamel J, Dufour S, Fortin D (1993) Case study methods. Sage Publications, Newbury Park

Harden BG (2011) Nihilism and cultural cohesion:(Re) considerations of Jean Baudrillard. Int J Baudrillard Stud 8:2

Hulme M (2008) The conquering of climate: discourses of fear and their dissolution. Geogr J 174:5–16
Hulme M (2009) Why we disagree about climate change: understanding controversy, inaction and opportunity. Cambridge University Press, Cambridge, UK
Joffe H (2008) The power of visual material: persuasion, emotion and identification. Diogenes 55:84–93
Kahneman D, Slovic P, Tversky A (1982) Judgement under uncertainty: heuristics and biases. Cambridge University Press, Cambridge, UK
Kenney K, Scott LM (2003) A review of the visual rhetoric literature. In: Scott LM, Batra R (eds) Persuasive imagery: a consumer response perspective. Lawrence Erlbaum Associates Inc., New Jersey, USA, pp 17–57
Kurtz DV (1996) Hegemony and anthropology: Gramsci, exegeses, reinterpretations. Critique Anthropol 16:103–135
Lakoff G (2010) Why it matters how we frame the environment. Environ Commun J Nat Cult 4:70–81
Leiserowitz A, Maibach E, Roser-Renouf C, et al (2015) Climate change in the American mind: March, 2015. Yale University and George Mason University. New Haven, CT, USA: Yale Project on Climate Change Communication
Leviston Z, Price J, Malkin S et al (2014) Fourth annual survey of Australian attitudes to climate change: Interim report. CSIRO, Perth, Australia
Lewandowsky S, Gignac GE, Vaughan S (2013) The pivotal role of perceived scientific consensus in acceptance of science. Nat Clim Change 3:399–404
Loesberg J (1993) Bourdieu and the sociology of aesthetics. ELH 60:1033–1056
Lorenzoni I, Nicholson-Cole S, Whitmarsh L (2007) Barriers perceived to engaging with climate change among the UK public and their policy implications. Glob Environ Change 17:445–459
McKie D, Galloway C (2007) Climate change after denial: global reach, global responsibilities, and public relations. Public Relat Rev 33:368–376
Miles MB, Huberman AM, Saldaña J (2014) Qualitative data analysis: a methods sourcebook. SAGE Publications Inc, Thousand Oaks, Califorinia, USA
Morris E, Sayler S (2014) The pensive photograph as agent: what can non-illustrative images do to galvanize public support for climate change action? Image Polit Clim Change Visual Imaginations Documentations 55:299
Nerlich B, Koteyko N, Brown B (2010) Theory and language of climate change communication. Wiley Interdisc Rev Clim Change 1:97–110
Nicholson-Cole SA (2005) Representing climate change futures: a critique on the use of images for visual communication. Comput Environ Urban Syst 29:255–273
O'Neill S, Nicholson-Cole S (2009) Fear won't do it, promoting positive engagement with climate change through visual and iconic representations. Sci Commun 30:355–379
Ragin CC (2006) How to lure analytic social science out of the doldrums: some lessons from comparative research. Int Sociol 21:633–646
Regniez G, Custead S (2011) The role and effectiveness of governmental and non-governmental communications in engaging the public with climate change. In: Whitmarsh L, O'Neill S, Lorenzoni I (eds) Engaging the public with climate change. Earthscan, London, UK, pp 200–216
Rose G (2011) Visual methodologies: an introduction to researching with visual materials. Sage Publications Limited, Thousand Oaks, California, USA
Schatzki TR (2003) A new societist social ontology. Philos Soc Sci 33:174–202
Scheufele DA, Tewksbury D (2007) Framing, agenda setting, and priming: the evolution of three media effects models. J Commun 57:9
Schirato T (1997) Between practice and structure: cultural literacy and Bourdieu's notion of habitus. South Rev Commun Polit Cult 30:259–267
Schirato T (1998) Commoditising communication skills. Aust J Commun 25:87–96
Scott LM (1994) Images in advertising: the need for a theory of visual rhetoric. J Consum Res 21:252–273

Scott LM, Batra R (2003) Persuasive imagery: a consumer response perspective. Lawrence Erlbaum Associates Mahwah, New Jersey, USA
Smith NW, Joffe H (2009) Climate change in the British press: the role of the visual. J Risk Res 12:647–663
Snelders D, Mugge R, Huinink M (2014) Using social distinctions in taste for analysing design styles across product categories. Int J Des 8:23–34
Thoyre A (2011) Social capital as a facilitator of pro-environmental actions in the USA: a preliminary examination of mechanisms. Local Environ 16:37–49
Tonkinwise C (2011) I ♥ Sustainability (because necessity no longer has agency). Des Philos Pap 2
Tversky A, Kahneman D (1974) Judgment under uncertainty: heuristics and biases. Science 185:1124–1131
Williamson J (1978) Decoding advertisements: ideology and meaning in advertising. Marion Boyars, London, UK

# Montreal and Kyoto: Needs in Inter-protocol Communications

**Aliaksandr Krasouski and Siarhei Zenchanka**

## 1 Introduction: Brief History of Montreal Protocol and Kyoto Protocol

The Vienna Convention for the Protection of the Ozone Layer (Convention 1985) and the United Nations Framework Convention on Climate Change (UNFCCC) (Convention 1992) are a multilateral environmental agreement defining human activity for prevention some negative processes influencing on the human health and the environment. 197 States joined to Vienna Convention and 197 States are the Parties of UNFCCC now.

The Montreal Protocol (Handbook 1991) on Substances that Deplete the Ozone Layer is a protocol to the Vienna Convention defining substances that are responsible for ozone depletion. The protocol pointed "that worldwide emissions of certain substances can significantly deplete and otherwise modify the ozone layer in a manner that is likely to result in adverse effects on human health and the environment". There were eight revisions of this protocol last of which was in Montreal in 2007. This protocol defined 1986 year as the year of comparison in calculating the levels of ozone-depleting substances (ODS) and contain the recommendations to the Parties on using, transferring, measuring and utilizing these substances. Some countries use 1990 as year of comparison. It should be noted that all the ODS defined in the protocol are synthetic, i.e. not presented in the natural environment.

A. Krasouski (✉)
Belarussian State University, 4, Nezavisimosti Avenue,
220030 Minsk, The Republic of Belarus
e-mail: krasovsky@bsu.by

S. Zenchanka
Minsk Branch of Plekhanov, Russian University of Economics,
127-2, Mayakovskogo Str., 220028 Minsk, The Republic of Belarus
e-mail: szenchenko@mesi.ru



W. Leal Filho et al. (eds.), *Handbook of Climate Change Communication: Vol. 1*,
Climate Change Management, https://doi.org/10.1007/978-3-319-69838-0_6

The Handbook (1991) for the Montreal Protocol on Substances that Deplete the Ozone Layer is an invaluable reference source for the decisions which the Parties have made in the process of developing the ozone regime. It contains the text of the Protocol with adjustments and amendments, all the decisions of the Meeting of the Parties starting since 1989, and supporting information.

ODS are used in a wide range of different applications such as refrigerators, air conditioners and fire extinguishers. They are also used as aerosol propellants, solvents and blowing agents for insulation foams.

The Kyoto Protocol (1998) extends UNCCC and defined the reducing the greenhouse gas emissions in the period 2008–2012. In accordance with article 2 of the Protocol the GHG concentration in an atmosphere must be at "a level that would prevent dangerous anthropogenic interference with the climate system". It outlined the basic rules, but did not provide details on their application. In accordance with this protocol the developed countries and countries with economies in transition took commitments to reduce or stabilize the emissions of greenhouse gases. Developing countries, including India and China didn't take any commitments.

The Kyoto Protocol includes flexibility mechanisms of implementation:

- International emissions trading,
- Clean development mechanism (CDM), and
- Joint implementation (JI).

The Kyoto Protocol was adopted until 2012 and then a new document should be signed. At a conference in Montreal (Stone 2006) a working group was established to discuss the items of protocol for the next period (2013–2017). The Cancun agreement (2010) defined that future global warming should be limited to below 2 °C. Doha Amendment (2012), which defines a second commitment period of the Kyoto Protocol, was agreed in 2012. Not all Parties of Protocol signed it and it didn't enter into force. Paris Agreement (2015) was adopted in December of 2015 by representatives of 195 countries. It defined a limiting global warming to less that 2 °C and the parties will endeavor to reach a limit the temperature increase to 1.5 °C. It was opened for signature on 22 April 2016. The agreement will enter into force after 55 countries that account for at least 55% of global emissions have deposited their instruments of ratification. G7 Ise-Shima Leaders' Declaration (2016) welcomes Paris Agreement on Climate Change and the decision in Dubai by the Montreal Protocol parties to work to address HFCs under the Montreal Protocol. The Kyoto Protocol, unlike the Montreal Protocol, refers to both natural and synthetic substances.

As the Vienne Convention as the UNCCC content demands on communication with their environment and between their Parties. These communications should be realized periodically. Communications help in disseminating knowledge on a community level and making decisions on a professional level. These communications do not often contain enough information on the mechanism of "ozone depletion—climate change" interaction and require additional research

## 2 Ozone Depletion and Climate Change

Ozone depletion and global warming are environmental challenges and their interaction and mutual influence are discussed at Vienna Convention, UNFCCC, Montreal and Kyoto protocols, scientific literature and different reports.

Article 2 of the Vienna Convention (1985) underlines the cooperation "with competent international bodies to implement effectively this Convention and protocols to which they are party". This cooperation means "systematic observations, research and information exchange in order to better understand and assess the effects of human activities on the ozone layer and the effects on human health and the environment from modification of the ozone layer" Article 3 say about the conducting of researches and scientific assessments not only the physical and chemical processes that may affect the ozone layer but "climatic effects deriving from any modifications of the ozone layer".

Vienna convention supports "cooperation with competent international bodies to implement effectively this Convention and protocols to which they are party".

The Montreal Protocol (2007) to the Vienna convention establishes demands to ODS. At the same time Article 7 of the Montreal Protocol supposes that "Each Party shall provide to the Secretariat statistical data on its production, imports and exports of each of the controlled substances"

As mentioned in a publication "The Montreal Protocol" (Handbook 1991) "The international response embodied in the Montreal Protocol has been widely regarded as the most successful environmental protection agreement ever reached to date. The Protocol has contributed significantly to reversing a problem with grave implications for life on earth".

UNFCCC and Kyoto Protocol (articles 2, 5, 7 and 10) implement and/or further elaborate policies and measures in accordance with its national circumstances, such as:

(i) Enhancement of energy efficiency in relevant sectors of the national economy;
(ii) Protection and enhancement of sinks and reservoirs of greenhouse gases not controlled by the Montreal Protocol, taking into account its commitments under relevant international environmental agreements; promotion of sustainable forest management practices, afforestation and reforestation;
(iii) Encouragement of appropriate reforms in relevant sectors aimed at promoting policies and measures which limit or reduce emissions of greenhouse gases not controlled by the Montreal Protocol;
(iv) Measures to limit and/or reduce emissions of greenhouse gases not controlled by the Montreal Protocol in the transport sector.

The Kyoto protocol considers GHG not included in Montreal protocol but which can be ODS. Some of these gases (HFCs) are considered as possible ODS. In April 2015 the EU has therefore proposed to amend the Montreal Protocol and phase down such gases at global level.

Most man-made ODS are very potent greenhouse gases. Eliminating these substances therefore has significant contribution to the fight against climate change.

Discussing climate models Nowack et al. (2015) pointed that "some processes, such as atmospheric chemical feedbacks, are still computationally expensive and are often ignored in climate simulations" and presented "evidence that the representation of stratospheric ozone in climate models can have a first-order impact on estimates of effective climate sensitivity". It was found that "using a comprehensive atmosphere–ocean chemistry–climate model an increase in a global mean surface warming of around 1 °C ( $\sim$20%) after 75 years when ozone is prescribed at pre-industrial levels compared with when it is allowed to evolve self-consistently in response to an abrupt 4 $\times$ $CO_2$ forcing".

UNEP (2010) report underlined that "There are strong interactions between ozone depletion and changes in climate induced by increasing greenhouse gases (GHGs). Ozone depletion affects climate, and climate change affects ozone". It pointed that "Model calculations indicate that the climate protection already achieved through the Montreal Protocol alone is larger than the reduction target of the first commitment period of the Kyoto Protocol (ending 2012)… It has been estimated that the cooling effect on the Earth's surface due to the phase-out of ODSs is approximately equal to the warming effect due to increasing GHGs at least through the first two decades of the 21st century. Thus, the Montreal Protocol has succeeded not only in reducing the content of chlorine in the atmosphere, but has also reduced global warming".

The Kyoto Protocol has similarities with the Montreal Protocol, but, addressing climate change, it is much more complicated than the phase-out of ODSs. If the Montreal Protocol refers mainly to consumers of ODSs, the Kyoto Protocol applies to all organizations that produce products and provide services, which is accompanied by the release of GHG.

The IPCC/TEAP (2005) "Special Report on Safeguarding the Ozone Layer and the Global Climate System" of the IPCC and the Montreal Protocol's Technology and Economic Assessment Panel (TEAP) contents a scientific assessment of the interrelations between the ozone layer and climate change. It contents different information on the ozone influence on climate change and vice versa. This report underlined that "Apart from their direct effect on climate, some gases have an indirect effect either from radiative forcing caused by their degradation products or through their influences on atmospheric chemistry".

Article (2015) "Ozone depletion and climate change" calls attention on important linkage between ozone depletion and climate change which involves the way that ozone-depleting substances and greenhouse gases alter certain processes in the atmosphere so as to enhance both global warming and stratospheric ozone depletion. These changes result in a warming of the troposphere and a cooling of the stratosphere. Stratospheric cooling is a key factor in the development of ozone holes over the poles.

Considering the benefits of the Montreal Protocol, UNEP (2011) report pointed that "The phase-out of ODSs under the Montreal Protocol avoided an estimated 10 Gt $CO_2$eq per year emissions in 2010, taking into account the ODS emissions that could have occurred in the absence of the Montreal Protocol, the offsets by ozone depletion and increases in HFC emissions. This figure for 2010 is about five times greater than the annual emissions reduction target for the first commitment period (2008–2012) of the Kyoto Protocol, estimated at 2 Gt $CO_2$eq per year".

# 3 Ozone Protection and Climate Change Communications

Climate change is a global issue, hence citizens, governments and businesses need accurate information about the problem and its causes. Communication, education, and outreach are powerful tools that government agencies, private organizations, and nonprofits can use to dispel misconceptions and to bring climate impacts and hazards to the attention of the public.

In accordance with Article 5 of the Vienna Convention "The Parties shall transmit, through the secretariat, to the Conference of the Parties established under article 6 information on the measures adopted by them in implementation of this Convention and of protocols to which they are party in such form and at such intervals as the meetings of the parties to the relevant instruments may determine". Scientific assessments of ozone depletion were published in 1989, 1991, 1994, 1998, 2002, 2006, 2010 and 2014. (https://www.wmo.int/pages/prog/arep/gaw/ozone_reports.html). Last "Assessment for Decision-Makers: Scientific Assessment of Ozone Depletion" was published in December 2014 (Assessment 2014), while the EEAP assessment report for 2014 was published in January 2015 (EEAP 2015).

These scientific assessments are based on national communications and expert reports. EEAP (2015) assessment underlined that "the previous SAP reports have served well the Parties to the Montreal Protocol, the scientific community, and the managers who deal with the research activities. However, the Montreal Protocol process has matured significantly and its needs have evolved". Proposed structure consists of two parts—Assessment Foundation from the Scientific Community which is a basis for Assessment for the Decision-making Community.

Article 7 of the Montreal Protocol obligates the Parties of Protocol "to provide statistical data to the Secretariat on its annual production (with separate data on amounts destroyed by technologies to be approved by the Parties), imports, and exports to Parties and non-Parties, respectively, of such substances for the year during which it becomes a Party and for each year thereafter. It shall forward the data no later than nine months after the end of the year to which the data relate."

UNEP Strategy (2010) underlines that "Information, communication and education (ICE) plays a vital and essential role as a strategic element of the Montreal Protocol implementation at national, regional and international levels by reinforcing investment activities, national policies and programs, and motivating changes in behavior among different targeted stakeholders".

UNFCCC considers communications as an important part of its realization. Parties of the Convention must present their Communications regular. These communications must content information about GHG inventory, Climate Change impact and adaptation, financial resources and technology transfer, education, etc.

Harvey et al. (2012) considered communication as a critical element in promoting effective and successful adaptation and mitigation strategies in Climate Change. They pointed that "originally presented as a complex and an abstract scientific problem, climate change information is increasingly being shared and discussed across disciplines and stakeholder groups at a range of scales". They consider communication as a continuum: "inform and educate individuals about climate change"—"achieve come type and level of social engagement/actions"—bring about changes in social norms and cultural values". There are different groups of aims for different social groups. Hence it is necessary to define principles of communication

As a part of the UK Climate Change Communications Strategy the principles of climate change communication were created by FUTERRA (2005) for the Climate Change Communications Working Group. 20 principles are groped in 6 themes:

1. Blowing Away Myths
2. A New Way of Thinking
3. Linking Policy and Communications
4. Audience Principles
5. Style Principles
6. Effective Management.

Ten strategies of climate change communication were suggested by Schweizer et al. (2009) using brainstorming exercise between scientists and local government:

1. Know your audience and select a believable messenger for that audience
2. Know why your message is appropriate for your audience
3. Connect your message to cultural values and beliefs
4. Make the message meaningful by appealing to values that are meaningful to your audience
5. Make the message, empowering with specific actions so your audience feels they can make a difference
6. Encourage your audience to think about the relationships and connections between climate, resources and themselves
7. Partner with other organizations
8. Start from the inside and involve your organization's leaders before communicating
9. Communicate about actions and use events as a mode of communication
10. Situate the issue in a specific location and place.

The IPCC Strategy (2012) has two main communications goals:

- To communicate its assessment findings and methodologies, by providing clear and balanced information on climate change 2, including scientific uncertainties, without compromising accuracy;

- To explain the way the IPCC works, selects its authors and reviewers and produces its reports and other products. This will promote the understanding of the reports and underpin its reputation as a credible, transparent, balanced and authoritative scientific body.

Communication plan as described by Raucher et al. (2014) can include:

1. Identify goals and objectives
2. Review guiding principle
3. Know your audience
4. Identify project impacts
5. Develop messages and create communication products
6. Evaluate the message.

It is seen that there are different approaches in development of climate change strategy and climate change plan. But Connor (2014) pointed that "Research has identified a number of psychological barriers that can prevent people from believing in or acting on messages about climate change". Hence, to overcome these barriers the communication strategy should be started from audience definitions and include next steps:

1. Choose audience. The volume of information that must be presented in communication is defined by the audience. Most full information is presented to science audience. Next group—decision making people. Last level—society, including higher education institutions. IPCC (Strategy 2012) defines its audiences as truly global in extent and therefore very diverse. In its communications and outreach activities, the IPCC will take the specific context of different countries into account, which may require tailor-made outreach activities. As stated in the Guide (2014) "understanding one's audience, where its members are coming from, and how they arrived there is the first key to unlocking success as a climate communicator".
2. Define goals and targets of communication. Strategy of communication is based on its aims. Goals and targets should be exactly defined for the specific audience. When goals and targets are known the audience can be defined.
3. Ground the strategy. There are many different approaches to ground the strategy and most of them are oriented to a specific audience.
4. Develop the plan. This plan must include responsibility, time-table, resources and methods of results dissemination.
5. Get a feedback. It is an essential part of a strategy which helps to develop and improve strategy and plan. As a strategy as plan should be presented to audience for discussing.

## 4 Inter-protocol Communications

Article 5 of the Vienne Convention (1985) underlines the importance of the transmission of information: The Parties shall transmit, through the secretariat, to the Conference of the Parties established under article 6, information on the measures adopted by them in implementation of this Convention and of protocols to which they are party in such form and at such intervals as the meetings of the parties to the relevant instruments may determine.

In accordance with Article 5 of the UNFCCC (Convention 1992) the Parties of Convention shall: "Support and further develop, as appropriate, international and intergovernmental programs and networks or organizations aimed at defining, conducting, assessing and financing research, data collection and systematic observation, taking into account the need to minimize duplication of effort".

The results presented in the last few years reports (Assessment 2014; EEAP 2015; Guide 2014; IPCC/TEAP 2005; Strategy 2010, 2012; UNDP 2014: UNEP 2010, 2011, 2015) shows the increasing importance of information exchange between the conventions and protocols.

These results underline the needs in further investigations of the interaction between changes in ozone concentration and climate change, in the development of physical models of such interaction and transfer these results to different grouped of specialists. These results are particularly important for decision-makers.

## 5 Some Results on Study of "Total Ozone—Climate Change" Interaction

Results of scientific researchers are an essential part of communication. Interaction "climate depletion—climate change" can work in both directions: changes in ozone can induce changes in climate, and vice versa. Conventions (1985, 1992) noted the importance of scientific researches for understanding of impact mechanisms of ozone depletion and climate change.

National Ozone Monitoring Research and Educational Center (NOMREC) was founded in 1997. Its main tasks are the monitoring of the ozone layer and other atmospheric components, measurement of surface ultraviolet radiation, the implementation of scientific research on the physics of the atmosphere, etc. It is responsible for overseeing the stratospheric ozone layer, the measurement of ground-level UV radiation and concentrations of ground-level ozone on the territory of Belarus. The observations are carried out in accordance with the commitments undertaken by the Republic of Belarus under the Vienna Convention for the Protection of the Ozone Layer.

Research results were included in National Communications of the Republic of Belarus on Vienna Convention and were reported on different conferences and symposium such as "Quadrennial Ozone Symposium of the International Ozone

Commission, 2016" (http://www.ozone-symposium-2016.org/QOS2016_programme_book.pdf).

Based on the results obtained in the center (for example, Turyshev et al. 2005; Krasouski and Turyshev 2011) the following mechanism of interaction "ozone depletion—climate change" can be formulated:

1. The ozone layer of the atmosphere, which is formed by the photophysic mechanism of ozone creation in the upper stratosphere by means of the hard UV radiation of the sun, is a major "transmission link" of impact of processes developing in the upper atmosphere on the troposphere. The solar activity modulates the state of the mesosphere while changing the ionic composition of the layer and the characteristics of atmospheric electric fields that affect the process of the formation of ozone in the upper stratosphere. Ozone actively distributes through the turbulent diffusion and gravitational settling, both vertically falling down as well as horizontally. The accumulation of ozone leads to heating of the stratosphere and the "sinking" of the tropopause layer. Many ozone—low tropopause (the polar zone).
2. Field of total ozone is formed by the competition of two mechanisms: the "radiation" mechanism which is responsible for the formation of ozone in the upper stratosphere and lowers the height of the tropopause, the other operates in the troposphere and called "thermodynamic". It creates conditions for the dry and moist instability and thereby raises the height of the tropopause and decreases the total ozone content.
3. "Thermodynamic mechanism", which consists in heating the underlying surface and the air heating in the unstable troposphere leads to "raise" the tropopause layer and increases the area of "negative altitude temperature gradient. The probability of destruction of ozone molecules by molecules of water vapor, hydroxyl ions and other destroyers, listed in the annex to the Vienna Convention for the Protection of the Ozone Layer", increases dramatically. Higher tropopause—a little ozone (tropical zone).
4. Mutual influence and the relative role of tropospheric ("thermodynamic") and stratospheric ("radiation") mechanisms vary according to region and season. Thermodynamic processes (convection) prevail in the tropical region. "Radiation" mechanism prevails in the unstable troposphere and in the polar latitudes.

Considering the past and present effects of ODSs and their substitutes on the Earth's climate and the ozone layer, IPCC/TEAP (2005) report make a conclusion that coincides with this concept: "Warming due to ODSs and cooling associated with ozone depletion are two distinct climates forcing mechanisms that do not simply offset one another". As was mentioned in UNEP (2015) report, "the phase-out of ODS, which are also greenhouse gases, will also have a positive effect on climate by avoiding their continued emissions into the atmosphere".

## 6 Conclusion

Presented results permit made next conclusions.

1. Montreal Protocols defines substances responsible for ozone depletion (ODS). All the ODS defined in the protocol are synthetic, i.e. not presented in the natural environment.
2. The Kyoto Protocol has similarities with the Montreal Protocol, but, addressing climate change, it is much more complicated than the phase-out of ODSs. The Kyoto protocol considers GHG not included in Montreal protocol but which can be ODS. Some of these gases (HFCs) are considered as possible ODS.
3. If the Montreal Protocol refers mainly to consumers of ODSs, the Kyoto Protocol applies to all organizations that produce products and provide services, which is accompanied by the release of GHG.
4. As the Vienna Convention for the Protection of the Ozone Layer, as the United Nations Framework Convention on Climate Change are directed to prevention of anthropogenic changes of environment and should transmit information to the Parties concerned.
5. In recent years, more and more publications are devoted to the elucidation of the mechanism of interaction of climate change, ozone change. But these mechanisms need to be clarified.
6. The analysis showed the role of communication and the need to develop communications strategies for various audiences, which should begin with the definition of audience.
7. The results of experimental studies on the mutual influence of changes in the concentration of ozone and climate are presented which can be summarized as follows:

   - While unstable thermodynamic processes in the troposphere (vertical convection) tend to raise the tropopause above, the impact of the ozone in the stratosphere mechanism (absorption of solar radiation and radiation heating) reduces its level.
   - Processes in the stratosphere, essentially associated with the processes in the mesosphere and ionosphere, changing the height of the tropopause in a given region, have an impact on the position of the stationary atmospheric fronts, separating global circulation cells.

These results further emphasize the need in inter-protocol communication for better understanding of the processes of ozone layer depletion and climate change.

## References

Article (2015) "Ozone depletion and climate change" Environment and Climate Change Canada https://www.ec.gc.ca/ozone/default.asp?lang=En&n=1662EC02-1/. Available June 2016

Assessment for Decision-Makers: Scientific Assessment of Ozone Depletion (2014) World Meteorological Organization, Global Ozone Research and Monitoring Project—Report No. 56, Geneva, Switzerland, 2014

Cancun Agreement (2010) http://cancun.unfccc.int/. Available June 2016

Connor P (2014) Climate change communication: key psychological research findings (and why you haven't heard about them yet. http://www.climatecodered.org/2014/04/climate-change-communication-key.html. Available September 2016

Convention (1985) The Vienna convention for the protection of the ozone layer. https://treaties.un.org/doc/Treaties/1988/09/19880922%2003-14%20AM/Ch_XXVII_02p.pdf. Available September 2016

Convention (1992) UNFCCC United Nations Framework Convention on Climate Change. FCCC/INFORMAL/84 GE.05-62220 (E) 200705. http://unfccc.int/resource/docs/convkp/conveng.pdf/. Available June 2016

Declaration (2016) G7 Ise-Shima Leaders' Declaration. http://www.mofa.go.jp/files/000160266.pdf/. Available June 2016

Doha Amendment to the Kyoto Protocol (2012) https://treaties.un.org/doc/Publication/CN/2012/CN.718.2012-Eng.pdf. Available June 2016

EEAP (2015) Protecting the ozone layer while also preventing climate change. Progress Report 2015. http://ozone.unep.org/en/assessment-panels/environmental-effects-assessment-panel. Available June 2016 United Nations Environment Programme, Nairobi, Kenya

FUTERRA (2005) The rules of the game principles of climate change communications. FUTERRA Sustainability Communications Ltd. http://www.stuffit.org/carbon/pdf-research/behaviourchange/ccc-rulesofthegame.pdf. Available June 2016

Guide (2014) Center for research on environmental decisions and ecoAmerica. Connecting on climate: a guide to effective climate change communication. New York and Washington, D.C

Handbook for the Montreal Protocol on Substances that Deplete the Ozone Layer (1991) http://ozone.unep.org/en/handbook-montreal-protocol-substances-deplete-ozone-layer/25411. Available June 2016

Harvey B, Ensor J, Carlile L, Garside B, Patterson Z, Naess LO (2012) Climate change communication and social learning–review and strategy development for CCAFS. CCAFS Working Paper No. 22. CGIAR Research Program on Climate Change, Agriculture and Food Security (CCAFS), Copenhagen, Denmark. Available online at www.ccafs.cgiar.org/. Available June 2016

IPCC/TEAP (2005) Special report safeguarding the ozone layer and the global climate system: issues related to hydrofluorocarbons and perfluorocarbons. In: Metz B, Kuijpers L, Solomon S, Andersen SO, Davidson O, Pons J, de Jager D, Kestin T, Manning M, Meyer L (eds) Cambridge University Press, UK, pp 478

Krasouski A, Turyshev L (2011) Interaction of "ozone-climate". Sci Innov 4:14–16 (in Russian)

Kyoto Protocol (1998) The Kyoto protocol to the United Nation framework convention on climate change. http://unfccc.int/resource/docs/convkp/kpeng.pdf. Available June 2016

Montreal Protocol (2007) European Commission, Montreal Protocol. Office for Official Publications of the European Communities, Luxembourg

Nowack PJ et al (2015) A large ozone-circulation feedback and its implications for global warming assessments. Nat Clim Change 5:41–45

Paris Agreement (2015) https://assets.documentcloud.org/documents/2646002/cop21-Final-Draft.pdf. Available June 2016

Raucher R, Raucher K, Leiserowitz A, Conrad S, Milan M, Dugan D (2014) Effective climate change communication for water utilities. Water Research Foundation, Denver

Schweizer S, Thompson JL, Teel T, Bruyere B (2009) Strategies for communicating about climate change impacts on public lands. Sci Commun 31(2):266–274
Stone SJ (2006) Comment on COP 11 to UNFCCC. Sustainable development law and policy. Winter, pp 45, 46, 67
Strategy (2010) OzonAction communication strategy for global compliance with the Montreal Protocol 2010. UNEP. http://www.unep.fr/ozonaction/information/mmcfiles/7440-e-Communication_Strategy.pdf. Available June 2016
Strategy (2012) Communication strategy decision taken with respect to the review of IPCC processes and procedures. IPCC 35th Session, 6–9 June 2012, Geneva, Switzerland
Turyshev L, Krasovski A, Zhuchkevich V, Barkouski N, Atrasheuski J, Tsvirko M, Sikorski V (2005) The development of the methods of short-time prognosis of the UV index near Minsk. Nat Resour 1:15–25 (in Russian)
UNDP (2014) Protecting the ozone layer and reducing global warming . www.undp.org/ozone/. Available June 2016
UNEP (2010) Environmental effects of ozone depletion: 2010 assessment. United Nations Environment Programme, Nairobi
UNEP (2011) HFCs: a critical link in protecting climate and the ozone layer. United Nations Environment Programme (UNEP), Nairobi
UNEP (2015) Synthetic of the 2014 report on the scientific, environmental effect, and technology & economic assessment panels of the Montreal Protocol. UNEP, Nairobi

## Author Biographies

**Aliaksandr Krasouski** is Associate Professor of Geography Departament of Belorusian State University, and has a Ph.D. in Physics. He is a founder of National Ozone Monitoring Research and Education Center of the Belarusian State University. His research focus is in the area of ozone monitoring, meteorology and ozone depletion-climate change interaction.

**Siarhei Zenchanka** is Professor of Economics Chair of Minsk Branch of Plekhanov Russian University of Economics, and has a Ph.D. in Applied Science. His research focus is in the areas of sustainable development, education for sustainable development. He is an auditor in environmental management.

# Communicating Climate Change: Theories and Perspectives

Henri-Count Evans, Lauren Dyll and Ruth Teer-Tomaselli

## 1 Introduction

Climate change has wide and severe negative impacts on humans, plants and the physical environment (IPCC 2013, Zakarya et al. 2015). Without serious global cuts in greenhouse gas emissions, climate change "poses great risks to human health, global food security, economic development and to the natural world on which much of our prosperity depends" (Zakarya et al. 2015). The Global South stands to be the worst affected by climatic changes—Africa and the small island nations (for example Indonesia and the Philippines) in particular have been seen to be the most vulnerable and exposed to climate change impacts "because of the range of projected impacts, multiple stresses and low adaptive capacity" (IPCC 2007). There is need for society to "take measures to adapt to these unavoidable impacts while taking action to cut greenhouse gas emissions" (Zakarya et al. 2015). There are generally two ways to respond to climate change.

The first, decarbonisation, is an attempt to reduce the quantity of greenhouse gas emissions to pre-1990 levels (IPCC 2007, 2013), essential if the globe is to try to prevent temperature increases beyond 1.5 °C by 2050 (IPCC 2013). The Intergovernmental Panel on Climate Change (IPCC) has recognised that failure to limit the temperature increase would exacerbate the already fragile conditions of the atmosphere marked by global warming, increased sea temperatures, extreme weather events, droughts etc. These impacts affect both the humanitarian and

H. -C. Evans (✉) · L. Dyll · R. Teer-Tomaselli
The Centre for Communication, Media and Society,
University of KwaZulu-Natal, Durban, South Africa
e-mail: henricount@gmail.com

L. Dyll
e-mail: dyll@ukzn.ac.za

R. Teer-Tomaselli
e-mail: teertoma@ukzn.ac.za

 
W. Leal Filho et al. (eds.), *Handbook of Climate Change Communication: Vol. 1*,
Climate Change Management, https://doi.org/10.1007/978-3-319-69838-0_7

ecological aspects of life with severe consequences in food security in the Global South, increased resource conflicts, threat to tourism and the preservation of biodiversity, and forces the diversion of budgetary resources from social services such as health and education to climate change-induced problems.

While Global South governments, often justifiably, have previously argued that they need not worry about reducing their emissions because the Global North countries are primarily responsible for the current environmental problems caused by industrialisation, recent data shows that this argument can no longer be sustained (Beck 1996, Cabello 2009). The developing Global South countries contribute significantly to overall global emissions, with the BRICS countries (Brazil, Russia, India, China and South Africa) together with the United States and the European Union, contributing the most towards global greenhouse gas emissions (RenataVidart Klafke et al 2015). This scenario makes relevant the call for global responsibility for the Global South countries to cut their emissions as they pursue economic development. The cuts will need to be supported by deliberate choices on the use of renewable energy, prioritising research and capacity into the renewable energy sources, investments in renewable energy technology and encouraging society-wide cuts in carbon footprints. The cuts in greenhouse gas emissions in the Global South have to be pursued but without stunting development projects. There is need for governments and citizens to change attitudes and behaviour, adopt environmentally friendly policies at individual, local, national and global levels by shifting from the use of fossil fuels to the use of renewable energy such as solar, hydro and wind.

The second response is to capacitate communities in adapting to climate change by adopting sustainable agricultural practices such as conservation agriculture, crop diversification, and the use of diverse drought resistant seed varieties.

In order for these two objectives to be achievable, there needs to be structural changes in society that allow for easier access to participatory forms of communication that enable the ordinary citizens, governments and the business sector to discuss and debate issues pertaining to climate change. Climate change communication is central to effective and sustainable mainstreaming of climate change in development policies, mitigation and adaptation policies, collective behavioural change, and more specifically attitudes towards climate change mitigation for improved efforts towards reducing greenhouse gas emissions. The communication media shape and construct "discourses and imaginaries circulating in cultural and political contexts" (Boykoff 2008a, b). This is made all the more pervasive with the advent of new media, and in particular the Internet, which promotes the multiplication of these spheres (Gitlin 1998), a point to be pursued later in this chapter. The communication media are actively involved in the construction of climate change as a social problem and building public perceptions and attitudes (Wilson 1993). Boykoff (2008a, b) strengthened this perception of climate change communication, seeing communication as a powerful "and heterogeneous set of non-nation state actors" where "discourses and imaginaries within spaces of cultural politics and geopolitics" are circulated, discussed and debated.

The global news media have become the global public sphere where issues in the international interests are discussed and debated. Climate change, being a worldwide challenge, has the potential to be addressed through an informed global citizenry. The news media, as argued by Todd Gitlin (1980), function as an authoritative version of everyday reality for everyday people and specialise in orchestrating consciousness for the people and thus are able to set the world climate-change agenda and influence global climate change decision-making processes and attitudes.

The media remain central to the provision of "public information" and in shaping people's perceptions, awareness and setting the public agenda regarding climate change (Carvalho 2010). Carvalho argued that the communication landscape should be diverse in terms of the participants, noting that participation in climate change communication needs to be equitable and to address the interests of diverse groups at all levels of society since only those voices that are active in the discussions get to shape action and policies relating to climate change. Thus far, attention to climate change has been elitist, with much emphasis on the global and mainstream transnational and national media, rather than equitable representation including diverse interests. Media coverage takes place in the "context of high-profile intergovernmental meetings and advance the notion that the global is the right political space for action" (Carvalho 2010). The global arena comprising a political and global elite alone cannot lead to collective social change for the environment: while mainstream media are central to the social construction of climate change, the construction "of climate change as a global political [problem creates a rift between] citizens and decision-makers" strengthening the idea that "climate change politics and action are reserved for the heads of the most powerful states" (Carvalho 2010).

The inclusion and participation of global, regional, national and local levels of society in climate change debate and action is critical to the collective sharing of global responsibility and empowerment. Thus our call for climate change communication that embraces participatory approaches, encourages citizen participation in discussions and decision-making and works towards a shared global climate-change action responsibility. The chapter does not document the pragmatic difficulties in the implementation of the participatory approach, but the authors are aware of their existence.

## 2 Communication Approaches Related to Climate Change Communication

Climate change communication has been theorised from varied paradigms. Approaches are adopted ranging from risk communication, development journalism, environmental communication, advocacy journalism and communication for social change and development. This section briefly discusses these approaches, citing their weaknesses when applied independently.

## 2.1 Risk Communication

Risk Communication is defined as the way in which "government agencies and organisations assess and manage risk and crisis situations, and how they communicate the nature of the crisis to stakeholders and members of the public" (PDE Office 2011, cited in Lie and Servaes 2015). John Sterman (2008) identified gaps between the "strong scientific consensus on the causes and risks of climate change" and the "widespread confusion and complacency among the public", viewing risk communication as "grounded in [a] deep understanding of the mental models of policy-makers and citizens". The success of risk communication relies "heavily on gaining cooperation from diverse countrywide entities" (Vaughan and Tinker 2009). The role of risk communication is similar to health communication as they both "instruct, inform, and motivate appropriate self-protective behaviour, update risk information, build trust in officials; and dispel rumours" (Vaughan and Tinker 2009). Highlighting the agentic nature of risk communication, they argued that:

> Ideally, pandemic communications maximise the public's capacity to act as an effective partner by encouraging prevention, promoting containment, and fostering resilience and recovery. Moreover, with an expanded role in pandemic planning and response, communication processes can prepare the public to adapt to changing circumstances or uncertainty during an emerging pandemic, educate public health planners about existing vulnerabilities [...] facilitate anticipation of surprising events, create dialogue between potentially affected populations and risk managers, and foster an environment of mutual trust.

Risk communication tends to be linear, from policy-makers and governments to the people who are affected. Instead of focussing on public understanding of risk, focus has been on the scientific expertise for climate change solutions (Sterman 2008). However, collective action is necessary to reduce emissions, and lobby governments to introduce and maintain policies that are designed to reduce greenhouse gas emissions and promote community capabilities for adaptation. Vaughan and Tinker (2009) recognised the need for the participation of the affected communities (though passively) when they noted that "the quality of the societal response depends partly on meeting the specific communication needs of all populations—especially those most vulnerable to the risks and most likely to experience communication gaps" (Vaughan and Tinker 2009). Thus, the risk communication approach, though useful in disaster situations, may not be as useful in the construction of collective knowledge and attitudes towards climate change, which is a long-term, sustained crisis. Climate change and social change communication need to build attitudes and perceptions that are socially shared, with emphasis on the participation of communities in climate change discussions and decision-making.

Health promotion and social marketing have been advanced as some of the ways to communicate risk information. However, these paradigms are criticised for their persuasive means that fail to consider serious contributors to health decision-making (Vaughan and Tinker 2009) and ignore the knowledge interests and power plays at stake in the context in which change is meant to occur (Thomas

2014). The problem with health and risk communication is that they focus more on the message producers and downplay the role of the audience, with whom meaning-production lies (Fiske 1989).

## 2.2 *Development Journalism*

Under the rubric of developmental journalism, Lie and Servaes (2015) saw journalists and international communication as useful contributors to sustainable development by delivering news, which is typically overlooked, to the public (Dixit 2010). Development journalism has its origins in postcolonial communication studies. The key ideas behind development journalism were conceived in the 1960s at a workshop of journalists in the Philippines (Xiaoge 2009). Development journalism was seen as an opportunity for journalists to evaluate the variations between planned development and the actual implementation (Aggarwala 1978), the idea being not to focus "on day-to-day news but on long-term development" (Kunczik 1988). It was linked to the calls for the New World Economic Order (NWECO) of the 1960s and 1970s and concerned with processes (not events) as opposed to the Western style of journalism, seeking cooperation between states/governments and the media.

Postcolonial governments in the Global South have tended to use the concept of development journalism to centralise power and to secure their authoritarian hegemony hiding under the notions of national development and the need to secure their nation-states from the perceived threats of cultural and media imperialism (see Galtung 1971, Schiller 1976). Robert White (1982) has described this as 'dissociation' theory.

## 2.3 *Advocacy Journalism*

Advocacy journalism positions journalists as the 'voice of the voiceless' speaking on behalf of those without "powerful spokesmen" (Janowitz 1975) and representing "specific interests, and motivated by the desire to redress power imbalances in society" (Waisbord 2009). The perspective of advocacy journalism has widened to include "organized groups that use the news media to influence reporting, and ultimately, affect public policies" (Waisbord 2009), which is different from the libertarian perspective role of journalism based on neutrality and objectivity.

Advocacy journalism shares close ties to political mobilisation. "[It]seeks to increase the power of people and groups, and to make institutions more responsive to human needs" attempting to "enlarge the range of choices that people have by increasing their power to define problems and solutions and participate in the broader social and policy arena" (Wallack 1993). Through social media platforms such as Twitter, Facebook and Whatsapp that facilitate citizen journalism, everyday man has been provided with the means to participate and contribute to the public debate as organised groups. Civic organisations, using the new media “aim to raise awareness, generate public debate, influence public opinion and key decision makers, and promote policy and programmatic change around specific issues” (Waisbord 2009).

On the negative side, due to specific groups having specific interests and access to the media (mass and interpersonal), advocacy journalism may promote partisan ideas and in the process subvert social change. Journalists and civic society organisers assume a privileged position of speaking for others, leading to the problems of people being ‘spoken about’ rather than ‘spoken with’ regarding their problems in a consultative manner—thereby entrenching the linear systems of communication. Civic organisations may have privileged access to the media, and climate change discussions, and decision-making platforms leading to lack of plurality of voices and the blurring of the distinctions between the journalist and the activist. In some cases, journalists may become pseudo-public relations practitioners of the civic movements and thus fail to play their watchdog role over both the government and the civil society.

## 2.4 Environmental Journalism

Environmental communication is a relatively new sub-discipline within communication sciences that includes the strategizing and analysis of “all interactions of human interaction with the environment” (Cox and DePoe 2015). The research on environmental communication arose first in the United States and Europe. Scholarship has moved through epochs of issues-attention, ideological representations, social constructions to an attempt to theorise climate change communication as a unique sub-discipline of communication and media studies. Cox (2007) defines environmental communication as “the pragmatic and constitutive vehicle for our understanding of the environment as well as our relationships to the natural world; it is a symbolic medium that we use in constructing environmental problems and negotiating society’s different responses to them”.

## 2.5 Communication for Development and Social Change

Communication for development and social change can be described according to two competing paradigms, firstly, diffusionist approaches (Rogers 1995) where communication is seen as the transfer of information focused on efficiency or effects, and secondly, the participatory paradigm that emphasizes bottom up initiatives (Servaes and Lie 2013; Tacchi and Lennie 2014; Thomas 2014; Tufte and Mefalopulos 2009; Waisbord 2014). In contemporary literature, ‘development’ is often replaced with ‘social change’ (Dutta 2011; Figueroa et al. 2002; Gumucio-Dagron and Tufte 2006; Servaes 2008). The term accounts more closely for the intangible processes and manifestations of changes that are called for within the participatory approach, such as voice, dialogue, agency and mobilization. This is opposed to the connotations of infrastructure, economic growth and authoritarian Western communication processes which are often automatically associated with the term ‘development’, that emanate from the dominant paradigm of which diffusionist thinking was a part (Dyll-Myklebust 2014).

The value of the participatory approach is that it “stresses the importance of cultural identity of local communities and of democratization and participation at all levels—global, international, national, local, and individual. It points to a strategy not merely inclusive of but largely emanating from the traditional receivers” (Servaes and Lie 2013). The central concept in multiplicity approaches is the need for dialogue and “actions that are essential for the process of conscientization, which leads to emancipatory social change” (Servaes and Lie 2013). Communication for development and social change considers change to be a “social process, which has its ultimate objective sustainable development at distinct levels of society. Communication media and ICTs are important tools in achieving this process but their use is not an end in itself. Interpersonal communication must also play a fundamental role” (Servaes 2007).

Servaes and Lie (2013) argue that the starting point should be the community where social issues are discussed. Environmental issues are one example. Mefalopulos (2005) argued that in social change, the key role of communication is to ensure people’s participation, noting that participation and empowerment remain the key “pillars of communication for sustainable development”. Freire (1997) saw empowerment as the ability of individuals to liberate themselves from relationships and structures of bondage. Narayan (2002) defined empowerment as “the expansion of assets and capabilities of poor people to participate in, negotiate with, influence, control, and hold accountable, institutions that affect their lives”.

Servaes and Lie (2013) call for the adoption of both participatory communication and communications for structural and sustainable social change (CSSC) as these are “more concerned about long-term sustained change at different levels of society”. The section below provides an overview of how communicating climate change could adopt a structural approach that appreciates scalar relations of existence; the global, the regional, the national and the local; and that foregrounds the ‘glocal’ with beneficial interdependency between the global and the local.

## 3 Towards a Normative Structural and Integrated Approach

The theoretical perspective adopted here views all forms of media and communication as central and interdependent, with emphasis placed on people’s participation and empowerment in decision-making and communication processes. The communication theories outlined above are relevant in various circumstances and have shortcomings in others.

Effective communication must embrace participation in order to promote inclusivity and fairness in climate change discussions and decision-making. The global level becomes a platform for collaboration, where the global media's ideal role is to inform sustainable global public policy through coverage of climate issues and response mechanisms from different regions in the world. At the regional level, the transnational media becomes a force for fostering regional policies compatible with regional aspects of sustainability (Thussu 2010) and cover issues that come from both the global and regional public spheres. National media informs national climate change policies and actions, while at the local level community communication affords local communities the opportunities to involve themselves in matters that deal with their local climatic conditions.

The media, while providing a platform for engagement and empowerment, also work as a watchdog, holding accountable the governments, policymakers and citizens at all levels of existence. The media evaluates, for example, the mitigation ambitions as proposed by different countries under the auspices of the United Nations Framework Convention on Climate Change (UNFCCC), comparing what is supposed to be done against what is actually happening. This role also involves making global political and business elites accountable to the electorates. The media is also instrumental in monitoring and exposing activities of industries that are responsible for carbon emissions, for example, oil companies and coal-based industries.

Borrowing from development journalism, climate change communication should engage and empower people in the process of climate change policy formulation, mitigation action and adaptation solutions. This may also be of benefit when local groups begin to apply indigenous knowledge systems to solving their climate change problems. Development journalism theorists make calls to empower the ordinary people (Xiaoge 2009). Servaes (1999) argued that for real social change to take place ordinary people should be collectively afforded the opportunity to make decisions concerning their lives and development at all levels of society (Fig. 1).

The theoretical underpinnings of participatory and integrated structural climate-change communication model are illustrated in Figure One above. The media, in all forms, remain central to the communication process; where the media act as the marketplace for ideas, provide a platform for engagement and debate. In this ideal context, the participants in the discourse including policymakers, ordinary citizens, experts/scientists, civil society and the media themselves engage equally and in a dialogic manner, based on principles of reciprocity. Ideas from the experts/scientists, policymakers, civil society and the ordinary people, enter the media for discussion with an equal opportunity for engagement and debate. The media become the meeting point of ideas from all participants without privileging one group at the expense of the other. This is the normatively ideal situation for any form of the media, be it mass media, interpersonal communication, community media or social media. The approach encourages equality and inclusivity of all ideas and perspectives on actions in relation to climate change.

No single model of communication can best be utilised in communication for development and social change (Servaes 1999). There is thus a need for a hybrid

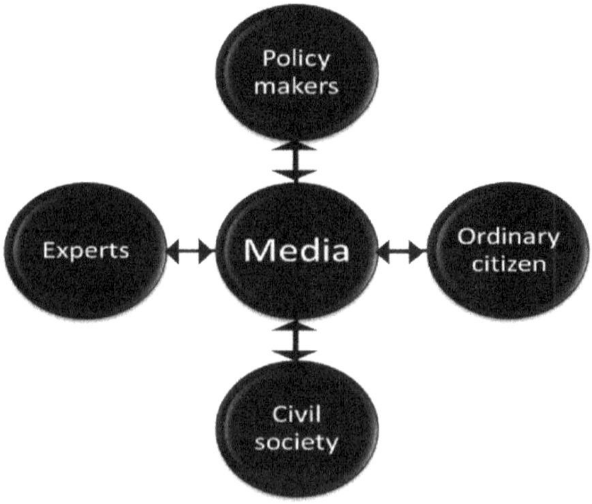


**Fig. 1** Normative integrated structural and participatory model for climate change communication (SPCCC). *Source* Developed by authors

climate-change communication model that builds on the advantages of both bottom-up and top-down processes. At the centre of this approach are three key provisions. Firstly, the approach must be participatory; secondly, it should be plural and lastly must embrace Information and Communication Technologies (ICTs), notably the Internet and the social media. Traditionally, the mass media such as television and newspapers have not enabled participation because of linear limitations; however, with the proliferation of new technologies, especially the Internet, opportunities exist for citizen participation in message production and consumption (Castells 2012).

Despite the evolution of communication for development and social change paradigms, the linearity of traditional media has made the media elitist, where the official/expert view is primarily considered, thus limiting ordinary people's participation in the communication process. This is true for climate change communication. The problem is exacerbated by the scientific nature of climate change. News reporters use scientific experts, business elites, politicians, and civic society organisations as news sources; that all talk about the problems affecting ordinary people from their point of view and agenda. Figure Two highlights the current trends in climate change communication, where the ordinary people are made passive recipients of media products. The communication process has remained a monopoly for the journalists, scientists, politicians and civic organisations, whose views have found prominence in the global climate change discourse. It can be argued that the poor reception and understanding of climate change among ordinary people, or their lack of climate action thereof, are linked to their non-involvement in the discourses that they should be active participants in and contribute to decision-making processes (Evans 2015).

We consider participation in climate change communication processes as central to the empowerment of ordinary people in the collective responsibility of influencing climate change decision-making processes locally and globally.

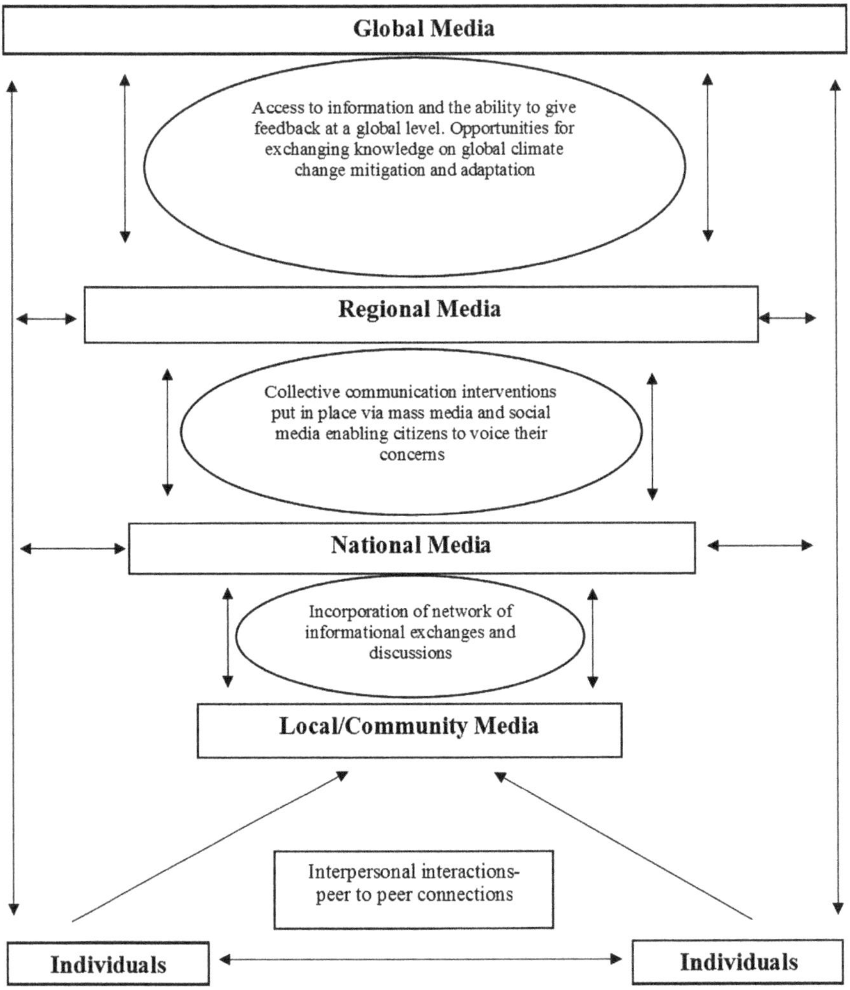


**Fig. 2** Schematic model for integrated structural and participatory model for climate change communication. *Source* Developed by authors

Figure 2 is cognisant of the different levels of existence and the contexts that influence individual and collective social action towards climate change. As illustrated above, the communication process is two-way allowing for the exchange of information and ideas, participation discussions and debates and the schematic model presented in Fig. 2 can be read from the bottom up or top down.

The need for an integrated structural approach is necessary with the realisation that no one single paradigm can lead to effective communication and no one explanation or methodology can be applied considering the complexity of the world today, issues of globalisation and interdependency, and the proliferation of ICTs across the globe (Servaes 2007, Castells 1996). Communicating climate change must cater for these complexities. While participation alone is not sufficient to achieve collective social change and urgent interventions are required for immediate crises, long term goals such as climate change need the active involvement of all sectors of the community.

The integrated approach builds on the potential of several approaches that incorporate interpersonal communication, mass communication, development journalism, and the new media of communications, especially social media. These distinct forms of communication when combined, ideally lead to participatory communication for structural and sustainable social change. A social ecological (Crepaz et al. 2006) or integrated structural approach takes a systems perspective to analysis that allows for a description of the complexity and interrelatedness of communication rather than just one isolated component. This may assist to establish the big picture of what is necessary to communicate around climate change.

## 4 Social Media for Climate Change Communication

Information and Communication Technologies (ICTs) have played a crucial role in support of communication for development and social change. Social media, though not an obvious route to large-scale development projects, is necessary in "encouraging small steps", noting that "digital initiatives may impart pro-environmental and other social values and help shape consumer preference for greener or alternative lifestyles. In other words, social media on their own will not necessarily save the planet" [however] "they might just encourage a more sustainable future" (Lie and Servaes 2015). Social media are useful vehicles for the sharing and exchange of information and to stimulate change at the global, regional, national, and local levels (Kavada 2014, 2012).

New media are the perfect examples of media convergence (text, audio and video), and this technological convergence has evolved into the convergence of interactions. It has enabled interactions between individuals and groups from different global regions, united by the same interests, and that these interactions have involved people from the Global North and those from the Global South on a horizontal level. To a small extent, therefore, these interactions have thus blurred the bi-polar distinctions of geopolitics, leading to the collapse of geographical boundaries, convergence of interactions, collapsing global and physical walls. This has resulted in dual communities that are both physical and virtual, participating in both the global and the local. Climate change communication needs to address these two communities by enabling participation, be it physical or virtual.

### 4.1 Interactions of the Global and the Local

Within the highly-globalised world, marked by social movements and conflicts, there is need for communications that focus on social change (Hemer and Tufte 2005). 'Development' and 'social change' are no longer a concern only for the 'developing Global South' but rather a phenomenon affecting all humanity (Pieterse 2001). The need to rethink the development-social change couplet as a "regional, transnational, global project" is thus necessary (Hemer and Tufte 2005). This global project is facilitated by the Internet, as "the backbone of the network society" (Castells 1996, 2001). Regardless of advances in technology and tele-density in the Global South, issues of access to technology are still pertinent. Through the ICT divide, members of the same locality (community—physically) are divided by their access or lack of it, in their participation in issues that affect and concern them. Distinctions between technology elites and 'have-nots' within the same countries or even the same communities inhibit full participation. These divides limit people's access to the platforms of community participation and engagement.

The proliferation of ICTs has resulted in a scattered diaspora "linked in transnational public spheres, which are undermining the 'imagined communities' of the national media" (Hemer and Tufte 2005). Borrowing from Arjun Appadurai (1996), Hemer and Tufte saw the new drivers of social change being what Appadurai called "grassroots globalization" that are characterized by "transnational advocacy networks" strengthening civil society and social activism. The transnational media thus have the capacity to "reinforce cultural (and national) identities, but transcultural processes are also a central feature of reflexive global modernity, expressed as 'creolization' or 'cultural hybridity" (Hemer and Tufte 2005).

Thomas Eriksen (2005) observed that fixed identities are now difficult, noting, "a fixed, stable, meaningful space, is becoming a scarce and flexible resource. Maintaining a predictable and secure group identity is hard work these days". He viewed "globalisation as all the socio-cultural processes that contribute to making distance irrelevant" (Eriksen 2005). Identity political problems in Europe are exemplified most recently by 'Brexit' (the 2016 British vote to exit the European Union) as well as in the United States, where nationalist/populist rhetoric has grown in recent years. Where proponents of the classical globalisation theory saw the global coming together, something independent of, but a result of globalisation has emerged with strong sentiments for the nation-state, parochialism and identity politics (Pieterse 2001).

## 5 Conclusion

This chapter set out theoretical perspectives that work towards climate change communication in the Global South by providing an array of communication approaches used in communicating climate change, and highlighting their strengths and weaknesses. No individual communication approach can be effective in communicating climate change, necessitating the use of an integrated and structural participatory approach that has as its premise people participation and empowerment. The authors are cognisant of the difficulties in the implementation of participatory approaches and offer a normative theoretical approach for communicating climate change in the Global South. Such an approach, appreciating scalar relations of existence and foregrounding the 'glocal', could advance a two-way exchange of knowledge and information on climate change.

This chapter endeavours to expand the scope of analysis and new research surrounding climate change communication, especially with reference to Global South. Seeing as there are rapid changes in the field of social change and development, the authors push forward the need for the movement towards stronger participatory communication approaches to capacitate communities in adapting to climate change. The chapter emphasises the role of climate change communication as a key pillar in influencing action and policy formulation that will meet the needs of diverse groups at all levels in society.

Some pertinent questions that have been identified as relevant for future research are: In what ways, can the media be practically equipped to enable participation both in the production and consumption of climate change related content? With the proliferation of social networking sites and the increased convergence of human interactions, how can communities at the grassroots levels be better incorporated into discussions about climate change in a way that breaks the barriers of the digital divide? In terms of theoretical grounding, how can the normative structural and integrated approach be fine-tuned to enable effective communication in various local and global regions in the world?

## References


Aggarwala NK (1978) A third world perspective on the news. Freedom at Issue, 13–20

Appadurai A (1996) Modernity at large: cultural dimensions of globalization vol 1. University of Minnesota Press

Beck U (1996) World risk society as cosmopolitan society? Ecological questions in a framework of manufactured uncertainties. Theory Cult Soc 13(4):1–32

Boykoff MT (2008a) The cultural politics of climate change discourse in UK tabloids. Polit Geogr 27(5):549–569

Boykoff MT (2008b) Lost in translation? United States television news coverage of anthropogenic climate change, 1995–2004. Clim Change 86(1–2):1–11

Cabello J (2009) The politics of the clean development mechanism: hiding capitalism under the green rug. In: Böhm S, Dabhi S (eds) Upsetting the offset: the political economy of carbon markets, pp 192–202

Carvalho A (2010) Media(ted) discourses and climate change: a focus on political subjectivity and (dis) engagement. Wiley Interdisc Rev Clim Change 1(2):172–179

Castells M (1996) The network society vol 1–3. Oxford: Blackwell

Castells M (2001) Virtual communities or network society. The Internet galaxy: reflections on the Internet, business, and society, 116–136

Castells M (2012) Networks of outrage and hope: social movements in the internet age. Cambridge, UK; Malden, MA: Polity Press

Cox R (2007) Nature's "crisis disciplines": Does environmental communication have an ethical duty? Environ Commun 1(1):5–20

Cox R (2012) Environmental communication and the public sphere. Sage publications

Crepaz N, Lyles CM, Wolitski RJ, Passin WF, Rama SM, Herbst JH, Purcell DW, Malow RM, Stall R (2006) Do prevention interventions reduce HIV risk behaviours among people living with HIV? a meta-analytic review of controlled studies. AIDS 20(2):143–157

Dixit K (2010) Dateline earth: journalism as if the planet mattered. IPI Asia-Pacific Foundation Inc, Bangkok

Dutta MJ (2011) Communicating social change: structure, culture, and agency. Routledge, New York

Dyll-Myklebust L (2014) Development narratives: the value of multiple voices and ontologies in Kalahari research. Crit Arts 28(3):521–538

Eriksen TH (2005) How can the global be local? Islam, the west and globalization of identity politics In: Hemer O, Tufte T (eds) Media and glocal change-rethinking communication for development, 26

Evans HC (2015) Climate-change awareness and online media in Zimbabwe: opportunities lost? In Halvorsen T, Ibsen H, M'kumbuzi VRP Knowledge for a sustainable world: a Southern African-Nordic contribution, 163

Figueroa ME, Kincaid DL, Rani M, Lewis G (2002) Communication for social change: an integrated model for measuring the process and its outcomes. The Communication for Social Change Working Paper Series No. 1. The Rockefeller Foundation and CFSC, New York.

Fiske J (1989) Reading popular culture. Hyman, London

Freire P (1997) A response. Mentoring the mentor: a critical dialogue with Paulo Freire, 60, 303–329

Galtung J (1971) A structural theory of imperialism. J Peace Res, 81–117

Gitlin T (1980) The whole world is watching: mass media in the making & unmaking of the new left. University of California Press

Gitlin T (1998) Public sphere or public sphericules? In: Liebes Tamar, Curran James, Katz Elihu (eds) Media, ritual and identity. Routledge, London, pp 168–174

Gumucio-Dagron A, Tufte T (2006) Communication for social change anthology: historical and contemporary readings. Communication for social change consortium New Jersey

Hemer O, Tufte T (2005) The challenge of the glocal. Glocal Times(1) http://biblioteca.clacso.edu.ar/clacso/coediciones/20100824061923/glocal.pdf

Intergovernmental Panel on Climate Change (2007) Summary for Policymakers. In: Climate change 2007: mitigation. contribution of working group III to the fourth assessment report of the Intergovernmental Panel on Climate Change. Cambridge University Press, Cambridge, United Kingdom and New York, NY, USA

Intergovernmental Panel on Climate Change (2013) Summary for policymakers. In: Climate change 2013: the physical science basis. Contribution of working group I to the fifth assessment report of the Intergovernmental Panel on Climate. Cambridge University Press, Cambridge, United Kingdom and New York, NY, USA

Janowitz M (1975) Professional models in journalism: The gatekeeper and the advocate. Journalism Mass Commun Q 52(4):618

Jeffrey JS (2008) Networking futures: the movements against corporate globalization. Experimental futures. Duke University Press, Durham, N.C

Kavada A (2012) Engagement, bonding, and identity across multiple platforms: avaaz on facebook, youtube, and myspace. MedieKultur—J Media Commun Res

Kavada A 2014 Transnational Civil Society and Social Movements. In: Wilkins KG, Tufte T, Obregon R (eds) The handbook of development communication and social change. Global handbooks in media and communication research. Chichester, West Sussex, UK; Malden, MA, USA: Wiley-Blackwell. http://alltitles.ebrary.com/Doc?id=10830394

Klafke RV, de Barros Ahrens R, da Silva RG, Pilatti LA, de Francisco AC (2015) Air pollution indicators in Brazil, Russia, India and China (BRIC) countries. Sci Res Essays 10(16):513–521
Kunczik M (1988) Journalismus als Beruf: Böhlau Köln, Wien
Lie R, Servaes J (2015) Disciplines in the field of communication for development and social change. Commun Theory 25(2):244–258
Mefalopulos P (2005) Communication for sustainable development: applications and challenges. Media and glocal change. Rethinking communication for development, 247–260
Narayan D (2002) Empowerment and poverty reduction. World Bank, Washington, DC
Pieterse JN (2001) Globalization and collective action Globalization and social movements pp 21–40. Springer
Rogers EM (1995) Lessons for guidelines from the diffusion of innovations. Jt Comm J Qual Improv 21(7):324–328
Schiller H (1976) Communication and cultural domination. International Arts and Sciences Press Inc, White Plains
Servaes J (1999) Communication for development: one world, multiple cultures. Hampton Press (NJ)
Servaes J (2007) Harnessing the UN system into a common approach on communication for development. Int Commun Gaz 69(6):483–507
Servaes J (2008) Communication for development and social change. Sage Publications India
Servaes J, Lie R (2013) Sustainable social change and communication. Communication Research Trends
Sterman JD (2008) Risk communication on climate: mental models and mass balance. Science 322 (5901):532–533
Tacchi J, Lennie J (2014) A participatory framework for researching and evaluating communication for development and social change. In: Wilkins KG, Tufte T, Obregon R (eds) The handbook of development communication and social change. Wiley, London, pp 298–320
Thomas PN (2014) Development communication and social change in historical context. In: Wilkins KG, Tufte T, Obregon R (eds) The handbook of development communication and social change. Global Handbooks in Media and Communication Research. Chichester, West Sussex, UK; Malden, MA, USA: Wiley-Blackwell
Thussu DK (2010) International communication: a reader. Routledge, Abingdon
Tufte T, Mefalopulos P (2009) Participatory communication: a practical guide. World Bank working paper, no. 170. The World Bank, Washington DC
Vaughan E, Tinker T (2009) Effective health risk communication about pandemic influenza for vulnerable populations. Am J Public Health 99(S2):S324–S332
Waisbord S (2009) Advocacy journalism in a global context. In: Wahl-Jørgensen K, Hanitzsch T (eds) The handbook of journalism studies. Routledge, New York & London 371
Waisbord SR (2014) The strategic politics of participatory communication. In: Wilkins KG, Tufte T, Obregon R (eds) The handbook of development communication and social change. Global Handbooks in Media and Communication Research. Wiley Blackwell, Chichester, West Sussex, UK Malden, MA
Wallack L (1993) Media advocacy and public health: power for prevention. Sage.
Wilson K (1993) Learning about Global Warming from the media: a study of print and electronic media as sources of geographic knowledge. University of Colorado, Colorado
White RA (1982) Contradictions in contemporary policies for democratic communication. Paper presented at IAMCR Conference, Paris, France
Xiaoge X (2009) Development Journalism. In: Wahl-Jørgensen K, Hanitzsch T (eds) The handbook of journalism studies. Routledge, New York & London, 357–370
Zakarya GY, Mostefa B, Abbes SM, Seghir GM (2015) Factors affecting CO2 emissions in the BRICS countries: a panel data analysis. Procedia Econ Financ 26:114–125

## Author Biographies

**Henri-Count Evans** is a Ph.D. Candidate at the University of KwaZulu-Natal in the Centre for Communication, Media and Society (CCMS). He has done research work pertaining to media practice and reporting of climate change issues and sustainable development. The research covered both the print, electronic media and the new media of communication in Zimbabwe, Southern Africa and the world at large. Henri-Count Evans has presented at several academic conferences and has published two book chapters and is currently working on upcoming peer-reviewed journal articles and conference proceedings.

**Dr. Lauren Dyll** is a Senior Lecturer in the Centre for Communication, Media and Society (CCMS) at the University of KwaZulu-Natal (UKZN) teaching graduate modules in research methodology, media theory, and social change and health communication. Her research interests and publication topics include: participatory communication (including health and environment), issues of identity, and critical indigenous qualitative research. She has been a key contributor to Prof Tomaselli's Rethinking Indigeneity project since 2002, and is a committee member and researcher in the international Consuming Environment research project between CCMS, UKZN and Högskolan i Gävle (HiG) (Sweden).

**Ruth Teer-Tomaselli** is a full-professor in the Centre for Communication, Media and Society (CCMS) and a University Research Fellow at the University of KwaZulu-Natal. She has published widely on South African media. Her research interests include research methodology and social theory in development; public service broadcasting; media history; and global media. Prof Teer-Tomaselli has been a Vice-President of the International Association for Media and Communication Research (IAMCR) and holds the UNESCO Chair in Communications for South Africa.

# Reconsidering Fictional Films for Communicating Climate Change Issues: An Analysis of the Filmmaking Strategies Behind Sustainable Energy Narratives

Michela Cortese

## 1 Introduction

In the last twenty years, there has been a growing recognition of the importance of visual representations across different academic areas. In particular, there has been an increasing effort among environmental scientists and engineers to collaborate with media scholars to share their knowledge and find new approaches and strategies for communicating issues of social relevance. The issues that have principally led, or even forced, these fields to come together are probably the ones related to climate change. Climate change has been the topic of numerous public, scientific and political debates and the media have demonstrably played a crucial role in framing climate change as a social, cultural and economic phenomenon (Anderson 2009).

The aim of this chapter is to encourage scholars to further analyse climate change films as part of the already existing literature on climate change visuals, and it will highlight the importance of reconsidering fictional films for effectively engaging people in the climate discourse. I explore the reasons why it would be a good idea to invest more in film productions as an instrument to communicate scientific findings regarding climate change. I underscore this through a review of the literature relating to the role of films and narrative from the perspective of cognitive and psychology studies, a very useful account for those who work in academia but also wish to undertake a project that involves the production of a film.

Besides acknowledging the fundamental role of films and noting their ability to raise emotions and initiate an active response among their audience, the last part of this chapter will provide a brief exposition of an environmental project that has been developed between members of the Department of Creative Studies and Media

M. Cortese (✉)
School of Creative Studies and Media, Bangor University, Bangor,
Gwynedd LL57 2DG, UK
e-mail: michela.cortese@bangor.ac.uk

 
W. Leal Filho et al. (eds.), *Handbook of Climate Change Communication: Vol. 1*,
Climate Change Management, https://doi.org/10.1007/978-3-319-69838-0_8

at Bangor University and the Department of Engineering at Swansea University. The project, part of a research grant from the *Sêr Cymru, Welsh National Research Network in Advanced Engineering and Materials* (*NRN*) and the *UK Engineering and Physical Sciences Research Council (EPSRC)*, involves the production of short fiction film, which will frame climate change through a metaphor on sustainability, rather than using a factual narrative or scientific data and statistics.

## 2 A Review of the Literature on Visual Representation of Climate Change in the Media

If we look specifically into the dimension of environmental communication and risk communication, textual analysis of media coverage of climate change demonstrates substantial progress in the last two decades but little research has been done on visual representations, especially on documentaries and fictional films (Hansen and Machin 2013; O'Neill and Smith 2014; Hansen and Cox 2015). Although there is a growing body of research concentrating on claiming the importance of images in environmental studies, less research has been done for investigating the consumption of climate change visuals, and there are even fewer studies dealing with how these images are constructed and produced (O'Neill and Smith 2014; Hansen and Machin 2013).

Visual representations of the environment in the media are all part of a cultural process that shapes our way of seeing. As receivers of the messages/images sent through the media every day, we assume that these are 'naturally' given. Certain environmental images are so largely distributed and rooted within our culture that we forget the fact that they are socially constructed. So, how specifically do scholars approach the issues around the production, distribution and consumption of those images which aim to represent the risks that our world would face due to climate change?

The body of research that analyses visual representations of the environment, has drawn on some iconic studies which try to tell us how visual climate discourse is represented in newspapers/magazines (Smith and Joffe 2009; DiFrancesco and Young 2011; O'Neill 2013; Nerlich and Jaspal 2013), television and films (Cottle 2000; Lester and Cottle 2009; Mellor 2009; Lester 2010; Brereton 2016) and advertising/campaigns (Hansen and Machin 2008; Linder 2006; Doyle 2007). The general view of some of these studies, mainly conducted through content and thematic analyses, is that the media tend to represent the environment "through the use of increasingly decontextualized global, symbolic and iconic images" (Hansen and Machin 2013: 156), rather than applying those elements which are identifiable according to their cultural and geographical features.

According to the set of studies that analyse the visual content of articles regarding climate change in the UK, US and Australian newspapers and magazines, there are specific groups of images which frequently occur across these different media items: (1) *impact of climate change (e.g. 'before' and 'after' pictures of retreating glaciers, polar bears struggling to swim to safety)*, (2) causes of climate change (e.g. smokestacks, deforestation, transport and fossil fuels), (3) *personification of climate change (e.g. politicians and celebrities)*, (4) *representation of climate change through graphs*. There are however some variations according to the country of publication. For example, O'Neill (2013) found that UK newspapers tend to visually frame climate change through the use of celebrities more than Australia and US, which in turn employs a large number of images of politicians. Furthermore, O'Neill found considerable visual coverage of climate change impacts in the UK compared to Australia and US.

According to Smith and Joffe's study, the use of images of melting polar ice caps and polar bears struggling in the Arctic would leave the audience with a sense of "desperation and futility" (Smith and Joffe 2009) more than engaging them with the issue. 'Members of an affected public' would play an equally important role as part of the personification frame and images of extreme weather events, for example those images related to Hurricane Katrina or flooding in British towns, "provide the mass media with an attractive hook with which document tangible examples of alleged climate change impacts" (Smith and Joffe 2009). Smith and Joffe also highlighted that broadsheet newspapers are more likely to include images of climate change impacts and graphs whereas tabloids are keener on pictures of people (O'Neill and Smith 2014). Another pattern which has been observed by O'Neill is that images of adaptation or mitigation are rarely featured, along with visuals on sustainable energy which are almost never employed.

In another study of images accompanying media reporting of the 2011 IPCC report, Nerlich and Jaspal (2013) also highlighted the importance of visual representations of extreme weather and their role as a means of demonstrating that climate change is happening. Images of extreme weather are extremely relevant in the analysis of visual environmental communication due to their power of evoking and "symbolising emotional responses, such as compassion, fear, guilt, vulnerability, helplessness, courage or resilience" (Nerlich and Jaspal 2013). In their analysis on various categories of environmental pictures (e.g. floods, heat, hurricanes, ice and sea level images), most of the images appeared to have largely negative emotional meanings and "might therefore not be conducive to engagement with the implicitly constructed antecedent of extreme weather, namely climate change" (Nerlich and Jaspal 2013).

Although most of the pioneer studies in risk and environmental communication concentrated on a visual content and thematic analysis of print and online press, they also reflect, to a certain extent, the same tendencies of visual representations of climate change in TV news and documentary films. For instance, in their investigation of TV news, Lester and Cottle (2009) remind us that climate change could have never become "the perceived 'global crisis' that it has in recent years without the help of media images symbolizing the harmful impacts of climate change on people, communities, and environments around the world" (Lester and Cottle 2009). Their study, which is based on an analysis of a two-week period of daily and high-rating news programs in six Anglophone countries and four satellite TV broadcasters, underlined three categories of news visuals that would match those presented in the aforementioned studies on newspaper visuals: iconic visuals, symbolic visuals (e.g. visuals of black smoke pouring out of factories and images of the globe), and spectacular visuals (e.g. visuals focusing on the destructive force of extreme weather events).

The results gathered from this study showed that more than 50% of TV news coverage of climate change relies on symbolic and spectacular visuals to bring the threat and reality of global climate change closer to the audience (Lester and Cottle 2009). Lester and Cottle also remark the significant role of the 'moving images' on TV, which are particularly constructed for reinforcing the connection between people's lives and nature. In particular, they refer to scenes of private domesticity within the threatened landscape. They observed that there is a very frequent use among TV news of images of families preparing meals while showing distant exterior shots, which would reinforce the idea of imminent risk, or family members looking at the pictures of the victims of a natural disaster. Moreover, according to this study, the constant and parallel use of the same visual sequences on the causes and impact of climate change across different TV news, seems to encourage the implicit connection for the viewers with climate change in the foreseeable future.

Perhaps the most significant contribution of Lester and Cottle's work, a point which was overlooked in the analysis of newspapers' visuals, is their argument on *trust* and *credibility*. Along with the debate on accuracy and authorial bias, these four elements can be particularly useful in the analysis of film and documentary and can open the door to further debate on whether fictional films would be a more suitable means of communication in some instances, as fictional narratives would reduce counter-arguing and issues related to accuracy.

In an analysis of Al Gore's 'An Inconvenient Truth' (AIT) 2006, and 'The Great Global Warming Swindle' (GGWS) 2007, Mellor (2009) introduces the problem of accuracy in the documentary genre by arguing that "the ambiguous figural status of the image in science documentaries makes accuracy an inadequate means by which to judge such films" and that accuracy should rather be considered "as a property related to the text" (Mellor 2009). The non-news factual genre has been underexplored by communication researchers although, whether distributed in cinemas or broadcasted on TV, the documentary film "can act as a prominent media product whose form and content become the subject of further public discourse" (Mellor 2009). In particular, an investigation into the Court arguments on the accuracy of these two productions showed that the judgment only brought to further problems and that "accuracy claims opened up room for more and more debate rather than closing down the debate into a final judgment", failing to resolve the dispute (Mellor 2009). According to Mellor, this kind of problems comes from a lack of attention to the rhetoric of the text when judging the film which is, in fact, the key element to assess the truthfulness of these productions.

The main issue with the environmental/scientific documentary is that we are facing two problems at the same time (1) the indexical nature of the images and (2) the scientific aspect of the argument. Both these features, especially in expository documentaries such as AIT and the GGWS (characterized by the voice-of-God narration, edited footages of interviews from the various experts, wallpaper shots etc.), significantly diminish the distance between reality and representation in the audience's perception of the 'story', and its indexical nature enables the audience to invest complete trust in it (Mellor 2009). In the case of AIT and the GGWS, a central role in this subtle manipulative process is also played by the use of graphical visuals, as "graphs are the most highly indexical of the representational techniques of science, tracing as they do data gathered from the material world" (Mellor 2009). All the aforementioned aspects of this genre will then represent an opportunity to reinforce the idea that what we are watching is the actual current state of events related to climate change, even though in most of these documentary productions we are dealing with the visualization of a set of natural processes that have not yet happened (Mellor 2009).

Another important element in the analysis of films is *ethics*. In *Environmental Ethics and Film* (2016), Brereton (2016) explores the role of fictional films in the construction of environmental ethics in contemporary societies and analyses how mass audience films can promote awareness and activate a response to environmental crises, through the use of cautionary tales. Brereton reminds us that there is a tendency in environmental analysis "to emphasise eco-cinema as first and foremost a cognitive, rather than an affective and emotional experience" (Brereton 2016). The present and the next sections aim to illustrate that the power of environmental films arises from both, intrinsically, cognitive and emotional responses, due to the numerous properties of the film as a text (e.g. music, images, costume design, characters). It is for this reason that the adoption of multimodal methods for the analysis of climate change films becomes necessary if one wants to have a complete view of their role and effects among the public.

## 3 Why Should We Examine the Use of Fictional Films for Communicating Climate Change?

The media have acted as translators of a complex vocabulary and scientific concepts for the *lay audience* since the major environmental disasters in the 1980s, such as Bophal (1984) and Chernobyl (1986), trying to represent the reality of science to its spectators, and research has confirmed this to be particularly true with climate change issues. For example, a survey of college students confirmed that the media are the main vehicle for bringing information about climate change to the masses, and more than a half of the students claimed to prefer television as a climate change source over printed media (Wilson 2000).

Another study from Lowe et al. (2006), following the sociological assumption that the media have a determining influence on the audience (Zinn and Taylor-Gooby 2006) showed that environmental films could have a significant impact on the public. Their research examines the impact of the disaster film *The Day After Tomorrow* (2004), and it does it through a survey of filmgoers in the UK. This study revealed that the movies changed people's perspective on climate change significantly, at least in the short-term, and reported that most of the viewers expressed a strong motivation to act after watching the film.

Even though most of the participants did not think that climate change would affect them significantly in their daily lives, they all confirmed that the film inspired them to find further information on climate change and 'what to do' in order to help with this global issue (Lowe et al. 2006). Only 5% of their sample did not see the necessity of taking any actions. Despite these findings, further research is needed to validate the length of the impact of climate change movies on people as the study above only referred to a film that belongs to the disaster genre. The way in which the disaster or the specific environmental issue is positioned in the text could make a substantial difference in this case.

Several factors can lead the consumer towards choosing 'audio-visual entertainment' sources over another kind of information based on numerical data and statistic models (e.g. degree of literacy, availability of certain documents, time available to the consumer after a day of work). This tendency among the lay public of preferring visual media forms has been recognised not only by experts in the social sciences but also by geographers such as James Balog who, in his documentary *Chasing Ice (*2012*)*, during an expedition in the Arctic, claimed:

> When I saw those [melting ice sheets and glaciers disappearing], the lights went off for me. I realised [that] the public doesn't want to hear about more statistical studies, more computer models, more projections, what they need is a believable, understandable piece of visual evidence, something that grabs them in the gut. - James Balog, min 9.41 (Chasing Ice 2012).

In the documentary, Balog also explains that using visual material is necessary not only for communicating with the lay public but also with an expert one. The use of images to convey the urgency of intervening on certain environmental issues has also been the focus for visual artists involved in the environmental debates. An example is the environmental photographer and filmmaker Edward Burtynsky, whose work has largely inspired governmental institutions, experts and the lay public. In his documentary, *Manufactured Landscapes* (2006), Burtynsky highlights the importance of sending messages through media forms that are distanced from 'politicised environments', without imposing on people what to see but rather 'let them judge for themselves':

> There are times when I've thought about my work and putting it into a more politicised environment. If I said that this is a terrible thing we're doing to our planet, then people would either agree or disagree. By not saying what you should see, that may allow them to look at something they have never looked at before, and to see their world a little differently.

Through his work, Burtynsky invites us to reflect on the difficult tasks that science and communication experts face when they need to send environmental messages to the public. Creative forms of communication in general, and films, in particular, can become a powerful tool for shaping people's mind but a crucial role is also played by claims coming from the leaders, the *stars*, of the films (Anderson 2011). For instance, Leonardo DiCaprio, while accepting his first Oscar at the

Academy Awards 2016, tailored his speech around the importance of recognising climate change as a *current real threat* rather than a *future risk*. Leonardo DiCaprio is one of the major Hollywood celebrities who has been acting as an ambassador for climate change for several years, through films and environmental campaigns, but his Oscar speech represented a new step. His speech was, in fact, an opportunity to remind the public of the significant impact that climate change is having on one very single aspect of our society, including the film industry.

The above citations are only a small sample of many other elements that we can individuate in order to demonstrate the pivotal role that films can have in communicating environmental risks to different audiences. However, another paramount element that is worth acknowledging, and which constitutes the core of every film, is the potential of the narrative (either fictional or factual) of an environmental story.

## 4 The Role of Narrative in Environmental Films

There are several ways in which one can communicate an idea or an emotional state, but 'narrative' is the fundamental mode of human interaction and is also the most efficient way of acquiring and sharing information (Hinyard and Kreuter 2007), especially if the information has social relevance. This applies in particular if we look at fictional narrative. In fact, several studies have proven that messages coming from fiction have a particular power to influence beliefs about (and within) social groups and social issues (Slater and Rouner 2002). Hinyard and Kreuter (2007) well explain a narrative as "any cohesive and coherent story with an identifiable beginning, middle, and end that provides information about scenes, characters, and conflict; raises unanswered questions or unresolved conflict, and provides resolution".

Compared with more expository communicative forms, narrative can captivate the audience, provide motivation to continue the story until a resolution is reached, and influence the audience to the point that can lead to behavioural change. These are valuable characteristics if there is a need to share scientific knowledge with people at risk for certain threats (Downs 2014). One of the biggest challenges in science is to communicate new empirical findings, and this may first require people to discard long-held beliefs or familiar structures for understanding this new information. A narrative can help facilitate this process by providing elucidations and explicit examples to which people can relate to more easily (Downs 2014). Among their many benefits, films can be particularly functional if used to engage members of the audience who have little interest in learning new information (e.g. adolescents, sceptics) (Downs 2014).

Furthermore, some studies suggest that there are better chances to engage an audience by including positive messages in film narratives about environmental risks, especially climate change, rather than adopting the strategy of fear, which would instead distance and disengage them, making them feel overwhelmed and helpless. Although shocking, catastrophic representations "may well act as an initial hook for people's attention and concern, they clearly do not motivate a sense of personal engagement with the issue and indeed may act to trigger barriers to engagement such as denial" and other negative or passive responses (O'Neill and Nicholson-Cole 2009). Fear appeals, "focus attention primarily on the threat, leaving less cognitive processing capacity available to continue taking in and making sense of further information" (Downs 2014).

Narrative is the core of mass media but, as we know, there are paramount differences among them. "Films and television, for example, provide rich visual imagery to viewers, whereas readers have to create mental images for themselves" (Green et al. 2004). Because of their narrative structure, films and other entertainment-education narratives can provide a strong emotional experience allowing the viewer to become involved with the characters of the story. This involvement can help to overcome various forms of resistance to persuasion (Moyer-Gusé 2008).

According to the *Elaboration-likelihood Model* (Slater and Rouner 2002), *Entertainment-Persuasion* (Moyer-Gusé 2008), *Social Cognitive Theory*, and *Transportation-imagery theory* (Green et al. 2004), the success of narratives is due to three important properties: memory retrieval processes, transportation and identification. Narratives create and reinforce memory traces for better recall and application over time (Downs 2014), especially when the experience of watching or reading has been particularly enjoyable (Green et al. 2004). Enjoyment leads to transportation, explained by the theorists as "a convergent process, where all mental systems and capacities become focused on events occurring in the narrative" (Green and Brock 2000). As the individual is swept into the story, he or she is also less keen or less motivated to counter-argue story points. Thus, through transportation, the acceptance of the messages, and the subsequent decrease of counter-arguing, there are better chances for persuasion to be successful (Green and Brock 2000; Slater and Rouner 2002; Moyer-Gusé 2008).

Like transportation, another element that increases absorption and reduces counter-arguing is identification with the characters in the story (Finnegan 2016). Identification is defined as a process "in which an individual perceives another person as similar or at least as a person with whom they might have a social relationship" (Slater and Rouner 2002; Moyer-Gusé 2008). Identification also refers to an emotional and cognitive process where "the viewer forgets about his or her own reality and temporarily becomes the character, taking on the character's perspective [...] and temporarily replaces his or her own identity with the character" (Cohen 2001). Identification involves four dimensions: "empathic (shared feelings with the character); cognitive (sharing the character's perspective); motivational (internalising the character's goals); and absorption (the loss of self-awareness during exposure)" (Moyer-Gusé 2008).

A set of studies which involved factual ("report of real events") or fiction ("resemblance to real persons or places is purely coincidental") stories suggested that individuals may be transported equally well into factual and fictional worlds

(Green and Brock 2000). However, "when individuals seek out media programming for entertainment, they often gravitate toward fiction" (Green et al. 2004).

## 5 Making the Environmental Film: A Collaboration Between Arts and Science

As mentioned earlier, a practice-led mode of research was also introduced as a method to understand certain dynamics connected to the film industry and film-making in general. An opportunity for collaboration between the creative arts and the sciences presented itself in January 2015 when I was introduced to Dr. Enrico Andreoli, senior lecturer in engineering at Swansea University, during a meeting on sustainable energy at Bangor University. The goal was primarily to produce a film in collaboration with somebody from the scientific field to understand and study the process of cooperation between the humanities and science and also understand the limitations and the obstacles in such partnerships. Dr. Andreoli expressed an interest in collaborating since the very beginning, as he was finding himself in a similar position but from another perspective.

Dr. Andreoli is a Senior Lecturer at the Energy Safety Research Institute (ESRI) of Swansea University where he leads research in Carbon Dioxide Capture and Utilisation. He is interested in sustainability, and his work is focused on changing the role of carbon dioxide in the current environmental and energy scenario, from a molecule that is a threat to the environment to an opportunity for long-term energy sustainability. Dr. Andreoli's study of sustainable energy and $CO_2$ represented the perfect opportunity for me to approach visualisation of climate change from a different angle and on a different topic from the most commonly depicted environmental catastrophes (e.g. hurricanes, floods, drought) while engaging with a subject that relates to other factors such as politics, economics and policy-making. Sustainable energy is rarely depicted in films and even less is $CO_2$ which, if ever featured, is always represented as the 'enemy'. This was then a chance to try to transform a negative element into a positive one, in order to represent its possible benefits.

Following the premises outlined in part 2 of this chapter, as the first project in this collaboration, Dr. Andreoli and I decided to focus our attention on producing a short fictional film which can be used for the purposes of my research but that could also be used as an 'ice-breaker' for Dr. Andreoli to engage his audience during conferences where he would primarily present technical and numeral data. One of the first steps in the construction of this video was then to individuate a public narrative on which one could tailor the elements of the 'story' with the concept of sustainable energy, so as to allow the viewer to engage with the film. The aim was to adopt the use of visual metaphors for leading the viewer through a 'cinematic' experience, instead of acquiring the information through those visual tropes that usually belong to the documentary film format.

By firmly supporting the thesis on the benefits of using fictional narratives for communicating to a wider audience, both inside and outside the professional field, during summer 2015, we secured funds for a total of £1600 from the *Welsh National Research Network in Advanced Engineering* and Materials and the *UK Engineering and Physical Sciences Research Council* in order to sustain the costs of production of our film (Grant ref: EPSRC: EP/N009525/1. Polymer-promoted Cu-catalysed conversion of $CO_2$ to $CH_4$).

Several ideas were proposed for the content of this film, but the final decision was to concentrate on the concept of sustainability first and then slowly include the other elements which can refer to advanced material and techniques, carbon dioxide and renewable energy. Among these ideas, we decided to use the concept of *tradition* as a metaphor to talk about *sustainability*. We elided the notion of *sustainability* with *tradition* as these two words are significantly linked. Tradition, like sustainability, involves a long-term commitment, a long-term action which survives during the time. Moreover, the two concepts suggest cognitive frames related to 'positive' events, as both tradition and sustainability relates to 'something good' and very rarely to a negative thought.

To have a film running the length of four minutes was also present in the minds of the production team throughout the development of the concept. Research and statistics on attention span show that today the average length that people watch a single video, either distributed on the internet or presented during private events, is actually under 5 min. Therefore, it was important that our creative decisions did not affect our aim of this running length. We also intentionally kept out any 'visually' direct link to engineering as our goal was to let the viewers make the link and let them elaborate the message presented between the images and the text in the film.

Even if the narrative (as a text) of the film has the implicit role of guiding the audience through the film, it is the construction of images themselves that will lead the viewer towards the final message. Dialogues were not used for this film. Instead, the visual-based information is supported by motion graphics, as well as appropriate music which helps reinforce the connection with the possible positive aspect of $CO_2$. Finally, the film ends with the slogan "Be sustainable, be tradition", as well as the research council's logos, ensuring that the thematic message is communicated effectively.

More specifically, the film deals with the story/evolution of a child's fascination with cooking and growing into adulthood where he becomes a chef. The story is set in Italy and spans a period from the 1970s to present day. It presents a child learning how to make pizza next to his grandmother, using products from their garden and utensils and pots which were handed down from generation to generation. These elements were included to suggest the importance of using materials which can survive through time. The Italian products from the vegetable garden also link to the idea of sustainability, and the concept of energy is transfer through the cooking itself.

The following tab shows the connection between the elements included in the film:

| Main concept | Metaphor |
|---|---|
| Sustainability | Tradition |
| Energy use | Cooking |
| Engineering techniques | Cooking technique |
| Evolution and development | Child growing up |
| Carbon Dioxide | Food from the vegetable garden |
| Advanced material | Ceramic and terracotta pots |

As previously mentioned, to explicitly talk about the scientific argument through a technical vocabulary was not our first concern. However, the film fits into the discussion as it will be understood in the context when presented to its public. In the end, we chose to entitle our film "Be tradition", and it can be found online through YouTube, Vimeo, Swansea and Bangor University websites.

Due to several factors, mostly related to the difference between a scientific and an artistic mind and way of working in these areas, it took us one and half years of brainstorming and communication through email and Skype in order to find a suitable idea, one that could serve both our needs in terms of research, and initiate the making of the film. During this process, we also collaborated with other experts in the field of film studies to develop the narrative of the film, but to agree on one idea was nonetheless very complicated. Interdisciplinary collaborations are a true challenge and, in an ideal scenario, it would have been beneficial to have a chance to work on previsualization and story-boarding, by collaborating with an artist for example, in order to help all the parties involved understand what they wanted through experimenting how the film will look like in advance. Unfortunately, in our case, the limited budget would not allow such an initial step.

Another significant issue comes from regulations related to the research council's funds, especially the ones coming from science. Most of the funding bodies require that those employed for a specific project must be registered as sole traders in order to be paid for their work. When dealing with a creative project where the aim is to hire the best artist, or the most suitable person for the job in general, the rule of employing sole traders becomes a limitation immediately for the project. While in scientific projects the focus may be on buying materials or technologies which can be supplied by a registered company, this is not always the case with artists. Very few artists (e.g. painters, film directors, designers) are usually recognised as businesses. Therefore, the producers are sometimes forced to choose collaborators according to their status as a company rather than for their skills, creativity, and excellence of their work.

However, even when the funds for the project are secured, in a multidisciplinary venture that involves the production of a creative medium for the public communication of a scientific topic, there is a general lack of understanding of the status of the media at the moment or, more specifically, there is a lack of knowledge about the potential of alternative media (e.g. video games, fictional films, creative narratives) to achieve the same objectives. With this reflection in mind, at the end of my project, I also identified a need for developing a new framework, guidelines and methodologies which will facilitate multidisciplinary projects and allow a follow-up analysis. Practice-based research is at the moment one of the most successful ways for investigating interdisciplinary processes of collaboration between humanities and science.

To conclude, some advice and solutions for those who wish to embark a similar journey into multidisciplinary projects for the production of creative pieces of work: (1) be open minded and understand where your colleague is coming from and try to compromise and respect each other's expertise and knowledge; (2) keep consistent communication with the participants of the project and avoid extended periods of silence (for creative purposes); (3) when budgeting for a creative project, especially films, make sure a right amount of the funds is kept and dedicated to the creation of pre-visualisation images which will help in understanding the final product (the importance of pre-visualisation work is a very obvious point for those working in the creatives arts but may be less evident for science experts); (4) besides using research councils funds, try to secure funds coming from other sources as well, which allow a more versatile use of the money in case of emergency; (5) define the role of each participant very carefully from the beginning in order to avoid future issues with the creative development of the work. Moreover, last, make sure you and your group of collaborators increase the conversation and spend a good amount of time exploring what other experts and artists have already achieved in similar areas of study.

All the above considerations derive from my own practice and relate to my experience in interdisciplinary academic partnerships. They may not be applicable to other projects, but I believe they represent a common trend in such projects and a chance for us all to reflect on how to meliorate in this increasing demand for interdisciplinary collaboration among scholars and other experts.

## 6 Conclusions

Media have the potential to generate emotional responses, which can, in turn, lead the audience to take action about certain risks, and this is particularly true if we consider fictional films. Due to their construction of narratives (textual and visual) and other elements such as music or special effects, films can engage the spectator in something that can be more than a mere moment of entertainment: they can be a true experience that can influence our ideas about nature (Cox 2013). Successful communication and following engagement of the public in momentous environmental issues can certainly be achieved through cross-disciplinary cooperation between media studies, social sciences and environmental science; a cooperation which is essential for properly addressing questions and resolving issues related to climate change and its effects on our society.

The film that has been produced as part of a project between arts and science, although developed while considering different types of public, was primarily conceived as a creative product to engage and interest policy makers and stakeholders. The aforementioned short film can be found through online platforms

(YouTube, Vimeo, universities websites), and it will be hopefully used as an instrument for further research. It is important to remember that this is only the first of a series of creative projects that will be undertaken in this collaboration, with the aim of effectively contributing with understandings of the processes and mechanisms of climate change communication, and ultimately encourage departments such as Engineering and Environmental Science to seek further partnerships with Creative Arts and Humanities.

Apart from the differences in reasoning between academic and professional disciplines, another issue which is worth mentioning for those who wish to undertake a similar project, are the limitations that come from receiving funding from science and engineering research councils since this type of funding does not necessarily meet the needs coming from a creative development such as the production of a film. Despite the various obstacles, more effort must be put into the analysis of the production of environmental fictional narratives, not only for the benefits of research itself but also to find better ways of reaching the audience and remark the urgency of climate change on a global scale.

## References


Anderson A (2009) Media, politics and climate change: towards a new research Agenda. Sociol Compass 3:166–182

Anderson A (2011) Sources, media, and modes of climate change communication: the role of celebrities. WIREs Clim Change 2:535–546

Brereton P (2016) Environmental ethics and film. Earth scan from Routledge

Cohen J (2001) Defining identification: a theoretical look at the identification audiences with media characters. Mass Commun Soc 4:245–264

Cottle S (2000) TV news, lay voices and the visualisation of environmental risks. In: Allan S, Adam B, Carter C (eds) Environmental risks and the media. Routledge, London, pp 29–44

Cox R (2013) Environmental Communication and the public sphere. 3rd edn

Di Francesco DA, Young N (2011) Seeing climate change: the visual construction of global warming in Canadian national print media. Cult Geogr 18(4):517–536

Downs JS (2014) Prescriptive scientific narratives for communicating usable science. PNAS 111 (4):13627–13633 (Sept 16 2014)

Doyle J (2007) Picturing the clima(c)tic: greenpeace and the representational politics of climate change communication. Sci Cult 16(2):129–150

Durkin, M (2007) The great global warming swindle, UK, 1hr 14 mins

Emmerich, R (2004) The day after tomorrow, USA, 2hr 4 mins

Finnegan J (2016) The screenplay and the spectator: exploring audience identification in narrative structure'. J Screenwriting 7(3):319–330. doi:10.1286/jocs.7.3.319_1

Green MC, Brock TC (2000) The role of transportation in the persuasiveness of public narratives. J Pers Soc Psychol 79:701–721

Green MC, Brock TC, Kaufman GF (2004) Understanding media enjoyment: the role of transportation into narrative worlds. Commun Theory 14(4):311–327

Guggenheim D (2006) An inconvenient truth. USA 1hr 36 mins

Hansen A, Cox R (2015) The Routledge handbook of environmental communication. Routledge Handbooks

Hansen A, Machin D (2008) Visually branding the environment: climate change as a marketing opportunity. Discourse Stud 10(6):777–794
Hansen A, Machin D (2013) Researching visual environmental communication. Environ Commun 7(2):151–168
Hinyard LJ, Kreuter MW (2007) Using narrative communication as a tool for health behavior change: a conceptual, theoretical, and empirical overview. Health Edu Behav XX (X):1–16
Lester, L (2010) Media and the environment. Polity
Lester, L and Cottle, S (2009) Visualising climate change: television news and ecological citizenship. Int J Commun 3:920–936
Linder SH (2006) Cashing-in on risk claims: on the for-profit inversion of signifiers for "global warming". Soc Semiot 16:103–132
Lowe T, Brown K, Dessai S, De Franca Doria M, Haynes K, Vincent K (2006) Does tomorrow even come? Disaster narrative and public perception of climate change. In: Public understanding of science. Sage Publications, pp 435–457
Mellor F (2009) The politics of accuracy in judging global warming films. Environ Commun J Nat Cult 3(2):134–150
Moyer-Gusé e (2008) Toward a theory of entertainment persuasion: explaining the persuasive effects of entertainment-education messages. Commun Theory 18(3):407–425
Nerlich B, Jaspal R (2013) Images of extreme weather: symbolising human responses to climate change. In: Science as culture, pp 1–24
O'Neill S (2013) Image matters: climate change imagery in US, UK and Australian newspapers. Geoforum 49:10–19
O'Neill S, Nicholson-Cole S (2009) Fear won't do it: promoting positive engagement with climate change through visual and iconic representations. Sci Commun 30(3):355–379
O'Neill S, Smith N (2014) Climate change and visual imagery. WIREs Clim Change 2014(5):73–87
Orlowski J (2012) Chasing ice UK, 1hr 15 mins
Slater MD, Rouner D (2002) Entertainment-education and elaboration likelihood: understanding the processing of narrative persuasion. Commun Theory 12(2):173–191
Smith N, Joffe H (2009) Climate change news in the British press: the role of the visual. J Risk Res 12:647–663
Wilson KM (2000) Communicating climate change through the media: predictions, politics and perceptions of risk. In: Allan S, Adam B, Carter C (eds) Environmental risk and the media. Routledge, London
Zinn JO, Taylor-Gooby P (2006) The challenge on (managing) new risks. In: Taylor-Gooby P, Jens O Z (eds) Risk in social science. Oxford University Press, New York

## Author Biography

**Michela Cortese** is a Ph.D. candidate and associate lecturer in Media Studies at Bangor University. She is an expert in risk communication and environmental communication, and she is a member and advisor of the Media and Persuasive Communication Network at Bangor. Her research focuses on the visual communication of climate change through films and television from a sociological perspective. She is currently the co-leader and producer of *Visually Capturing* $CO_2$, a series of projects that involves the production of environmental visual representations in collaboration with the Department of Engineering at Swansea University

# Role of Emotions in Climate Change Communication

**Sefat Salama and Khalil Aboukoura**

## 1 Introduction: Why Emotions Matter

Emotions are a powerful motivator for human behaviours (see e.g., Damasio 1994; Izard 2002; Kahneman 2003; LeDoux 2012; Levine and Leven 2014). Damasio argues that emotions are a necessary ingredient to almost all decisions. Damasio's view is based on his studies of people whose connections between the "thinking" and "emotional" areas of the brain had been damaged. They were capable of rationally processing information about alternative choices but were unable to make decisions because they lacked any sense of how they felt about the options (Damasio1994, 2005; Damasio and Carvalho 2013). Myers (2004) defined emotion as a complex state of feeling that results in physical and psychological changes that influence thought and behaviour. Emotion is a state of feeling, but it encompasses physiological, cognitive and behavioural components (Solomon 2008). For example, emotion can be a response to a situation—fear when exposed to a threat; it can motivate behaviour—anger can motivate one to act aggressively; or it can be a goal in itself (Rathus 2012). Another study shows that emotion can also have a significant, direct impact on judgment and choice (Loewenstein and Lerner 2003). Emotions serve as affective prompts for engagement with an issue and lead to forming predispositions for action when a relevant situation arises (Baumeister et al. 2007). Furthermore, emotions themselves can serve as frames by which future information is interpreted (Nabi 2003). Recent studies have also pointed to the importance of emotions, in addition to cognition, in public perceptions of risk and associated decision making (see e.g., Kahneman 2011; Stanovich and West 2008;

S. Salama (✉) · K. Aboukoura
Cardiff University, Cardiff, UK
e-mail: salamas@cardiff.ac.uk

K. Aboukoura
e-mail: khaleelk13@yahoo.co.uk



W. Leal Filho et al. (eds.), *Handbook of Climate Change Communication: Vol. 1*,
Climate Change Management, https://doi.org/10.1007/978-3-319-69838-0_9

Slovic et al. 2005; Kahneman and Frederick 2002). Positive emotions can also play an important role in message response (Monahan 1995). Also, emotional responses to hazards and events shape people's perception of risk, sometimes leading to reactions that differ from what would be expected based on what is known, or communicated, about a risk (Rakow et al. 2015). According to broaden-and-build theory, formulated by Fredrickson (1998), positive emotions often initiate a cycle of more positive emotions. Specifically, positive emotions can facilitate the development of skills, networks, resources, and capacities which in turn promote wellbeing and fulfilment. Broaden-and-Build theory of positive emotions is used to demonstrate how cultivated positive emotions can expand individual awareness of their connections to Earth's living systems, increase their capacity to creatively and effectively address environmental problems. It also helps them recognize that wellbeing and environmental health go hand in hand (Fredrickson 1998, 2001, 2004).

Negative emotions, such as anger or fear, motivate us to the need to be on "high-alert" and to seek more information about an issue (Baron et al. 1994). They can also lead to greater risk perceptions and greater policy supports (Leiserowitz 2006, 2005). Also negative emotions have the potential to transform apathy or indifference into perceived importance and behavioural action (Thomas et al. 2009). Leiserowitz (2006) and Smith and Leiserowitz (2012) found that negative affect and imagery toward climate change were the strongest predictors of risk perception. Hazards that evoke negative emotions tend to be perceived as more risky and requiring mitigation. Public risk perceptions tend to consider hazards more risky when they are relatively more unfamiliar and evoke more negative emotions due to their catastrophic potential (Slovic 1987). However intense negative emotions can have a counterproductive effect on risk protection behaviour, with fear and anxiety leading to avoidant behaviours and defensive denial (Witte and Allen 2000; Taylor et al. 2012). This type of emotions was frequently described by some researchers as obstacles or barriers which hamper individuals' engagement with the climate change issue.

The concept of barriers is increasingly used to describe the obstacles that hinder the planning and implementation of climate change adaptation (Klaus et al. 2014). Moser and Ekstrom (2010) define barriers as obstacles that make adaptation less efficient, less effective or may require changes that lead to missed opportunities or higher costs. Moser (2009) argued that there are internal psychological and cognitive processes that may prevent an individual from engaging on climate change issue, as well as social, political, economic, and other structural external barriers to such engagement. Processing of climate change information may undermine the motivation of individuals to engage on the issue if they have emotional responses to it that are demotivating.

Emotions are to be seen as cultural-cognitive products related to values and social norms of a society and that social cognitions or representations held by individuals as well as of textual and visual objects carry emotions as an inseparable part of them. They are furthermore ethically important for informing behaviour (Höijer 2010). Recently, researchers have begun to explore the role of emotions in influencing how individuals process and react to climate change information (see e.g., Chadwick 2015; Lu and Schuldt 2015, 2016).

This paper aims to address the important role of emotions as internal or individual barriers to communicate issues surrounding climate change through reviewing available literature which discuss the impact of emotions in climate change communication.

## 2 Emotions and Cimate Change

Climate change is an urgent issue that effects our environment and natural resources and our everyday life in many ways of present and future generations. Nevertheless, people are unwilling to change their behaviours. Ockwell et al. (2009) propose that significant reductions in greenhouse gas emissions can only be achieved through understanding the emotional aspects of the public's responses to climate change which may cause irrational beliefs or behaviours even in the face of scientific evidence. Hornsey et al. (2016) highlight two major ideas about the public's feelings on climate change. First, the analysis suggests that out of all the personal characteristics examined by scientists so far, political affiliations, worldviews and values were the most significant predictors of a person's beliefs about climate change. Second - and perhaps somewhat disheartening—a person's belief in climate change doesn't necessarily translate into big support for climate-friendly action. Research by Norgaard (2011) on the role played by emotions in shaping attitudes in responses to climate change raises major questions for how we go about climate communication. She shares the now widely-argued view that a lack of information about climate change is not the problem, observing that "information deficit models" fail to account for the behaviour of those who believe in global warming and express concern about it. Norgaard maintains that understanding emotions is central to understanding the social processes that shape responses to climate change. In addition, emotional engagement also leads to a higher degree of motivation than a detached, rational stance on climate change (Weber 2006). Emotions are also related to a perception of climate change (see e.g., Smith and Leiserowitz 2014; Van der Linden 2014; Myers et al. 2012; Roeser 2012; Doherty and Clayton 2011; Siegrist et al. 2007). In a large scale survey of Australians' perceptions and understandings of climate change in Australia, Reser et al. (2012) reported that 20% of people show appreciable distress about climate change. A similar survey from the Yale Climate Project in America reported that a large percentage of people surveyed about climate change report feeling disgusted, hopeful, helpless, sad, depressed or guilty about the issue and a quarter report feeling depressed or guilty (Maibach et al. 2008). A study by Feinberg and Willer (2010) suggested that the conflict between the negative impacts of climate change and their belief in a just world led to the message being ignored—and even used as evidence that climate change was not occurring.

Emotions can be seen as a necessary source of reflection and insight concerning the moral impact of climate change. Yang and Kahlor (2013) found those who had negative feelings toward climate change such as depression, anxiety actively sought more information about climate change. Those driven by a positive affect toward climate change such as hopefulness, excitement, happiness, actively avoided exposure to additional information on the issue. Smith and Leiserowitz (2014) also found that some discrete emotions such as worry, interest, and hope have a large influence on how individuals support climate change policies, while fear found to be a relatively weak indicator of positive support for preventative climate change action. Smith and Leiserowitz explain this result by pointing out that worry is related to long-term problems such as career, health, and family issues and acts as a motivating agent to help people find solutions to these problems, while fear prepares people for immediate threats and is related with mainly short-term issues (Smith and Leiserowitz 2014).

Debate exists about whether negative or positive emotions are more conducive to encouraging behaviour that mitigates climate change (Leviston et al. 2014). Höijer (2010) examined how the Swedish media communicated emotions in the social construction of global warming risk and found that hope and compassion were used as emotional anchors to help people understand projected climate impacts. Höijer argued that emotional representations of climate change may on the one hand enhance public engagement in the issue, but on the other hand may draw attention away from climate change (Höijer 2010). Lombardi and Sinatra (2013) suggest that teacher-held emotions may affect instructional practice. The researchers found that In-service secondary teachers who do not currently teach about climate change exhibited greater anger and decisiveness than preservice elementary teachers and in-service secondary teachers who do teach about climate change. Teachers' topic emotions (anger and hopelessness) were significant predictors of plausibility perceptions with more anger associated with lesser plausibility and greater hopelessness associated with higher plausibility perceptions.

## 3 Climate Change Communication: Role of Emotions

The complex nature of climate change issue poses many challenges for scientists, decision makers and communicators interested in engaging diverse public audiences who don't pay more attention to the issue. There is a gap between scientific knowledge about climate change and public motivation to respond. Barriers to effective climate change communication have been well documented (Markowitz and Shariff 2012; Moser 2010; Van der Linden et al. 2015). Psychologists are identifying countless psychological 'barriers' that obstruct behaviour change despite knowledge about anthropogenic ecological degradation, that include perceptual, cognitive, emotional, interpersonal and group processes (Gifford 2011). Among them are the perception that climate change is a distant issue and the fact that different segments of the public have divergent opinions regarding its existence and origins (Bliuc et al. 2015; Spence et al 2012).

To overcome the barriers to engagement on climate change, communicators must be critically aware of the essential role of emotions in behaviour change. Climate change communicators have examined the ways that evoking such emotions can lead to engagement or disengagement with climate change. O'Neill and Nicholson-Cole (2009) argued that there is a lack of clarity in the literature about the impacts that fearful messages in climate change communication have on people's senses of engagement with the issue and associated implications for public engagement strategies.

While emotional appeals are often conceptualized as a particular type of framing and frequently promoted as an efficacious strategy for persuasion (Dillard and Nabi 2006; Nabi 2003), few studies have directly explored their potential in communicating climate change. However, accumulating evidence suggests that emotions may play an unappreciated role in this domain (Lu and Schuldt 2016).

Literature in cognitive psychology has shown that increased emotional intensity of an event/object can reduce its perceived distance (Cole et al. 2013; Van Boven et al. 2010). Also, previous psychology and communication research suggests that emotions can affect policy preferences that are typically strongly linked to political ideology (see e.g., Kühne and Schemer 2015; Small and Lerner 2008; Lu and Schuldt 2016).

Effective climate communicators understand that emotions play a key role in motivating action. But many communicators want to know which specific emotions they should try to evoke to motivate action on climate change. People's emotional reactions to messages about climate change are an important consideration, yet to date they have often been overlooked both in research and in communication efforts (Myers et al. 2012). Smith and Leiserowitz (2014) found that "worry" about climate change was the strongest predictor of support for policies to mitigate climate change. "Disgust," on the other hand, was the strongest predictor of opposition to these policies. Interestingly, researchers also found that "fear" did not strongly correlate with support for policies to mitigate climate change.

Thagard and Findlay (2011) argue that "encouraging belief change…about climate change requires dealing with emotional constraints as well as cognitive ones". Markowitz and Shariff (2012) identify positive emotional appeals as one of the most promising strategies for climate change communication. Indeed, emotional responses to novel information occur effortlessly and almost instantaneously, influencing subsequent cognitive processing of information (Kahneman 2011).

Feelings of hope related to climate change, for example, likely increase the probability that individuals will choose to engage with the issue and adopt beliefs and behaviours consistent with efforts to stem the problem (Markowitz and Shariff 2012; Snyder 2002; Swim et al. 2010). Hope is often aroused as perceptions of efficacy, the belief that an individual has the capacity to implement a proposed response to a threat and that the recommended action can effectively mitigate the threat upsurge (Bandura 1977). Feelings of hope and efficacy are strongly correlated with a willingness to engage in pro-environmental behaviours and to support climate change policies (Lorenzoni et al. 2007). Conversely, feelings of hopelessness and inefficacy related to climate change are linked with a tendency to ignore the problem or to rationalize inaction (Norgaard 2011).

Experimental evidence from psychology suggests that fear framing is indeed effective in motivating behaviour change across various different behaviours. In general, the greater the level of fear produced by a communication, the greater the level of attitude change found. However, Moser (2007) rightly cautions that fear framing cannot be used indiscriminately, and that a number of preconditions also need to be in place (a sense of personal vulnerability, self- efficacy, low response costs, and wider social support (see Spence and Pidgeon 2010). As people move toward contemplating taking action, stronger fear appeals can help form a behavioural intent. However such fear appeals must be coupled with constructive information and support to reduce the sense of danger (Moser 2007). Weber (2006) argues that risk communication strategies should explicitly appeal to emotions, but in a careful way. Similarly, Meijnders et al. (2001) argue that communication about climate change should appeal more directly to feelings such as fear. Now it might be argued that appeal to emotions and feelings such as fear is rather tricky, as feelings and emotions are generally seen as irrational and biased states that are inferior to rational forms of information processing (Roeser 2012).

Studies have shown that climate change communications which appeal to emotions such as fear and shock are not always effective in promoting deeper engagement. For example, with regard to the issue of a sustained uptake of low carbon behaviours, fear can be experienced as disempowering if people do not feel able to reduce the threat (see e.g., O'Neill and Nicholson-Cole 2009; Howell 2014). The principal problem with fear as the main message of climate change communication is that what grabs attention (dire predictions, extreme consequences) is often not what empowers action. Numerous studies have documented that audiences generally reject fear appeals (or their close cousin, guilt appeals) as manipulative (Moser 2007; O'Neill and Nicholson-Cole 2009, in Moser and Dilling 2011). O'Neill and Nicholson-Cole (2009) found that fear is not a useful tool for motivating genuine personal engagement with climate change and action. Research has shown that deliberate attempts to instil fear or guilt in people carry a considerable risk of backfiring. Dickinson et al. (2013) suggest that while some dire messages are ineffective, those evoking concern for target species of significance to the learners may be as successful as positive messages. In fact, the study suggests that mentioning dangers for humans did not increase participants' interest in taking personal action on climate change, but mentioning dangers for birds was highly effective. Highlighting the positive collective impacts of small behavioural changes also increased participants' interest in taking personal action.

Also, there has been increasing attention given to the images used in the context of climate change. There is now growing recognition of the importance of investigating how climate change is conveyed in visual images (see e.g., Nicholson-Cole 2005; O'Neill and Hulme 2009; Smith and Joffe 2009; Doyle 2011). A study conducted by Lester and Cottle (2009) in which the authors distinguished between human and natural impact images, underscored the fact that images are a useful adjunct in communicating climate change via news. Another study by Höijer (2010) in Sweden highlighted that the use of images in the press and television can trigger a range of emotions among people. Climate change visuals trigger instant reflexes and feelings engaging people's emotions (Sheppard 2005) and help them remember the designed messages (Bartels and Nelissen 2000). Visuals may also increase their empathy with other people making them more open to consider different perspectives in decision-making (Petheram et al. 2012). Nerlich and Jaspal (2014) found that images of extreme weather in the media create negative emotional meanings and might lead to disengagement with the issue of climate change.

The images symbolised fear, helplessness and vulnerability and, in some cases, guilt and compassion appealing to fear of disaster can lead to denial and paralysis rather than positive behaviour change. According to Harré (2011) and Burke (2015), we can be even more successful about promoting sustainable behaviours when we can find emotionally-charged stories that elicit positive emotions and have pro-social moral stories that can generate emotions that steer us towards behaviours that are good for others and away from behaviours that damage others. Burke stated that 'These stories make us feel good and make us want to pass them on to others' (2015). Harré adds that to be really useful for conversations about climate mitigation and adaptation, 'stories need to provide the full motivational and detailed how-to-do-it package that we human imitators respond to so well' (Harré 2011, p. 54). Images of climate impacts can prompt a desire to respond, but because they are emotionally powerful, they can also be overwhelming. Coupling images of climate impacts with a concrete behavioural 'action' for people to take can help overcome this (Corner et al. 2015). Markowitz and Shariff (2012) suggest six inter-related strategies for communicators to reiterate action on climate change as a moral imperative. Among them are emotions; the authors state that motivating through appeals to hope, pride, and gratitude rather than the negative emotions of guilt, shame, and anxiety is far more efficacious. It has even been demonstrated that pro-social behaviour arising from an appeal to these positive emotions "leads to increase in subjective well-being." In 2009, the Centre for Research on Environmental Decisions at Columbia University (CRED) put out a guide on how to communicate about climate change. The guide urges educators and scientists to avoid overusing emotional appeals and accentuating worst-case scenarios, noting that they may backfire.

However, Roeser (2012) writes that "emotions are generally excluded from communication and political decision-making about risky technologies and climate change, or they are used instrumentally to create support for a position". She adds, emotions may be the "missing link in effective communication," and concludes that emotions are necessary for understanding the moral impact of risks of climate change, and provide motivation for people to change. We cannot and should not divorce feeling from societal issues like this, otherwise we may never achieve any action.

Effective climate change communication has to target people's emotions. Policy should be directed to climate change communication tools with a focus on emotional engagement to encourage people to take personal responsibility in climate change and act, catalysing the desired behavioural change (Ramkissoon and Smith 2014).

We suggest that, by understanding peoples' emotional responses to climate change, we may be able to devise guidelines for effective communication and develop plans for mitigating and adapting to climate change.

## 4 Should Scientists Show Emotions While Discussing and Communicating About Climate Change?

According to The Guardian, a professor of ocean geology wept as she discussed the impact carbon dioxide ($CO_2$) emissions is having on the sea. She fears we are acidifying and heating the ocean so fast that her young daughters may no longer enjoy coral reefs and shellfish by the end of the century and her passion for the oceans triggered tears (Harrabin 2015).

Oreskes (2013) argues that scientists should express more alarm about climate change. She recalls a presentation at the American Geophysical Union conference where an audience member stood up and said, "You are telling us that we have a very serious problem, but you don't sound at all worried. You don't even sound upset!" Oreskes argues that expressing concern would help convey the seriousness of the issue that it's difficult to get excited about something when the experts themselves seem dispassionate (Kearns 2015).

Marshall (2014) believes that when it comes to climate change, the unwillingness of scientists to show emotion has been counterproductive in engaging the public.

Farbotko and Mcgregor (2010) explain that emotions need to be taken seriously in the context of climate change by drawing on sadness and discomfort triggered by a member of the delegation representing the Pacific Island nation of Tuvalu at Copenhagen Conference of Parties (COP)15 held in December 2009 and demonstrate a need for more research linking geography, emotion and science. Farbotko and McGregor (2010) argue that possibilities for change arise when emotions enter climate change negotiations, even though (or perhaps because) these are arenas that privilege rational exchange.

Emotions are necessary for understanding the moral impact of the risks instigated by climate change. They also paradigmatically provide for motivation. Emotions might be the missing link in effective communication about climate change (Roeser 2012). However, emotions are generally excluded from communication and political decision making about climate change, or they are used instrumentally to create support for a position (Roeser and Nihlen 2013).

## 5 Conclusion

Emotions play a key role in motivating actions on climate change. However, the essential question which many climate change communicators seek to illuminate is which specific emotions trigger a behavioural response to climate change. An excessive focus on the negative impacts of climate change (i.e. a severe 'diagnosis') without effective emphasis on solutions (a feasible 'treatment') typically disengages the audience (Moser and Dilling 2011). Emotional considerations may be the key to changing behaviour. Communication should be tailored to appeal to the morality of the audience, inciting emotions which trigger a behavioural response (Roeser 2012).

Extensive research is needed to analyse the emotional underpinnings of responses to different ways of framing about climate change. Effective climate change communication is not only to convey information and data, but it is also about people. We must go beyond the idea of simply communicating facts; we must acknowledge the primary role that emotions and feelings play in how we understand those facts. It is important to know what motivates the audience to behave in certain ways and what might inspire them to change that behaviour.

There is an urgent need to consider emotional aspects when we discuss the implications of climate change and how to communicate adaptation and mitigation to climate change. To overcome the challenges and barriers to engagement with climate change, communicators must be critically aware of the role emotions have in prompting behaviour, and should continue to develop an understanding of the ways in which people emotionally engage with climate change. Furthermore, climate change communicators have to carefully consider how to use both positive and negative emotions to connect and engage audiences. Provoking emotions toward climate change has to be handled with care as too much can go into the wrong direction. To conclude, more attention should be paid to how emotions and climate change are intertwined in climate change mitigation and adaptation strategies and policies.

Understanding the nature of emotions impact in climate change communication is essential to explore strategic ways of dealing with climate change barriers. This paper highlighted the paucity in empirical studies about the role of emotions in communicating issues related to climate change. It also demonstrated discrepancies in the outcomes of available research studies in this area. Future imperial research about emotions as internal individual barriers is much needed.

## References

Bandura A (1977) Self-efficacy: toward a unifying theory of behavioral change. Psychol Rev 34 (2):191–215

Baron RS, Logan H, Lilly J, Inman ML, Brennan M (1994) Negative emotion and message processing. J Exp Soc Psychol 30(2):181–201

Bartels G, Nelissen W (2000) Developing and marketing communication strategies with the aid of the FCB grid. In: Bartels G, Nelissen W (eds) Marketing for sustainability. Towards transactional policy-making, IOS Press, Amsterdam

Baumeister RF, Vohs KD, DeWall CN, Zhang L (2007) How emotion shapes behaviour: feedback, anticipation, and reflection, rather than direct causation. Pers Soc Psychol Rev 11 (2):167–203

Bliuc AM, McGarty C, Thomas EF, Lala G, Berndsen M, Misajon R (2015) Public division about climate change rooted in conflicting socio-political identities. Nat Clim Change 5(3):226–229

Burke S (2015) Communicating about climate change. In: Walker R, Mason W (eds) Climate change adaptation for health and social services. CSIRO Publishing, pp 51–69

Chadwick AE (2015) Toward a theory of persuasive hope: effects of cognitive appraisals, hope appeals, and hope in the context of climate change. Health Commun 30(6):598–611

Cole S, Balcetis E, Dunning D (2013) Affective signals of threat increase perceived proximity. Psychol Sci 24(1):34–40

Corner A, Webster R, Teriete C (2015) Climate visuals: seven principles for visual climate change communication (based on international social research). Climate Outreach, Oxford

CRED—Center for Research on Environmental Decisions (2009) The psychology of climate change communication. A guide for scientists, journalists, educators, political aides, and the interested public, Columbia University, New York

Damasio AR (1994) Descartes'error: emotion, reason, and the human brain. Grosset/Putnam, New York

Damasio AR (2005) Descartes'error: emotion, reason, and the human brain. Penguin Books, New York

Damasio AR, Carvalho GB (2013) The nature of feelings: evolutionary and neurobiological origins. Nat Rev Neurosci 14(2):143–152

Dickinson JL, Crain RL, Yalowitz S, Cherry T (2013) How framing climate change influences citizen scientists' intentions to do something about it. J Environ Educ 44(3):145–158

Dillard JP, Nabi RL (2006) The persuasive influence of emotion in cancer prevention and detection messages. J Commun 56(Suppl. 1):S123–S139

Doherty TJ, Clayton SD (2011) The psychological impacts of global climate change. Am Psychol 66(4):265–276

Doyle J (2011) Mediating climate change. Ashgate, UK

Farbotko C, McGregor HV (2010) Copenhagen, climate science and the emotional geographies of climate change. Aust Geogr 41(2):159–166

Feinberg M, Willer R (2010) Apocalypse soon? Dire messages reduce belief in global warming by contradicting just-world beliefs. Psychol Sci 22(1):34–38

Fredrickson BL (1998) What good are positive emotions? Rev Gen Psychol 2(3):300–319

Fredrickson BL (2001) The role of positive emotions in Positive Psychology: The broaden and build theory of positive emotions. Am Psychol 56(3):218–226

Fredrickson BL (2004) The broaden and build theory of positive emotions. Philos Trans R Soc B 359:1367–1377

Gifford R (2011) The dragons of inaction: Psychological barriers that limit climate change mitigation and adaptation. Am Psychol 66(4):290–302

Harrabin R (2015) Is it ok for scientists to weep over climate change? The guardian, 9 July 2015 at: http://www.theguardian.com/environment/2015/jul/09/is-it-ok-scientists-weep-over-climate-change. Accessed 28 June 2016

Harré N (2011) Psychology for a better world: strategies to inspire sustainability. University of Auckland, New Zealand

Höijer B (2010) Emotional anchoring and objectification in the media reporting on climate change. Public Underst Sci 19(6):717–731

Hornsey Matthew J, Harris Emily A, Bain Paul G, Fielding Kelly S (2016) Meta-analyses of the determinants and outcomes of belief in climate change. Nat Clim Change 6(6):622–626

Howell RA (2014) Investigating the long-term impacts of climate change communications on individuals' attitudes and behavior. Environ Behav 46(1):70–101

Izard CE (2002) Translating emotion theory and research into preventive interventions. Psychol Bull 128(5):796–824
Kahneman D (2003) Maps of bounded rationality: psychology for behavioral economics. Am Econ Rev 93(5):1449–1475
Kahneman D (2011) Thinking fast and slow. Farrar, Straus, and Giroux, New York
Kahneman D, Frederick S (2002) Representativeness revisited: attribute substitution in intuitive judgment. In: Gilovich T, Griffin D, Kahneman D (eds) Heuristics and biases: the psychology of intuitive judgment. Cambridge University Press, Cambridge
Kearns F (2015) Scientists have feelings too. Hippo reads (2015) at: http://read.hipporeads.com/scientists-have-feelings-too/. Accessed 28 June 2016
Klaus E, Susanne CM, Esther H, Richard K, Christoph O, Anna P, Maja R, Catrien T (2014) Explaining and overcoming barriers to climate change adaptation. Nat Clim Change 4:867–872
Kühne R, Schemer C (2015) The emotional effects of news frames on information processing and opinion formation. Commun Res 42(3):387–407
LeDoux J (2012) Rethinking the emotional brain. Neuron 73(4):653–676
Leiserowitz AA (2005) American risk perceptions: is climate change dangerous? Risk Anal 25 (6):1433–1442
Leiserowitz A (2006) Climate change risk perception and policy preferences: the role of affect, imagery, and values. Clim Change 77(1):45–72
Lester L, Cottle S (2009) Visualizing climate change: television news and ecological citizenship. Inter J Commun 3:920–936
Levine DS, Leven SJ (2014) Motivation, emotion, and goal direction in neural networks, 2nd edn. Psychology Press, New York
Leviston Z, Price J, Bishop B (2014) Imagining climate change: the role of implicit associations and affective psychological distancing in climate change responses. Eur J Soc Psychol 44 (5):441–454
Loewenstein G, Lerner JS (2003) The role of affect in decision making. In: Davidson R, Goldsmith H, Scherer K (eds) Handbook of affective science. Oxford University Press, New York
Lombardi D, Sinatra GM (2013) Emotions about teaching about human-induced climate change. Inter J Sci Educ 35(1):167–191
Lorenzoni I, Nicholson-Cole S, Whitmarsh L (2007) Barriers perceived to engaging with climate change among the UK public and their policy implications. Glob Environ Change 17(3–4):445–459
Lu H, Schuldt JP (2015) Exploring the role of incidental emotions in support for climate change policy. Clim Change 131(4):719–726
Lu H, Schuldt JP (2016) Compassion for climate change victims and support for mitigation policy. J Environ Psychol 45:192–200
Maibach EW, Roser-Renouf C, Leiserowitz A (2008) Communication and marketing as climate change-intervention assets a public health perspective. Am J Prev Med 35(5):488–500
Markowitz EM, Shariff AF (2012) Climate change and moral judgement. Nat Clim Change 2 (4):243–247
Marshall G (2014) Don't even think about it: why our brains are wired to ignore climate change. Bloomsbury, USA
Meijnders AL, Midden CJH, Wilke HAM (2001) Role of negative emotion in communication about CO2 risks. Risk Anal 21(5):955–966
Monahan JL (1995) Thinking positively: using positive affect when designing health messages. In: Maibach E, Parrott RL (eds) Designing health messages: approaches from communication theory and public health practice. Sage, London
Moser SC (2007) More bad news: the risk of neglecting emotional responses. In: Moser SC, Dilling L (eds) Creating a climate for change; communicating climate change and facilitating social change. Cambridge University Press, Cambridge, UK

Moser SC (2009) Communicating climate change and motivating civic action: renewing, activating, and building democracies. In: Selin H, VanDeveer S (eds) Changing climates in north american politics: institutions, policymaking and multilevel governance. The MIT Press, Cambridge, MA
Moser SC (2010) Communicating climate change: history, challenges, process and future directions. Wiley Interdisc Rev Clim Change 1(1):31–53
Moser SC, Dilling L (2011) Communicating climate change: closing the science-action gap. In: Norgaard R, Schlosberg D, Dryzek J (eds) The Oxford handbook of climate change and society. Oxford University Press, New York
Moser SC, Ekstrom JA (2010) A framework to diagnose barriers to climate change adaptation. Proc Natl Acad Sci USA 107(51):22026–22031
Myers DG (2004) Theories of emotion. Psychology. Worth Publishers, New York
Myers TA, Nisbet MC, Maibach EW, Leiserowitz AA (2012) A public health frame arouses hopeful emotions about climate change. Clim Change 113(3–4):1105–1112
Nabi RL (2003) Exploring the framing effects of emotion do discrete emotions differentially influence information accessibility, information seeking, and policy preference? Commun Res 30(2):224–247
Nerlich B, Jaspal R (2014) Images of extreme weather: symbolising human responses to climate change. Sci Cult 23(2):253–276
Nicholson-Cole SA (2005) Representing climate change futures: A critique of the use of images for visual communication. Comput Environ Urban Syst 29(3):255–273
Norgaard KM (2011) Living in denial: climate change, emotions, and everyday life. The MIT Press, Cambridge, MA
O'Neill S, Nicholson-Cole S (2009) 'Fear won't do it': promoting positive engagement with climate change through visual and iconic representations. Sci Commun 30(3):355–379
Ockwell D, Whitmarsh L, O'Neill S (2009) Reorienting climate change communication for effective mitigation. Forcing people to be green or fostering grass-roots engagement? Sci Commun 30(3):305–323
O'Neill S, Hulme M (2009) An iconic approach for representing climate change. Glob Environ Change 19(4):402–410
Oreskes N (2013) The scientist as sentinel. Limn Mag 3 at: http://limn.it/the-scientist-as-sentinel/. Accessed 28 June 2016
Petheram L, Stacey N, Campbell B, High C (2012) Using visual products derived from community research to inform natural resource management policy. Land use Policy 29(1):1–10
Rakow T, Heard CL, Newell BR (2015) Meeting three challenges in risk communication: phenomena, numbers and emotions. Policy Insights Behav Brain Sci 2(1):147–156
Ramkissoon H, Smith L (2014) The relationship between environmental worldviews, emotions and personal efficacy in climate change. Inter J Arts Sci 7(1):91–109
Rathus SA (2012) Psychology: concepts & connections (Brief version, 9th edn. Wadsworth Publishing, Belmont, CA
Reser JP, Bradley GL, Glendon AI, Ellul MC, Callaghan R (2012) Public risk perceptions, understandings, and responses to climate change and natural disasters in Australia and Great Britain. National Climate Change Adaptation Research Facility, Gold Coast, Australia
Roeser S (2012) Risk communication, public engagement, and climate change: a role for emotions. Risk Anal 23(6):1033–1040
Roeser S, Nihlen J (2013) Risk Communication and Moral Emotions. In: Arvai J, Rivers III L (eds) Effective risk communication, Routledge
Sheppard S (2005) Landscape visualisation and climate change: the potential for influencing perceptions and behaviour. Environ Sci Policy 8:637–654
Siegrist M, Cousin ME, Kastenholz H, Wiek A (2007) Public acceptance of nanotechnology foods and food packaging: the influence of affect and trust. Appetite 49(2):459–466
Slovic P (1987) Perception of risk. Science 236(4799):280–285
Slovic P, Peters E, Finucane ML, MacGregor DG (2005) Affect, risk, and decision making. Health Psychol 24 (Suppl. 4):S35–S40

Small DA, Lerner JS (2008) Emotional policy: personal sadness and anger shape judgments about a welfare case. Polit Psychol 29(2):149–168
Smith N, Joffe H (2009) Climate change in the British press: the role of the visual. J Risk Res 12 (5):647–663
Smith N, Leiserowitz A (2012) The rise of global warming skepticism: exploring affective image associations in the United States over time. Risk Anal 32(6):1021–1032
Smith N, Leiserowitz A (2014) The role of emotion in global warming policy support and opposition. Risk Anal 34(5):937–948
Snyder CR (2002) Hope theory: rainbows in the mind. Psychol Inq 13(4):249–275
Solomon RC (2008) The philosophy of emotions. In: Lewis M, Haviland-Jones JM, Barrett LF (eds) Handbook of emotions. Guilford Press, New York
Spence A, Pidgeon N (2010) Framing and communicating climate change: the effects of distance and outcome frame manipulations. Glob Environ Change 20(4):656–667
Spence A, Poortinga W, Pidgeon N (2012) The psychological distance of climate change. Risk Anal 32(6):957–972
Stanovich KE, West RF (2008) On the relative independence of thinking biases and cognitive ability. J Pers Soc Psychol 94(4):672–695
Swim J, Clayton S, Doherty T, Gifford R, Howard G, Reser J, Stern P, Weber E (2010) Psychology and global climate change: addressing a multi-faceted phenomenon and set of challenges. A report by the American Psychological Association's task force on the interface between psychology and global climate change at: http://www.apa.org/science/about/publications/climate-change-booklet.pdf. Accessed 20 July 2016
Taylor AL, Dessai S, Bruine de Bruin W (2012) Public perception of climate risk and adaptation in the UK: a review of the literature. Clim Risk Manag 4–5:1–16
Thagard P, Findlay S (2011) Changing minds about climate change: belief revision, coherence, and emotion. In: Olsson EJ, Enqvist S (eds) Belief revision meets philosophy of science. Springer, Berlin
Thomas EF, McGarty C, Mavor KI (2009) Transforming "apathy into movement": the role of prosocial emotions in motivating action for social change. Pers Soc Psychol Rev 13(4): 310–333
Van Boven L, Kane J, McGraw AP, Dale J (2010) Feeling close: emotional intensity reduces perceived psychological distance. J Pers Soc Psychol 98(6):872–885
Van Der Linden S (2014) On the relationship between personal experience, affect and risk perception: the case of climate change. Eur J Soc Psychol 44(5):430–440
Van der Linden S, Maibach E, Leiserowitz A (2015) Improving public engagement with climate change five "best practice" insights from psychological science. Perspect Psychol Sci 10 (6):758–763
Weber EU (2006) Experience-based and description-based perceptions of long-term risk: why global warming does not scare us (yet). Clim Change 77(1):103–120
Witte K, Allen M (2000) A meta-analysis of fear appeals: implications for effective public health campaigns. Health Educ Behav 27(5):591–615
Yang ZJ, Kahlor L (2013) What, me worry? The role of affect in information seeking and avoidance. Sci Commun 35(2):189–212

## Author Biographies

**Sefat Salama** Researcher, lecturer, science communicator and journalist, writing mainly about science, technology and environment related issues. She authored a number of books in Arabic including her most read and cited book Nanotechnology: Tiny World and Big Future, Foreword by Professor Munir Nayfeh, 2009, The Ethics and Challenges of the Era of Robots, 2014 (co-author), Science Fiction and Developing Creativity, foreword by professor Farouk El-Baz, 2006 (co-author) and Climate Change: The Missing Aspects (in press).

**Khalil Aboukoura** Freelance psychologist, speaker and author. His research interests include the psychology of climate change, science and technology risk communication, and media and climate change.

# Climate Change Communication in Australia: The Politics, Mainstream Media and Fossil Fuel Industry Nexus

**David Holmes and Cassandra Star**

## 1 Introduction

The Australian story of climate change communication began with great promise in the late 1980s, with politicians and the media raising awareness of climate science. In 1990 emissions targets to cut greenhouse gases by 20% by 2005 were announced. After the first IPCC (Intergovernmental Panel on Climate Change) report news outlets were already reporting the growing consensus amongst scientists. But by the mid-1990s, this bipartisan political and media consensus began to change. Over the next two decades the national interest, political elites and the public narrative were overtaken by corporate interests.

This chapter argues that it is not possible to understand this change or climate change communication in Australia without first looking at the political economy of the fossil-fuel industry, politics and mainstream media nexus. The economy, mostly in the Australian public imagination, is strongly dependent upon natural resource extraction and export, especially of fossil fuels. This ensures a strong compact between political elites and the interests of fossil fuel companies, which exhibit high concentrations of ownership and control. Coal companies are the largest donors to both major political parties, federally and in the state of Queensland.

Parallels can also be drawn to Australian media corporations with 72% of capital city daily newspapers being owned by just one company (News Corporation). This chapter demonstrates the editorially divided nature of the Australian press on

D. Holmes
School of Media, Film and Journalism, Monash University,
PO Box 197 Caulfield East VIC 3145, Clayton, Australia
e-mail: david.holmes@monash.edu

C. Star (✉)
Flinders Government, College of Business, Government and Law,
Flinders University, GPO Box 2100, Adelaide, SA 5001, Australia
e-mail: casssandra.star@flinders.edu.au



W. Leal Filho et al. (eds.), *Handbook of Climate Change Communication: Vol. 1*,
Climate Change Management, https://doi.org/10.1007/978-3-319-69838-0_10

climate change. Since the 1990s, climate change has become highly politicised, with News Corporation outlets being stridently hostile to climate science and effective carbon abatement policy. On the other side of the divide are Australia's national broadcaster, the ABC, Fairfax Press and start-up open-web news outlets that have focused on raising the level of climate literacy in Australia. The potential role of social media in raising awareness of dangerous climate change, is limited given its demonstrated tendency to simply amplify mainstream media messages.

Nevertheless, there has been almost no research in Australia exploring the link between climate change reporting and social attitudes toward climate, and the absence of such research presents a limitation for this paper. Public concern for taking action on climate change has remained high in recent years. Mainstream news reporting has not fueled this concern, which leaves unanswered the sources of such concern. Moreover political parties have largely ignored public opinion, and consistently neglected to take the action that publics have demanded.

## 2 The 'Value Gap' in Communicating Climate Change

Globally, the problem of effectively communicating climate change stands between climate science and climate action. Understanding this gap where action has for so long been catching up to science, is first and foremost the task of media and communications scholars, journalists and social and political analysts. This gap between values and action, can be understood by recognising the diverse audiences that consume climate messages. Mike Hulme has illustrated this problem in terms of the different ways in which successive IPCC findings have been 'framed' by different kinds of news outlets catering to divergent audiences:

> Climate change is framed in a multitude of different ways, either informed by the world-views of those communicating or filtered by the intuitive world-views of those listening. Different climate change discourses use different linguistic repertoires, often related to the specific goals of the discourse coalition involved. (Hulme 2009: 245)

To address the problem of communicating the science, more than just climate scientists are needed to effectively convey and contextualise the science. In a recent essay, one of Australia's leading public intellectuals Robert Manne, marvels at the success of the IPCC and how it has provided policymakers and global publics with rigorous but cautious summaries of the findings of the world's climate scientists (Manne 2015):

> Interestingly, however, social scientists other than economists – sociologists, psychologists, political scientists, students of international relations – have not been invited to contribute to the IPCC reports nor have they participated in the global conversation on climate change. This is seriously strange. For no less important than the impact of climate change on the Earth and its creatures is the question of why human beings – international society, governments of nation-states, communities, individuals – have so far failed so comprehensively to rise to its challenge.

In this chapter, such a challenge is exemplified in the case of Australia. Australia offers a lesson on how a dysfunctional relationship between a country's media, its political culture and its scientific institutions can seriously impede progress on climate action. We look at how inaction came about in a nation which is arguably one of the most exposed to climate change impacts in the southern hemisphere.

The sheer range of extreme weather events: chronic drought, floods, cyclones, heatwaves and bushfires have, in only the last 10 years managed to touch so many lives in Australia as to guarantee ever broader demand for the Australian government to do something to ameliorate climate change. Some of the most notable extreme weather events include the Black Saturday firestorm in Victoria on February 7 2009, which killed 173 people, injured 414, and destroyed 2100 homes. The fire front, fuelled by eucalyptus oil in the native trees reached temperatures of 700 °C. Enough heat was released to equate to 1500 Hiroshima bombs and the post-fire inquiry showed that the fire was capable of liquefying the roofs of cars. Such fire behaviour was repeated in a series of fires in the summer of 2015–16, starting with a fire in Pinery, South Australia (November 25, 91 homes destroyed) and on Christmas day 120 homes were incinerated at Wye River and Separation Creek in Victoria. This was followed by a fire in Western Australia on 7 January 2016 which destroyed 143 buildings that constituted the town of Yarloop.

The heatwaves in south-east Australia in 2013 and 2014 were also the worst on record resulting in almost 1000 deaths. Combined with floods like those in the Hunter Valley in January 2016 and the Brisbane Queensland Floods (climate change attribution studies discussed below) in January 2011, many of the most populated areas of Australia have felt the full force of climate-fuelled extreme weather, which has worsened since the 1990s.

Yet, despite these impacts, successive Australian governments have delayed, diminished and even reversed action on climate change. As we outline in detail below, such failures go back to the 1980s on both the Labor and the Liberal-National sides of politics. In conjunction with such inaction, the broad shape of climate reporting in Australian media has not held politicians accountable. A polarisation of reporting has emerged since the early 2000s. Most progressive press has accepted the science and run stories about the impacts of, and solutions to, climate change. However the tabloid press and commercial television news have either failed to report climate change stories or have been stridently hostile. This is particularly true of Newscorp newspapers, who have arguably spearheaded a culture of denialism in Australia (Bacon and Jegan 2013). Given the extent of Newscorp's influence on Australian climate politics, such a denialist stance has politicised climate change policy to a surprising degree. This has provoked a 'debate' that has not moved on from questioning the science itself, whilst the impacts grow year on year.

In this chapter we examine the historical, economic and mediatized contexts of Australia's climate politics and how these have given rise to the kinds of debates explored by research scholars in Australia. We argue that understanding the relationship between the developmentalist political economy of Australia's resources industry and the concentrated nature of Australia's mainstream media, makes clear the immediate context and constraints of climate change communication in Australia.

## 3 Political Economy of Climate Change in Australia

Critical to understanding the evolution, current state and effectiveness of climate communication in the Australian context is recognition of the landscape of climate politics. While many studies of climate communication focus solely on media content messages, the receptiveness of citizens to media messages is a much more complex process (Olausson 2011). Therefore, it is essential to examine not just media messages, but the broader societal context in which different audiences are located. Key to understanding Australia's current climate politics are two main factors—the history of climate politics and the political economy—both have significant impacts and important lessons for effective climate communication. These factors are also intertwined with the media industry in the Australian context, explored further below.

Australia inhabits an interesting position in the global politics of climate change. Australia is a small Western nation, with aspirations to "middle power" status. It is a relatively insignificant player economically and in terms of overall greenhouse gas emissions, but, it has played a pivotal role as a "spoiler" in international negotiations since the mid-1990s, often significantly influencing and cooperating with more powerful friends, including until recently, the United States.

To understand the motivations and enduring concerns behind Australia's position at international climate change negotiations and domestically, two key factors need to be considered: Australia's emissions profile and the political economy of development in the Australian domestic context.

While Australia's domestic greenhouse gas emissions may not seem significant in global terms—1.5% in 2005 (Garnaut 2008) of global emissions—the country has one of the highest per capita emissions profiles internationally. The reason for this is linked not to energy intensity of GDP or high incomes, but to the high emissions intensity of energy generation and use in Australia (Garnaut 2008). This intensity of energy use is inextricably linked to the commitment and policies of successive Australian governments, on both sides of politics, to a particular political economy of development centred on the extraction and export of high-emissions primary products, especially coal (Pearse 2009), but more recently, the developing coal seam gas industry.

Australian governments have a strong history of encouraging and supporting primary resource development for export, especially coal. Australia is the fifth largest coal producing nation globally. Resources and energy comprise four of the top seven Australian exports by value. It is valued at $AUD 174 bn (Australian Mining 2015), but this represented an 11% decline in 2015. In recent years, supply of coal exports on the global market has increased and demand softened, resulting in lower prices at a time when Australian production capacity is increasing (Australian Government, Department of the Treasury 2014). Commitment to this approach can be linked to Australia's history as a resource-rich colony for the British Empire, exporting much needed raw material to the mother country (for example, see Walker 1994). The lack of evolution in Australia's economic development approach has been referred to as a "Quarry Vision" (Pearse 2009):

> It's our "natural competitive advantage endowed by providence", the "engine room of economic growth", "the backbone of our economy", or so we hear constantly from a political, business and economic chorus. Debate rages about virtually everything else, but there's perfect harmony on the importance of the Quarry… the rest of the economy is a mere sideshow (pp. 1–2).

In Australia, resources, including minerals, coal and gas, remain the property of the Crown (government), rather than the property of individual landholders. In the case of onshore resources, licensing and regulation of these resources falls to individual state governments. Potential developments are subject to regulatory regimes including licensing to explore and/or develop, environmental and planning regulations and eventual taxation and royalties (St. John 2014). The law and regulatory regimes are strongly in favour of exploration companies.

The current financial position of Australian state governments is a motivator for enthusiastic resource development. Regardless of political persuasion, Australian state governments are unashamedly developmentalist. This can be attributed to two main issues. Firstly, historically, Australian governments, state and federal, have been particularly wedded to primary resource developmentalism, and this has not changed. Australia, like the United States, is a "younger" country, forged of a "frontier" mentality that is still evident. This is pronounced, especially due to the continued Australian dependence on primary production, mostly resource extractive industries, for economic activity (Dyster and Meredith 2012). Secondly, Australian state governments do not raise income tax. All income tax is collected by the Federal government, and there is a process of dispersement to the states. There is a pronounced vertical fiscal imbalance in the Australian federation (Hunter 1974). Additionally, state governments' key sources of raising discretionary income not tied to Federal government programs and priorities is limited and shrinking (Parkin and Anderson 2007). One area however where state governments can raise revenue is via royalties for extractive activities (Eccelston and Wooley 2014). In addition, while materially these revenues may be comparatively modest, the economic activity and employment they purport to generate is especially prized.

Despite analyses that the resources era or long boom as it has been known in Australia, is at an end (Sheehan and Gregory 2013), deep commitment to this approach to economic development remains. Estimates indicate that the Federal government in Australia spent $AUD 4.5 bn on subsidies to resource extracting industries in 2012, not including tax concessions (Duxfield 2013). This has made state governments, of both major political persuasions, relatively hostile to actors and activities seen as anti-development, including campaigns to constrain climate change emissions.

This has substantial impacts for environmental policy making, and in particular for climate policy in the Australian context—often seeing it play out as a trade-off between climate action and economic growth. This situation has obvious impacts for effective climate communication.

## 4 Climate Change Politics in Australia

Australian politics is a two-party parliamentary democracy, with a significant third force in the Australian Greens. The Australia Labor Party (ALP) is the centre-left progressive party, with the other key party a centre-right coalition with the Liberal Party as the senior partner and the National-Country Party as the junior member of declining importance. As a parliamentary democracy, Australian government responses to climate change are largely driven by social attitudes and public commitment to climate change, which are in turn influenced by media coverage and elite attitudes and debate. Since 2007, successive Australian governments have responded to public attitudes around climate change even if policy initiatives have been ineffective. There has been some success in the area of renewable energy, but also numerous policy failures, reflecting the fact that governments have been pre-occupied with impression management more than facing the hard challenges of emissions reduction.

The emergence of global climate change as an issue in the late 1980s coincided with the government of Bob Hawke in Australia. The Hawke (ALP) government is credited with a number of substantial environmental initiatives on issues of bio-diversity and sustainability (for example see: Staples 2012). Action on climate change has been on the policy agenda in Australia since that time. The Hawke government went to the 1990 election with a commitment to reducing emissions by 20% by 2005, with parallel policies being developed (and perhaps exceeding these commitments) by the Liberal Party. Recently released cabinet papers from this time also show that the Hawke government considered a carbon tax (Hudson 2015a, b).

After the election and Australia's commitment to an international target at the Rio conference, the Hawke government faced internal and external pressures on proposed climate actions that remain largely unchanged. Internally pressure came from resource and economic ministers in the government (primarily the Treasurer Paul Keating) and external pressure came from resource extractive industries. The power and influence of these industries in Australian politics, on climate policy and on climate communication in particular is well documented (for example see Pearse 2009). In 1991 Keating took over from Hawke as leader of the ALP and as Prime Minister, and began stepping back from Hawke's environmental commitments, including those on climate change. So, Australia's internationally committed target was largely ignored, as emissions and coal exports increased. The economic and political power of the resource extractive industries was further entrenched. With the election of the Howard Liberal government in 1996, public commitment to climate change action declined further and bipartisan support ended. This period peaked with John Howard's position not to ratify the Kyoto Protocol, insisting he would not sacrifice economic growth to do so, drawing on the familiar framing of a conflict between economic growth and climate action. Only in the dying days of the government in 2006 did he indicate an intention to undertake policy change when confronted with the overwhelming popularity of Kevin Rudd's (ALP) call to action on climate change as the "great moral, environmental and economic challenge of our age". With increased public support for action and a re-emergence of bipartisanship, conditions were ripe for politicians to declare immediate and effective efforts to combat climate change.

What followed was a significant election victory for Kevin Rudd's ALP, after more than a decade in opposition. Ratification of the Kyoto Protocol was an early action of the new government, much to the celebration of the Australian environment movement. The movement had actively supported the Rudd/ALP election campaign and looked forward to meaningful climate policy change. By 2010, despite avoiding the global financial crisis (GFC), the Rudd government faced strong internal division and falling popularity. Many link this to Rudd's inability to negotiate a climate policy package acceptable to the Australian Greens, who held the balance of power required to pass the measure in the Australian Senate (for example see: Rootes 2011; Marr 2010; Hudson 2015a). The bipartisanship on climate change also receded with the displacing of Malcolm Turnbull by Tony Abbott as leader of the opposition Liberal Party. In the wake of this, Rudd lost the leadership of the ALP to Julia Gillard, who immediately set out to negotiate an alternative to Rudd's failed *Carbon Pollution Reduction Scheme* (CPRS).

After the failure of the CPRS legislation in 2009 and 2010, the environment movement strongly supported the Gillard government's price on carbon with the *Clean Energy Act 2011*. The *Say Yes* campaign was formed and active in securing public support for the Act. The *Say Yes* campaign and the *Clean Energy Act 2011* secured a modest price on carbon under the Gillard government. During the Gillard government and through to the Kevin Rudd led election campaign in 2013,[1] the government sustained significant attacks over climate policy, especially the carbon price, and it became a key divisive issue leading up to the election, with the opposition framing their rhetoric around the familiar clash between climate policy and economic prosperity. This framing resonated particularly with those on fixed incomes, especially pensioners, blue-collar workers and small business owners. The opposition attacks on the carbon price and Prime Minister Gillard, during which she was referred to as "Ju-liar", and chants of "burn the witch" were heard at anti-carbon price rallies, were ferocious.

Nevertheless, the legislation was in place successfully and achieving emissions reduction targets until the election of the Abbott government in 2013. The

[1]Kevin Rudd won the leadership of the ALP back from Julia Gillard in a challenge on 26 June 2013.

Abbott-led Liberal Party came to government on an election promise of an immediate end to the carbon price—and these promises came to fruition in July 2014, with the dismantling of a number of relevant pieces of legislation.

Key lessons about the Australian context for climate communication are evident by examining both the political economy and the history of climate change policy. The Australian political elite are significantly constrained by the historical dependence of the economy on resource-led development. Despite this, significant evolution in climate policy can be achieved when there is bipartisanship on this issue. When bipartisanship on the issue is absent, public opinion can be polarised, not necessarily on belief in climate change, but on commitment to policy action. These lessons have a significant impact on the ability to effectively communicate climate change in the public sphere, further complicated by factors within Australia's media industry, as explored below.

## 5 Australian Public Attitudes to Climate Change

Australian public attitudes to climate change have been examined in the literature in a couple of key ways—belief in climate change, its causes and attitudes about action on climate change. Each of these approaches are investigated below.

Since 2007, the majority of Australians believe climate change is real and is happening, though the actual percentages fluctuate slightly over time. Research undertaken between 2010 and 2014 demonstrates that a majority of Australians accept climate change (almost 80%). The number that thought human activity influenced it changed alongside political debates, though remained at over 60% (Leviston et al. 2015). Where we see more inconsistent commitment is in the number of Australians that would translate this into policy action and how. While prior to the 2007 election 68% agreed that steps should be taken, even if significant costs were incurred, by 2012 this had dropped to 36% before picking up again to 56%, according to a number of surveys just prior to the 2016 federal election (Holmes 2016).

Additionally, attitudes to climate change action in Australia are strongly divided along partisan party lines, as in many other Western nations (see Poortinga et al. 2011; Whitmarsh 2011; Eurobarometer 2009; Tranter 2011), with a significant element being the influence of leaders' stances. Political research over time has identified the impact and influence of the attitudes of leaders (see Miller et al. 1976 for an early example of this literature). This is of particular interest in the Australian landscape between 2006 and 2015 given the personal attitudes of the leaders of the major parties. This is particularly true of Tony Abbott's leadership of the Liberal Party (discussed further below). His personal views have had a strong influence on recent Australian debate.

In 2009, Abbott described the science on human-caused climate change as "crap" and referred to comments linking Australian bushfires to climate change as "hogwash". In his political autobiography, *Battlelines*, discussion of the issue follows a number of climate denialist talking points and argues against policy action:

> It sounds like common sense to minimise human impact… It doesn't make sense, though, to impose certain and substantial costs on the economy now in order to avoid unknown and perhaps even benign changes in the future (Abbott 2009: 170–1).

Abbott's view, backing away from John Howard's acceptance of the need for action in the lead up to the 2007 election, has contributed to changes in attitudes to policy interventions in the Australian electorate. This aligns with arguments that divided elites have a strong polarising impact (Sciarini et al. 2007). Thus, we can see that political elites, particularly when bipartisanship is absent, can have a strong impact on effective climate communication and therefore on public attitudes. This influence in turn has driven issue-attention cycles highlighted further at the end of this chapter.

## 5.1 *Five Australias?*

However, it can be argued that the polarisation of climate action in Australia has itself obscured a more nuanced reality of public attitudes toward climate changes. An analysis follows by the Yale Centre for Climate Change communication in the United States, which identifies 'Six Americas' (see Leiserowitz et al. 2009). The work divides American society into six stances on climate: Alarmed, Concerned, Cautious, Disengaged, Doubtful and Dismissive. These groups not only differed in their worldview on climate, but also in the way they sought information. For example, Alarmed Americans would rely much more on media sources, whilst those in the doubtful category would look to the views of family and friends. Since that initial study, and its updates (Leiserowitz et al. 2013a), it has been extended to India (Leiserowitz et al. 2013b) and replicated in Germany (Metag et al. 2015) where 'Five Germanys' were found.

In Australia, work to replicate this study has been done by Morrison et al. (2013) and Sherley et al. (2014). However, the most comprehensive analysis was conducted by Hine et al. in (2013) by the National Climate Change Adaptation Research Facility in Australia. This study sought to identify 'climate change interpretive communities', which share similar concerns about climate change. A sample of 3096 Australians was surveyed in 2010 to see what kinds of attitudinal segmentation would result. The researchers in this initial phase of the project identified audience segmentation comprised of five distinct interpretive communities: *Alarmed* (26%), *Concerned* (39%), *Uncertain* (14%), *Doubtful* (12%), and *Dismissive* (9%).

The authors observed that:

> The five interpretive communities also differed in their responses to climate change. Self-reported frequency of engaging in several psychological (e.g. thinking about climate change) and behavioural (e.g. using less electricity) responses increased incrementally across

> the five groups; ranging from very low average levels reported by *Dismissive* Australians to very high average levels reported by *Alarmed* respondents (Hine et al. 2013: 25).

However two years later, the survey was repeated, this time with a much greater proportion of male respondents and respondents from urban settings. The researchers came up with only three Australias: Alarmed, 34%, Uncommitted 45% and Dismissive 20%. Notwithstanding, both samples revealed conceptually equivalent 'anchor segments' (Dismissive and Alarmed). This suggests that the more enduring and 'stable' groups in the analysis are ones that are opposed, mirroring the politicisation of climate change Australia media.

As with the studies in the US, Germany and India, the takeaway message from the study was that:

> Communication specialists should take this heterogeneity in the Australian population into account when developing climate change messaging strategies. In short, the content and structure of climate change adaptation messages matter, and our results suggest that it is important to match message attributes with audience characteristics to increase their impact (p. 59).

However, as the authors also point out, understanding the Five Australias will not, by itself, be able to guide climate change communicators. Rather, changes in the media landscape and the new kinds of audiences that result, as well as changes in climate change itself, will impact on the kind of messages that people in the different groups will accept or not accept. In addition, the impact of the political context is not to be discounted.

## 6 Climate Change and the Media in Australia

For over four decades, Australia has had, and continues to have one of the highest concentrations of media ownership in the industrialised world (Jones and Pusey 2008). In newspaper ownership, Australia has a situation in which nearly all major capital newspapers are owned and controlled by just two companies: Rupert Murdoch's NewsCorp (with 72% circulation of capital city dailies) and Fairfax (with 22%). Commercial free-to-air television stations, the number one source from which most Australians get their news, also enjoy an acute oligopoly, with just three television networks controlling all of the capital city spectrum in Australia (Australian Government, Department of Broadband and Communications 2012 p. 9).

Whilst it is true that Australians prefer to get their news from the packaged format of evening news bulletins, the influence of the morning newspapers should not be under-estimated. The tabloid papers in particular, which outsell the broadsheets 40 to 1, set the agenda for the day's news in the hierarchy of stories, treatments and topics, which the evening news uses as its reference for its own agendas and sense of newsworthiness.

Thus many of the same narratives emerge in relation to particular topics that are adapted between print and screen, the major difference being that television leads with footage of news events that newspapers can only accommodate in their online versions.

## 7 Climate Change Reporting in Australia

When climate change first became an issue in the 1980s, almost all of the Australian press coverage was led by scientific sources. The role of human activity was not contested at all until climate change became politicised in the 1990s by interest groups, politicians and an adversarial "debate" between climate change deniers and an IPCC-led science of consensus.

This 'debate' between climate change deniers (Washington and Cook 2011) and consensus-scientists has been quite one-sided, with deniers getting a disproportionate degree of news media access compared to scientists in the Australia mass media. Disproportionate, in this instance relates to studies showing there to be a 97% consensus amongst climate scientists that anthropogenic climate change is occurring (see Cook et al. 2013; Holmes 2013a). Since around 2001 Australia, Newscorp news outlets, in Australia, have overwhelming favoured a select number of scientists (both in the climate area and other areas of science) in their reporting who have had fallings out with the IPCC (Bacon and Jegan 2013). At the same time, the number of environmental journalists reporting in Australian newspapers has declined, leaving the public broadcaster, The Australian Broadcasting Corporation, The Guardian (Australian Edition), and pockets of Fairfax Press as the only mainstream outlets to engage climate change politics and science. To some degree, the decline in mainstream climate reporting has been compensated for by start-up open websites like reneweconomy and The Conversation (see Australian Government, Department of Broadband and Communications 2012). The fact that the large mainstream audiences are missing out on climate reporting is a significant problem in Australia that has broad implications for effective climate communication and policy innovation in Australia.

A study of Australian newspapers conducted by the Australian Centre for Independent Journalism by Wendy Bacon in 2011 and 2012, found that one-third of all news articles surveyed rejected climate science (Bacon and Jegan 2011, 2013). A sample of 602 articles over February and April in 2011 and 2012 evaluated whether each article accepted or denied the fact of anthropogenic global warming. There were four categories that each of the articles were sorted into: accepted, rejected, suggested doubt, and unable to discern. The study found that more than 60% of the articles in the two largest circulating tabloids of News Corp, Sydney's *Daily Telegraph*, and Melbourne's *Herald Sun*, rejected anthropogenic global warming. Fairfax and Seven West's Perth newspaper published very few skeptical pieces and were overall accepting of the science. When it came to columnists, the percentages were even higher for News Corp, with 97% of comment articles in the

*Herald Sun* being skeptical. The study also found that paradoxically News Corp columns dealt with climate much more often that other publications in order to discredit the science.

As a host of other analysts have shown, across the nation, the Murdoch press, with its control of 72% of capital city dailies, has led the way in actively creating doubt about climate change (Chubb and Nash 2012; Manne 2011). For NewsCorp scholar David McKnight, News Ltd papers were tolerant of views urging action on climate in the 1990s. This shifted from around 2002 onwards to an editorial stance of ridiculing climate science and climate action (McKnight 2010). The masthead of the Murdoch Press, *The Australian* newspaper, sets the agenda for the news around the country and has been run at a loss for many years because it is used by News Ltd to influence political elites and even other newspapers (Manne 2011). As Robert Manne has observed, '*The Australian* is ruthless in pursuit of those who oppose its worldview' which includes market fundamentalism and minimal action on climate change (Manne 2011: 2).

As Manne (2011) found in a content analysis of *The Australian* from April 2003 – April 2011, articles that denied any need to take action on climate change outnumbered articles supportive of climate change action by 4 to 1. Access given to opinion columns is even more unbalanced, with the overwhelming majority of columns written by employees of lobby groups for business or the fossil-fuel industry, who have no expertise in climate science. According to McKnight, in *The Australian* "The climate deniers and sceptics, regardless of their lack of evidence and (in most cases) scientific qualifications, were elevated to the status of brave dissidents against an oppressive set of beliefs" (McKnight 2010, p. 703).

It is fair to say that Australian climate scientists are also reluctant to appear in the news media. The climate denier industry is so well organised in Australia that many climate scientists don't want to handle the large volume of emails and Freedom of Information requests whenever they publicise their research. Moreover, when their research is picked up online, they are inundated by the online version of climate deniers or 'trolls'. Trolls are online skeptics who vigorously "criticize any evidence that supports man-made global warming and yet embrace any argument, op-ed, blog or study that purports to refute global warming" (Skeptical Science.com).

Despite this, Australia has seen the emergence of some excellent climate communicators, who have valiantly tried to raise the level of climate literacy across the nation in the face of much hostile press. Foremost amongst these are David Karoly (University of Melbourne), Will Steffen and Andrew Glickson from the Australian National University, as well as Andy Pittman and Sarah Perkins from the University of New South Wales. The climate science acceptor community have also cultivated communicators that do not have a climatology background, such as Clive Hamilton (Hamilton 2007, 2009, 2010 and 2013) and Tim Flannery (Flannery 2005, 2008, 2009a, 2009b, 2010 and 2015) who have written books communicating the climate crisis whilst also having success in diverse media platforms.

But these figures rarely make it into the tabloid and mainstream news feeds, which is where they would attain greater audiences with people that may be searching for an explanation for the extreme weather events they experience around them.

## 8 Case Study of Climate Change Reporting of the Brisbane Floods 2011

Following a very wet December in 2010, one of the most unprecedented rainfall events to affect Queensland, culminated in the flooding of Brisbane between January 10 and 16. During this time, an area five times the size of Germany was underwater. Two hundred towns were flooded along with over 200,000 homes. Upstream from Brisbane, in the Lockyer Valley, high above sea level, the rain was so intense as to be described by many news outlets as an "inland tsunami" of water that killed 35 people.

From a climate science standpoint the 2011 Brisbane floods were unique. Unlike the cyclone-generated floods of 1893 (Cyclone Wanda) and 1974 (Cyclone Oswald) there was no cyclone driving the floods. Instead, unprecedented evaporation and rain had characterised the months before the flooding in and around Brisbane and several detection and attribution studies since, have linked the severity of the floods to climate change (see Evans and Boyer-Souchet 2012; Hendon et al. 2013; Ummenhofer et al. 2015). Such evaporation is directly linked to record sea surface temperatures along the Queensland Coast, which were fuelled by an exaggerated La Niña phenomenon. Climate science predictions drawn from the IPCC suggest that increased water vapour in tropical and sub-tropical regions will become increasingly dense, so that when it rains it will rain longer and much harder than ever before, increasing the frequency of events like the Brisbane floods.

In an analysis of the mainstream media coverage of the floods, Holmes found that of the 1433 news articles published in the Australian press during the six peak days of the floods, very few made a link to climate change; only 16 articles mentioned climate change and of these only 7 suggested there might be a link to climate change (Holmes 2013b). The newspaper database Factiva was used to examine all Australian printed news outlets during the six peak days of the floods and found 1433 articles using the search term 'Brisbane near4 floods' (this search terms finds every instance of these two words within 4 words of each other).

These articles were then searched for reference to 'climate change' OR 'global warming'. Given that only seven articles or just under 0.5% of all the articles retrieved, made a causative link to climate change suggests a missed opportunity for climate change reporting in Australia.

Very little of the news coverage of the Queensland floods made any connection between the floods and climate change and the focus was instead on the impending disaster. Of the 16 articles that mentioned climate change during this period: two were unequivocal that climate change was the major factor in the floods and five were in the 'maybe' category. The remaining nine articles were critical of climate change as a factor. Two posited La Niña as the decisive force in the floods and connected this to cyclical weather patterns that have a long history, whilst the other articles were dismissive of any suggestion climate change was real or had anything to do with the floods. These articles often appealed to a strong nature/culture dichotomy which re-determined climate (and indeed weather) as an external force that had nothing to do with human activities.

An editorial in *The Australian* entitled 'Sticking together in the face of the floods' ferocity' (12 January 2011. p. 13) was long enough to provide the background science needed to understand the events, but stated:

> It would be a mistake to jump to quick conclusions and attribute this disaster to climate change, inadequate urban planning or emergency service shortcomings: the brutal truth is that on days this summer, we have been overwhelmed by nature (2011: Editorial 1 p. 13).

Thus, mainstream media reportage of this event reinforces our earlier message about the lack of opportunity for climate communication in the Australian context.

## *8.1 Social Media and the Brisbane Floods*

The question of the audience impact of the mass media reporting of the Brisbane floods is something that can be explored by analysing social media. Traditional media has become the predominant content of social media. Users of social media feed off mass media. For example, the primary use of Twitter and Facebook is for media sharing, or linking to stories in traditional news media. Thus, the audience of mass media is made much more visible by social media and it is from social media that we are able to learn where audiences concentrate their attention in the finite spectrum of mass media.

Therefore it is worth looking at the retweet amplifications of mass media available from two studies conducted in 2011 (Bruns et al. 2012; O'Donnell et al. 2011) as a measure of the influence of mass media in public understandings of the event and then examining how much of this content is concerned with climate change.

In a sample of 40,000 tweets, 39% or 15,500 retweeted media available outside of twittersphere. The most popular kind of news media during the floods was image-based and this was mirrored in the twittersphere. The highest number of tweets and retweets fed off twitterpics. The second highest category of tweets and retweets was media sharing of radio and online newspaper content.

Here, Twitter itself, reinforces a media-reactive view of the world, and mirrored the insatiable demand for 'immediacy' that is expressed in the genre of 'liveness'. With such limited capacity to exchange analysis, it extended the reach of news media rather than provided an alternative to it (see also Kwak et al. 2010).

Of the remaining 24,500 tweets, (61%) tweets were used predominantly for organisational communication, to assist those affected in overcoming the floods. These could be categorised as:

- Advice
- Situational Information
- Request for Information

- Help and Fundraising
- Personal Narrative and Eyewitness reports
- Adjunctive Discussion
- Personal reaction
- Thanks
- Support
- Meta-discussion of the role of twitter in crises.

As Bruns et al. (2012) report, there was very little 'adjunctive discussion' (discussion of environmental politics or climate change) on Twitter at the time of the floods. This reflected the lead that was taken by news organisations in shaping not what audiences think, but what they think about. Thus, social media coverage of extreme weather events in the Australian context does not fill the climate communication gap, but instead, it amplifies the climate denial messages of the mainstream media.

## 9 Climate and 'Issue Attention' Drivers in Australia

Bruns et al's (2012) finding that there was little 'adjunctive discussion' about climate change during extreme weather events, needs to be connected back to the main drivers of 'climate issue attention cycles' in Australia. The Australian state and concentrated media industries in Australia act as a protectorate for mining and fossil-fuel interests, therefore climate change does not have the same persistence in issue attention cycles that it might otherwise have.

The exception to this in Australia is around election cycles and climate summits, where some research on the link between climate communication, voter intention and issue attention has been conducted (see Lester et al. 2015; Gurney 2014; Rootes 2011; Rowe 2011). In an article comparing issue attention on climate change in Australia, Germany and India between 1996 and 2010, Schäfer, Ivanova and Schmidt show that political elections are one of the main drivers of attention on climate in Australia (Schaefer et al. 2014).

Their study examined two leading newspapers in each country, collecting articles during the 1997–2010 time series which mention 'climate change', 'greenhouse effect' or 'global warming.' For the Australian case, they looked at *The Australian* and the *Sydney Morning Herald*. In Australia they found that 'climate change' coverage took up 1.42% of all articles published but found that attention had risen over the time series. Schaefer et al. (2014) found that political events (especially elections) increased media attention on climate change in Australia more than any other country in the analysis. UNFCCC Conferences of the Parties were political events that had their strongest influence in Australia. The study also found that climate change related feedback from political actors, such as statements about climate change by politicians (like those reviewed in detail in the above section), both in and out of parliament were moderately influential drivers, as were statements from environmental non-government organisation actors.

The study also confirmed that scientific activity, measured by the monthly output of articles on climate change in high-profile journals had negligent impact on issue attention in Australia. More significant were international factors such as the UN climate conferences (CoPS). However, the most important drivers of attention on climate were domestic political activities which were found to have the 'strongest influence on issue attention relative to the other two countries' (Schäfer et al. 2014, p. 76).

## 10 Conclusion

The fact that political events and political speech acts are the major driver of climate change communication has delimited the kinds of research conducted by Australian researchers. As the climate issue is so shaped by the interests of a fossil-fuel protectorate and an oligopoly of media power in Australia, most research on climate communication is focussed on elections and the nature of these power relations. Whilst Australia is highly exposed to climate impacts, as a 'middle power' Australia is not as vulnerable as developing countries to these impacts. Its capacity to adapt to, at least in the near-term, the impacts of climate change is high. If this were not the case, communicating climate change would be much more urgent than it is. Two exceptions to this are the high degree of newspaper coverage and corresponding research to both rural Australia (Quiggan et al. 2010; Waitt et al. 2012; Semujju 2015) and the Pacific (Ryan 2010; Harris 2013; Dreher and Voyer 2015), where issues of drought, sea-level rise, adaptation and climate justice have received moderate attention by Australian researchers.

However, many researchers in the field recognise that the relationship between media, the fossil-fuel industry and political elites is the primary context involved in Australia recognising and meeting its climate mitigation commitments. This has made this issue one that has become so heavily politicised and polarised as to paralyse constructive discussion of the climate challenges and appropriate policy responses that Australia has so long failed to address. Future research that investigates how the sustained high public concern for climate change policy action in the Australian context can fracture the media–fossil fuel industry–political elite roadblock that currently polarises the issue.

## References

Abbott T (2009) Battlelines. Melbourne University Press, Melbourne

Australian Government, Department of Broadband and Communications (2012) Convergence review: final report. AGPS, Canberra

Australian Government, Department of The Treasury (2014) Long run forecasts of Australia's terms of trade. AGPS, Canberra

Australian Mining (2015) Australia's export earnings in decline. Australian mining http://www.australianmining.com.au/features/australia-s-export-earnings-in-decline. Last accessed 2 Oct 2016

Bacon W, Jegan A (2011) Sceptical climate part 1: climate change policy. In: Bacon W (ed) Sceptical climate report. Australian Centre of Independent Journalism, Sydney
Bacon W, Jegan A (2013) Sceptical climate part 2: climate science in Australian newspapers. In: Bacon W (ed) Sceptical climate report. The Australian Centre for Independent Journalism, Sydney
Bruns A, Burgess J, Crawford K, Shaw F (2012) #qldfloods and @QPSMedia: crisis communication on Twitter in the 2011 South East Queensland floods. Queensland University of Technology, Brisbane, Centre for Creative Industries
Chubb P, Nash C (2012) The politics of reporting climate change at the Australian broadcasting corporation. Media Int Aust 144(1):37–48
Cook J, Nuccitelli D, Green SA, Richardson M, Winkler B, Painting R, Way R, Jacobs P, Skuce A (2013) Quantifying the consensus on anthropogenic global warming in the scientific literature. Environ Res Lett 8(2):1–7
Dreher T, Voyer M (2015) Climate Refugees or Migrants? Contesting media frames on Climate Justice in the Pacific. Environ Commun 9(1):58–76
Duxfield F (2013) Mining subsidies top $4.5 bn: Australia institute, ABC Rural http://www.abc.net.au/news/2013-06-25/nrn-dist-mining-subsidies/4778042. Last accessed 2 Oct 2016
Dyster B, Meredith D (2012) Australia in the global economy: continuity and change. Cambridge University Press, Cambridge
Eccleston R, Woolley T (2014) From Calgary to Canberra: resource taxation and fiscal federalism in Canada and Australia. Publius J Federalism 45:1–39
Editorial 1 (2011) The Australian entitled 'Sticking together in the face of the floods' ferocity' (12 January 2011. p. 13)
Eurobarometer (2009) Europeans' attitudes towards climate change. Special Eurobarometer 322/ Wave 72.1. European Commission, Brussels
Evans JP, Boyer-Souchet I (2012) Local sea surface temperatures add to extreme precipitation in northeast Australia during La Nina. Geophys Res Lett 39(10):L10803
Flannery T (2005) The weather makers: the history & future impact of climate change. Grove Press, New York
Flannery T (2008) Now or never: a sustainable future for Australia? Q Essay 31:1–66
Flannery T (2009a) Now or never: a sustainable future for Australia?. Black Inc, Melbourne
Flannery T (2009b) Now or never: why we need to act now for a sustainable future. HarperCollins, Toronto
Flannery T (2010) Here on earth. Text Publishing, Melbourne
Flannery T (2015) Atmosphere of hope: searching for solutions to the climate crisis. Atlantic Monthly Press, Boston
Garnaut R (2008) The garnaut climate change review. Cambridge University Press, Cambridge
Gurney M (2014) Missing in action? The 'non'-climate change debate of the 2013 Australian federal election. Glob Media J Aust Ed 8(2):1–21
Hamilton C (2007) Scorcher: the dirty politics of climate change. Black Inc, Melbourne
Hamilton C (2009, 16 October) How to deal with climate change grief. Crikey.com http://www.crikey.com.au/2009/10/16/hamilton-3-stages-of-climate-change-grief-denial-hope-angry-acceptance/. Last accessed 2 Oct 2016
Hamilton C (2010) Requiem for a species: why we resist the truth about climate change Allen & Unwin, Sydney
Hamilton C (2013) What history can teach us about climate change denial. In: Weintrobe S (ed) Engaging with climate change: psychoanalytic and interdisciplinary perspectives. Routledge, New York, pp 16–32
Harris U (2013) Using participatory media to assess climate change impact on vulnerable communities. Media Asia (Asian Media Information & Communication Centre) 40(4):321–326
Hendon H, Lim E, Arblaster JM, Anderson DLT (2013) Causes and predictability of the record wet in east Australian Spring 2010. Clim Dyn 42(5):1155–1174
Hine DW, Phillips WJ, Reser JP, Cooksey RW, Marks ADG, Nunn PD, Watt SE, Ellul MC (2013) Enhancing climate change communication: strategies for profiling and targeting Australian

interpretive communities. National Climate Change Adaptation Research Facility, Gold Coast. http://www.nccarf.edu.au/sites/default/files/attached_files_publications/Hine_2013_Enhancing_climate_change_communication.pdf. Last accessed 2 Oct 2016

Holmes D (2013a) Climate change and the politics of consensus. The Conversation http://theconversation.com/climate-change-and-the-politics-of-consensus-18940. Last accessed 2 Oct 2016

Holmes D (2013b) The tabloidization of climate change at four round conference: media engagement: East and West. Hong Kong Baptist University, Hong Kong 8–11 May

Holmes D (2016) Why has climate change disappeared from the Australian election radar? The Conversation http://theconversation.com/why-has-climate-change-disappeared-from-the-australian-election-radar-59809. Last accessed 2 Oct 2016

Hudson M (2015a) Carbon coups: from Hawke to Abbott, climate policy is never far away when leaders come a cropper. The Conversation https://theconversation.com/carbon-coups-from-hawke-to-abbott-climate-policy-is-never-far-away-when-leaders-come-a-cropper-47542. Last accessed 2 Oct 2016

Hudson M (2015b) 25 years ago the Australian government promised deep emissions cuts, and yet here we still are. The Conversation http://theconversation.com/25-years-ago-the-australian-government-promised-deep-emissions-cuts-and-yet-here-we-still-are-46805. Last accessed 2 Oct 2016

Hulme M (2009) Why we disagree about climate change: understanding controversy. Inaction and Opportunity Cambridge University Press, Cambridge

Hunter JSH (1974) Vertical intergovernmental financial imbalance: a framework for evaluation. Public Finance Analysis 2:481–492

Jones P, Pusey M (2008) Mediated political communication in Australia: leading issues, new evidence. Aust J Soc Issues 43(4):583–599

Kwak H, Lee C, Park H, Moon S (2010). What is Twitter, a social network or a news media? In: Proceedings of the 19th international conference on World Wide Web (WWW'10) ACM Press, New York

Leiserowitz A, Maibach E, Roser-Renouf C, Smith N (2009) Global warming six Americas 2009. Yale project on climate change and George Mason University Center for Climate Change Communication, New Haven, Connecticut http://environment.yale.edu/climate-communication/files/climatechange-6americas.pdf. Last accessed 2 Oct 2016

Leiserowitz A, Maibach E, Roser-Renouf C, Feinberg G, Howe P (2013a) Global warming's six Americas: September 2012. Yale project on climate change and George Mason University Center for Climate Change Communication, New Haven, Connecticut http://environment.yale.edu/climate-communication-OFF/files/Six-Americas-September-2012.pdf. Last accessed 2 Oct 2016

Leiserowitz A, Thaker J, Feinberg G, Cooper D (2013b) Global Warming's Six Indias. Yale project on climate change and George Mason University Center for Climate Change Communication, New Haven, Connecticut http://environment.yale.edu/climate-communication/article/Global-Warming-Six-Indias. Last accessed 2 Oct 2016

Lester L, McGaurr L, Tranter B (2015) The election that forgot the environment? issues, EMOs, and the Press in Australia. Int J Press/Politics 20(1):3–25

Leviston Z, Greenhill M, Walker I (2015) Australian attitudes to climate change: 2010–2014. CSIRO, Perth

Manne R (2011) Murdoch's Australian and the shaping of the nation. Q Essay 43:1–142

Manne R (2015) Diabolical: how we have failed to address climate change. The monthly: summer edition, December 2015–January 2016 http://www.themonthly.com.au/issue/2015/December/1448888400/robert-manne/diabolical. Last accessed 2 Oct 2016

Marr D (2010) Power trip: the political journey of Kevin Rudd. Q Essay 38:1–91

McKnight D (2010) A change in the climate? The journalism of opinion at news corporation. Journalism 11(6):693–706

Metag J, Füchslin T, Schäfer M (2015) Global warming's five Germanys: A typology of Germans' views on climate change and patterns of media use and information. Public Understanding of Science Published online before print July 3 2015 doi:10.1177/0963662515592558

Miller AH, Miller WE, Raine AS, Brown TA (1976) A majority party in disarray: policy polarization in the 1972 election. Am Polit Sci Rev 70(3):753–778
Morrison M, Duncan R, Sherley C, Parton K (2013) A comparison between attitudes to climate change in Australia and the United States. Aust J Environ Manage 20(2):87–100
O'Donnell K (2011) The architected ritual of twitter: personal users disaster communication in the 2011 Queensland floods. Honours Dissertation, Communications and Media Monash University, Melbourne
Olausson U (2011) "We're the ones to blame": citizens' representations of climate change and the role of the media. Environ Commun 5(3):281–299
Parkin A, Anderson G (2007) The howard government, regulatory federalism and the transformation of commonwealth-state relations. Aust J Polit Sci 42(2):295–314
Pearse G (2009) Quarry vision: coal, climate change and the end of the resources boom. Q Essay 33:1–122
Poortinga W, Spence A, Whitmarsh L, Capstick S, Pidgeon N (2011) Uncertain climate: an investigation into public scepticism about anthropogenic climate change. Glob Environ Change 21:1015–1024
Quiggin J, Adamson D, Chambers S, Schrobback P (2010) Climate change, mitigation and adaptation: the case of irrigated agriculture in the Murray-Darling basin in Australia. Can J Agric Econ 58(4):531–554
Rootes C (2011) Denied, deferred, triumphant? Climate change, carbon trading and the Greens in the Australian federal election of 2010. Environ Politics 20(3):410–417
Rowe D (2011) Comparing newspaper coverage of climate change during election campaigns in the United States, Canada and Australia. PhD Research thesis, Syracuse University, Syracuse http://surface.syr.edu/com_etd/87/. Last accessed 2 Oct 2016
Ryan Y (2010) COP 15 and Pacific Island states: a collective voice on climate change. Pacific J Rev 16(1):193–203
Schäfer MS, Ivanova A, Schmidt A (2014) What drives media attention for climate change? Explaining issue attention in Australian, German and Indian print media from 1996 to 2010. Int Commun Gaz 76(2):152–176
Sciarini P, Bornstein N, Lanz B (2007) The determinants of voting choices on environmental issues: a two-level analysis. In: de Vreese C (ed) The dynamics of referendum campaigns: an international perspective Palgrave Macmillan, New York
Semujju B (2015) frontline farmers, backline sources: women as a tertiary voice in climate change coverage. Feminist Media Stud 15(4):658–674
Sheehan P, Gregory RG (2013) The resources boom and economic policy in the long run. Aust Econ Rev 46(2):121–139
Sherley C, Morrison M, Duncan R, Parton K (2014) Using segmentation and prototyping in engaging politically-salient climate-change household segments. J Nonprofit Public Sect Mark 26(3):258–280
Skeptical Science Website http://www.skepticalscience.com
Staples J (2012) Non-government organisations and the Australian government: a dual strategy of public advocacy for NGOs. PhD Research thesis, UNSW, Sydney
St. John A (2014) Resource development and landholders' rights: a quick guide. Australian Parliamentary Library, Canberra http://www.aph.gov.au/About_Parliament/Parliamentary_Departments/Parliamentary_Library/pubs/rp/rp1314/QG/ResourceDevelopment. Last accessed 2 Oct 2016
Tranter B (2011) Political divisions over climate change and environmental issues in Australia. Environ Polit 20(1):78–96
Ummenhofer CC, Gupta AS, England MH, Taschetto AS, Briggs PR, Raupach MR (2015) How did ocean warming affect Australian rainfall extremes during the 2010/2011 La Niña event? Geophys Res Lett 42:9942–9951. doi:10.1002/2015GL065948
Wait G, Farbotko C, Criddle B (2012) Scalar politics of climate change: regions, emissions and responsibility. Media Int Aust 143(1):36–46

Walker KJ (1994) The political economy of environmental policy: an Australian introduction. UNSW Press, Sydney
Washington H, Cook J (2011) Climate change denial: heads in the sand. Earthscan, London
Whitmarsh L (2011) Scepticism and uncertainty about climate change: dimensions, determinants and change over time. Glob Environ Change 21(2):690–700

# Inclusion of Gender in Africa's Climate Change Policies and Strategies

**Mary Nyasimi, Ayansina Ayanlade, Catherine Mungai, Mercy Derkyi and Margaret O. Jegede**

## 1 Introduction: Overview of Climate Change in Africa and Gender-Related Adaptation

Africa is a diverse continent with varied geography and topology, resources, population sizes and cultures, and differences in developmental progress. African countries continue to face the global climate change challenge, which threatens their economic gains and prospects (McCarthy 2001; Shackleton et al. 2015). This is becoming increasingly evident as their economies are already experiencing increasing vulnerability due to their dependence on climate sensitive sectors such as agriculture and forestry for economic growth (Leal Filho et al. 2015; Midgley and Bond 2015; Souza et al. 2015). Since climate change requires global solutions, therefore, Africa and the rest of the world need to join in managing the effects of climate change on different sectors of the economy. The Intergovernmental Panel on Climate Change (IPCC) is a scientific body set up by the United Nations to engender scientific and technical information on climate change issues. The reports generated by IPCC are used by parties to United Nations Framework Convention

M. Nyasimi (✉) · A. Ayanlade · C. Mungai
CGIAR Research Program on Climate Change, Agriculture and Food Security (CCAFS), International Livestock Research Institute (ILRI), Nairobi, Kenya
e-mail: m.nyasimi@cgiar.org

A. Ayanlade
Department of Geography, Obafemi Awolowo University, Ile-Ife, Nigeria

M. Derkyi
Department of Forest Science, School of Natural Resources, University of Energy and Natural Resources (UENR), Sunyani, Ghana

M. O. Jegede
African Institute for Science Policy and Innovation (AISPI), Obafemi Awolowo University, Ile-Ife, Nigeria



W. Leal Filho et al. (eds.), *Handbook of Climate Change Communication: Vol. 1*, Climate Change Management, https://doi.org/10.1007/978-3-319-69838-0_11

on Climate Change (UNFCCC), which is the main international treaty on climate change, for negotiations (Ho-Lem et al. 2011; Madzwamuse 2011).

African countries participate in the global discussions on climate change through the African Group of Negotiators (AGN), Group of 77 plus China (G77) and Least Developing Countries (LDC). A major focus for Africa during negotiations is strengthening adaptation efforts due to the continent's low adaptive capacity and its negligible carbon footprint compared to the rest of the world. Africa also lobbies for adaptation finance, technology support and mechanisms to address the current loss and damage from extreme climatic events (Yamin and Depledge 2004; Roger and Belliethathan 2016). Nonetheless, it is equally important that African leaders are open to mitigation opportunities that will bring technology innovations much closer to their development agenda. The Paris Agreement (2015) that entered into force on 4th October, 2016 by being ratified by 81 parties representing more than 55% of global emissions, creates opportunities for African countries to draft and implement nationally relevant climate change policies and strategies. As of October 2017, 169 of the 197 Parties have ratified to the Convention. Notably are the international commitments such as the Intended Nationally Determined Contributions (INDCs), now refered to as National Determined Contributions (NDCs), national polices and strategies including National Adaptation Plans (NAPs), Nationally Appropriate Mitigation Actions (NAMAs). Implementation of these actions will commence prior to 2020. While progress has been made in developing polices and strategies for climate change adaptation and mitigation, it is critical to ensure that these do not lead to inequalities, especially gender and social inequality during implementation.

## 2 Gender and Climate Change Issues in Africa

Climate change and variability is exacerbating the drying up of already arid zones in Africa, decreasing in volumes of inland surface and subsurface water, causing shifts in cropping and farming systems, changes in pests and disease dynamics and creating competing demands for shrinking natural resources that have the potential for creating bilateral and multilateral conflicts (Buhaug et al. 2015; Mora et al. 2015; Rose 2015; Akresh 2016; Sonwa et al. 2016). There are quite a lot of studies on the challenges of climate change and its impacts in Africa, but there is a gap yet to be filled on women, gender and climate change. The question of who will be affected the most—male or female, old or the young, unfortunately, remains relatively underexplored in climate policy domain. Scattered research across the continent has yielded some gendered vulnerability to climate change. Some of the studies report that the risks of climate change are mostly felt by rural dwellers and especially the women and youth who are the most exposed to various ill-health which result from environmental changes (Bradshaw 2010). This is because they have low coping capacity and are highly vulnerable to the negative outcomes of climate change (Adger et al. 2009; Vincent et al. 2010).

Adger et al. (2009) reports that women and youth are vulnerable to weather related disasters due to socially and culturally determined gender roles. For example, women do not always learn to swim and so can be vulnerable to flooding. Concerns about their physical and psychosocial safety in emergency shelters and camps can also impede the willingness of women to leave their homes, even when at risk from weather disasters. There is often differential mortality between men and women after disasters, with one study of extreme events in 141 countries from 1981 to 2002 finding that more women are killed during extreme weather events than men (Neumayer and Plümper 2007). In some regions of Africa, climate change is causing rivers and lakes to dry up and, women and youth in the rural areas are seen with pots and other water carrying materials, walking long distances in order to get water for their domestic use. In Nigeria, persistent droughts and flooding have had serious drawbacks in the academics of the young children especially in the northern states (Adejuwon and Jegede 2011). According to this study, school enrolment in some part of Sokoto state, Nigeria was very low in the years of drought as compared to those of other years when drought was not recorded. The livelihoods of women dependent on fishing on this lakes has dwindled with decline in fish numbers (Iyun 1999).

Recognition of the above differential realities between women, men and the youth are thus an important prerequisite to ensuring that actions aimed at adaptation and mitigation are gender-responsive. Few countries have enacted climate change policies and plans however, the actions identified within these policies do not address the needs of the most vulnerable people, that is women and youth (Mburia 2015).

This chapter, therefore, reviews climate change related policies and strategies in East and West Africa through a gendered lens. The countries are Kenya, Uganda, Tanzania in East Africa and Ghana and Nigeria in West Africa. Drawing upon a gender inclusion criteria, the study assesses how countries have integrated gender in their climate change policies and strategies. We did a qualitative review of existing climate change policies and strategies using minimum principles for gender responsive climate change policies and strategies that comprise a set of the criteria shown in Table 1. These are some of the principles that countries can incorporate to address the bottlenecks that prevent the prioritization and institutionalization of gender at national levels. In this chapter, the term gender refers to women, youth and other vulnerable groups.

**Table 1** Principles and minimum criteria for a gender-responsive climate change policy or strategy

| Principles | Criteria |
|---|---|
| 1. Evidence-based | Analyze, whether the climate change policies/plans/strategies, did a gender analysis on climate risks, impacts, and vulnerability |
| 2. Positive social and gender norms | Analysis whether vulnerable people are identified, defined and targeted and their knowledge and capacity to address risk and vulnerability assessed |
| 3. Equal benefits | Analyze whether the design of initiatives and activities within the climate change policies/plans/strategies aim to ensure resulting benefits, systems and services are beneficial to women and youth |
| 4. Participation | Analyze whether the climate change policies/plans/strategies promote participation, voice and inclusion of all groups, especially women and youth. Analyze whether they recognize and strengthen the capacity and rights of vulnerable men, women and youth to continually participate in adaptation decision-making |
| 5. Needs assessment and analysis | Analyze whether the climate change policies/plans/strategies identified adaptation and mitigation actions that respond to needs of women, youth, and other vulnerable groups |
| 6. Strategic planning and implementation | Analyze whether the technology choices identified in the policies/plans/strategies respond to risk and vulnerability as well as anticipation of future climate and uncertainty at community and national level, and consider trade-offs and synergies between different vulnerable groups<br>Analyze whether the technology choices identified in the climate change policies/plans/strategies build on local knowledge, and consider availability, accessibility, affordability, relevance to women, youth and other vulnerable groups<br>Analyze the extent to which climate change policies/plans/strategies facilitate access to finances across gender and youth |
| 7. Monitoring and evaluation | Analyze whether the monitoring system incorporates reflection by women and youth and identifies feedback loops to inform modifications as the project evolves so that activities are successfully adapted to new learning and does not create/lead to more inequalities |

# 3 Climate Change Policies, Strategies, and Plans

## *3.1 Intended Nationally Determined Contributions*

During the 19th session of the UNFCCC Committee of Parties (COP 19) held in Warsaw, Poland, countries agreed to submit to the UNFCCC their INDCs, that are now referred to as National Determined Contributions (NDCs) after signing of the Paris climate agreement. It is important to point out that the Paris Agreement (2015) does not reinforce the commitments made with the INDCs. This therefore was a disappointment. Each country sets of target and actions will come into effect post-2020. By August 2016, a total of 181 countries had submitted their INDCs (Deprez et al. 2015; Huang 2015; Stavins 2015). A practical and comprehensive INDC should be written through a transparent process involving grassroots, domestic and international stakeholders. This will build trust and accountability and help stakeholders to track progress towards lowering a country's emissions. Specifically, the actions outlined within the INDC must be articulated clearly and measurable in order to show how the actions will contribute to emissions reductions and climate resilience in the future (Zhang and Pan 2016). An INDC must also be linked with a country's sustainable development goals and national climate change policies. The critical component missing in this is the inclusion of gender dimension as an indicator of a good INDC. Despite the lack of inclusion of dimensions of gender and social inclusion as an indicator for a good INDC, there has been gradual progress towards recognizing the need to address gender equality in climate change dialogues (e.g., Paris agreement), climate change policies and plans. To understand the manner in which the five African countries have included gender-responsive actions in their INDCs, we identified few variables to analyze the INDCs as outlined in Appendix I.

All the five countries submitted their INDCs and they are focused on both mitigation and adaptation actions to achieve the below 2 °C target. However, implementations of the mitigation and adaptation actions are split into two, unconditional (what a country can do without external financial support) and conditional (what a country can do with external financial support). It is important to note that the INDCs are not comprehensive documents, in that, they did not include the process on how they were written; it was not possible to apply all the principles identified in Table 1. Four of the five countries—Kenya, Uganda, Ghana, and Nigeria have mentioned some related concept of 'gender" in different sub-sections of their INDCs. This indicates a national consciousness of the critical gender and climate change inter-linkages within a country's development agenda. Tanzania does not mention gender or any other related dimension of gender.

As these countries are beginning to deliberate on their mitigation and adaptation policies and actions as they work towards post 2020, an obvious opportunity stands out for revisiting the INDCs and other national plans and strategies to ensure that that gender remains an integral component. Uganda and Nigeria provided a gender analysis and identified vulnerable groups. These are women, children, elderly, persons with disabilities or sick, the rural poor and those living in slums as the most vulnerable to the impacts of climate change. Nigeria and Uganda clearly outline the vulnerability of women, children, the elderly, and persons with disabilities or sick in terms of food insecurity, water shortage and fuel wood scarcity. Nigeria integrated the gender and social inclusion dimension to ensure that different groups of people based on 'their social status, gender, wealth, political power and access to and control over resources' are also targeted. Furthermore, Uganda's INDC stresses on the importance of targeting the rural poor who derive their livelihoods from agrarian activities, and the poor living in slums in peri-urban and urban areas. These categories of people are especially vulnerable as they have lower capacity to cope with and adapt to the impacts of the changing climate. Nigeria developed eight criteria to

**Table 2** Analysis of countries INDCs

| Characteristics of the INDC | Kenya | Uganda | Tanzania | Nigeria | Ghana |
|---|---|---|---|---|---|
| Percentage of emission reduction goal by 2030[a] | 30%[b] | 22%[c] | 10–20%[c] | 20%[d] (45% with external support) | 15%[d] (45% with external support) |
| Gender mainstreaming in policy | Yes | Yes | No | Yes | Yes |
| Identification of gender, vulnerable groups and youths | Yes | Yes | No | Yes | Yes |
| Specific actions | Strengthen the adaptive capacity of the most vulnerable groups and communities through social safety nets and insurance schemes | Respect for human rights and gender-responsive climate change actions<br>The protection of vulnerable groups, including women<br>Mainstreaming gender into development policies, plans and strategies as well as observance of human rights in all climate change adaptation and mitigation actions<br>Promote and encourage the mainstreaming of gender considerations in climate change issues | | | Resilience for gender and the vulnerable<br>Implementation of community led adaptation and livelihood diversification for vulnerable groups |

[a]Relative to a business-as-usual (BAU) scenario emission of 73.95MtCO2e 2 by 2030
[b]Relative to the BAU scenario of 143 MtCO2eq and with external support on finance, investment, technology development and transfer, and capacity building)
[c]The BAU project of 77.3 MtCO2eq/year in 2030
[d]Relative to the BAU scenario of 138–153 Million tonnes of carbon dioxide equivalent (MtCO2e)

assess potential mitigation actions and gender and social inclusion was one of them. Despite that, there is a lack of gender-based differential in identified priority actions for all the countries' INDCs (Table 2). Although the four countries indicated that "gender-responsive climate change actions' will be considered, there is a lack of gender-differential actions for all the different categories of people identified, that is, women, youth, disabled, the rural poor and the urban poor. It is important to point out that the measures included within Nigeria's INDCs were considered to be gender neutral and/or to enhance social inclusion. Therefore, as countries transition from INDCs to NDCs, more gender differential actions are needed.

In terms of financing for INDCs actions that target women and other vulnerable groups, only Ghana has made a commitment to implement community led adaptation diversification at the cost of USD 1023 million that can build resilience of women and other vulnerable groups. The Ghana National Climate Change Adaptation Strategy (NCCAS) document didn't do gender analysis on climate risks, impacts and vulnerability; rather, it looked at the impacts on the various sectors of the Ghanaian economy such as economic, social and infrastructural groups. The document further said the cumulative effects on these sectors determine the impacts and vulnerabilities of various livelihoods groups and places in the country. The document however mentioned poverty as a good indicator of resilience, while occupation and location determine sensitivity to climate change in Ghana. Poverty in Ghana is highly specialized such that rural areas harbor the bulk of the poor compared to urban areas. The Ghana National Climate Change Policy (NCCP) document however did gender analysis on climate risks, impacts and vulnerability. According to the policy document, climate change will hit women harder than men because of existing vulnerabilities and gender inequalities since women primarily constitute the majority of the poor and are more dependent on natural resources for their livelihood (Kalame et al. 2009; Essegbey et al. 2016).

Kenya and Uganda have mentioned mainstreaming gender in climate change policy, development policies, plans and strategies. None of the countries' INDCs mention a gender-sensitive process of Monitoring, Reporting and Verification (MRV). The MRV is a key pillar that a country can use to track the implementation and impact of climate change policies, plans and strategies. Therefore, if gender is lacking in the MRV, the five countries will lose out on reporting on the gendered impacts and outcomes of the proposed actions listed in their INDCs.

## 3.2 National Climate Change Policy and Strategies

In Ghana, gender issues have been given special prominence in the broad thematic areas and programmes in the National Climate Change Policy. Women and the girl child are often responsible for the very areas that are most vulnerable to climate change, including household chores, water, food and fuel supplies. The NCCP recognizes their particular vulnerability to climate change and ensures that they are part of the response (Sarpong and Anyidoho 2012; Essegbey et al. 2016). Ghana's

NCCP asserts that the impacts of climate change will hit women harder than men because of existing vulnerabilities and gender inequalities since women primarily constitute the majority of the poor and are more dependent on natural resources for their livelihood. It further encourages equal participation of women and men as well as allowing for the differentiated impacts on women and men from climate change and its response measures to be taken into account in line with various national policies on women's rights and gender equality. The NCCP intends to;

- Promote equal opportunities and affirmative action for women and vulnerable groups in climate change adaptation and mitigation through mainstreaming gender issues into national and sub-national climate-change-related policies.
- Increase knowledge and strengthen capacity at all levels on gender-responsive climate change policies, strategies and programmes.
- Emphasis on promotion of vocational training, especially for youth, in places with high likelihood of receiving in-migration.
- The Ministry of Gender, Children and Social Protection is represented in the NCCP implementation which is spearheading national drives to overcome the challenges of gender inequality.

The NCCP and NCCAS identified the disabled, aged, women, youth, and children as the most vulnerable to climate change impacts. Men are included as vulnerable. However, the degree of vulnerability is different for men, women and children. Additionally, Ghana's NCCP and NCCAS identified other vulnerable as people who are poorer and living in the rural areas or marginal environments and depending on natural resource-based activities for their livelihoods. The socio-economic groups affected most by climate change include small-scale food crop farmers, women small-scale farmers, livestock operators, fishermen and fishmongers, slum dwellers, and migrant farm workers and smallholder farmers in drier areas. The NCCAS targeted livelihoods and agriculture and the actions include; strengthening the relationship between scientific knowledge and traditional or indigenous knowledge; building and strengthening capacity of local farmers to increase agricultural productivity and awareness of climate issues; and enhance the living standards of vulnerable groups through acquisition of alternative livelihoods skills.

Ghana's climate change policy identified gender equality as important precondition for successful climate change adaptation and mitigation responses and equitable social development (Table 3).

Nigeria identified women, children, and resource-poor men as those vulnerable to climate change impacts. Uganda, identified the vulnerable groups to include the poor, people living with disability, youth, people living with HIV/AIDS, the elderly, orphans and vulnerable children, refugees, and marginalized communities, due to their limited adaptive capacity. Tanzania's NCCS, states that climate change impacts are differently affecting gender and vulnerable groups due to their different roles in the society and therefore gender perspective in development of adaptation and mitigation strategies is therefore crucial to narrow the gender gap and reducing

**Table 3** Inclusion of gender in national climate change policies and strategies (Republic of Uganda, 2012; Republic of Ghana, 2013; Government of Kenya, 2010; Federal Republic of Nigeria, 2011; United Republic of Tanzania, 2012)

| Characteristics of the climate change policy and strategy | Ghana | Uganda | Kenya | Nigeria | Tanzania |
|---|---|---|---|---|---|
| Document reviewed | NCC, 2013<br>NCCAS | NCCP, 2012 | NCCRS, 2010 | NASPA-CCN 2011 | NCCS, 2012 |
| Gender analysis on climate risks, impacts and vulnerability | Yes | Yes | Yes | Yes | Yes |
| Identification of gender, vulnerable groups and youth | | Yes | | Yes | |
| Direct benefits to women, youth and other vulnerable groups | Yes | No | No | No | Yes |
| Participation and empowerment of women and youth | Yes | Yes | No | No | Yes |
| Building on local knowledge | yes | No | No | No | No |
| Budget allocation for gender specific actions | Yes | No | No | No | Somewhat |
| Mainstreaming gender | Yes | Yes | Yes | Yes | Yes |
| Gender responsive monitoring and evaluation systems | Yes | No | No | No | No |
| Specific actions | Systematic gender analysis, collection and utilization of sex-disaggregated data, and the establishment of gender-sensitive benchmarks and indicators need to be undertaken<br>Promotion of vocational training, especially for youth | Strengthen the capacity of relevant stakeholders at national and local levels to promote gender-sensitive approaches to climate change adaptation | n/a | n/a | Promoting safeguards that will ensure gender and vulnerable groups' equity in benefit sharing<br>related to climate change initiatives<br>Generate gender disaggregated data on impacts and response |

impacts. The same is echoed in Kenya's NCCRS that calls for a Gender-based response strategy.

Uganda's policy document advocates for climate change response policies and activities to be gender sensitive. It also calls for strengthening the capacity of relevant stakeholders at national and local levels to promote gender-sensitive approaches to climate change adaptation. In Ghana, the policy document calls for policies that support gender equality in access, use and control over science and technology, formal and informal education and training to enhance the nation's capability in disaster reduction, mitigation and adaptation to climate change.

Only Ghana has plans to strengthen the relationship between scientific knowledge and traditional or indigenous knowledge and build and strengthens the capacity of local farmers on climate issues and awareness. Ghana emphases on budget allocation for gender issues and climate change. Ghana also intends to promote gender equitable financing as a means of responding to the differential impacts of climate change by gender. It requires establishing clear mechanisms for integrating a gender dimension into the design, implementation and monitoring of all climate funds. According to Ghana, the government is considering the potential for a national financing mechanism or facility that can meet the needs of different themes, such as gender issues. Tanzania's strategy did not specify any facilitation to access finance across gender, however, an integrated approach and coordinated working system was highly recommended to ensure that funds to address climate change are used to achieve the objectives presented in the strategy. Ghana's NCCP outlines the development of relevant gender-sensitive indicators as part of Ghana's monitoring and reporting system. Both Uganda and Tanzania do outline a monitoring and evaluation system that will be linked to the planned outcomes and outputs of the costed Implementation Strategy. The results and findings shall guide in revising the Strategy.

Ghana, Uganda and Tanzania advocate for enhancing equitable representation of women and vulnerable groups at all levels in planning, decision making and implementation of adaptation and mitigation actions. The documents also emphasize equal access to training and capacity-building programmes to ensure women's full participation in climate change initiatives. Uganda's policy intends to mainstream gender issues in climate change adaptation and mitigation approaches in order to reduce the vulnerability of women and children to the impacts of climate change. The policy has the priority to ensure that communities are empowered and that both men and women participate meaningfully in planning, testing and rolling out adaptation and mitigation activities in rural and urban areas.

Finally, Ghana advocates for linking climate change to the Ministry of Gender, Children and Social Protection, which is spearheading national drives to overcome the challenges of gender inequality.

## 4 The Gender Dimensions of Climate Change Communication for Policy Development and Implementation

As countries continue to develop and implement climate change policies, programmes, actions and strategies, it is essential to support effective communication and information sharing from policy makers to implementers and the community at large. This should also include feedback loops to ensure that information from the local level is integrated into policy development at national, regional and international level. Communicating climate change is complex, given that the phenomenon of climate change itself is neither 'black' nor 'white'. The causes and impacts of climate change are difficult to discern given that climate change is not a stand-alone phenomenon. Indeed one may wonder how communicating climate change is different from communicating other environmental problems, economic challenges, risks, health issues, policy dilemmas, or the need for political engagement and behavior change (Moser 2010). Moser further noted that:

> Climate change—no matter how certain and urgent to experts—for now, and maybe for some time, is fundamentally a mediated, ambiguous problem for most audiences and easily trumped by more direct experiences. This means that lay audiences need to receive ample, clear, sufficiently strong, and consistent signals that support the necessary changes. This clear signaling task cannot be underestimated given the hurdles of cognition, the human-nature disconnect, climate and societal system lags, and other matters competing for constant attention. (Moser 2010).

While communicating climate change itself is not complex enough in itself, there is also the need to consider the gender dimensions of communicating climate change. Taking into cognizance the differential impacts of climate change and the respective response strategies informed by gender and social differentiation characteristics. Climate change communication in the African context needs to take into account cultural diversity, communities understanding of issues as well as how women and youth (specifically), in these contexts express their hopes and aspirations. Other key issues that determine the effectiveness of communication include the conditions under which the message is crafted, possible blocks or hindrances, feedback from recipients and the effects of the message sent (Röhr 2009). There are different tools which can be used to communicate climate change. It is critical to understand the strengths and weaknesses of the different tools using a gender lens. Traditional communication tools such as radio, theater, video and face-to-face interactions are the most effective for women, particularly in instances where new technologies are not accessible. Indeed radio has emerged as the most accessible communication tool for grassroots women because it is cheap, easy to use and does not require literacy. Also, it doesn't not interfere with their work and allows information to reach communities immediately and over large areas (Tall et al. 2014).

With the advent of modern ICT tools, other innovative communication channels, such as mobile phones, the internet, and interactive voice-response systems, could bridge the gap and spur a wider impact on the capacity of farmers to managing risks in agriculture. The number of people owning mobile phones in Africa is expected to increase in the next few decades (Hampshire et al. 2015). This platform could support information dissemination on expected local weather and climatic conditions, local market prices, and other relevant information. The platform may also enable farmers, irrespective of gender, to make inquiries (in all languages) specific to their crops, and receive personalized replies from agricultural extension experts on their phones (AGRA 2014). Focusing on agriculture in Africa, where women contribute larger percentage of the labor force (Cleaver 1993), several strategies have been used to communicate weather and climate information and their respective derived agro-advisories. Innovative communication strategies could complement community level efforts to enhance social cohesion through stronger community networks and local institutions and gender equality focused on improving farmers' capabilities in managing climate-related risks. This could be particularly important in communities where farmers are involved in groups that motivate and shape proactive and participatory decision-making (AGRA 2014).

## 5 Conclusion

Gender equality is essential to the effective initiation and implementation of climate change policies and strategies in Africa. Women, youth and other vulnerable groups are not just helpless victims of climate change, but can be powerful agents of change if their knowledge and leadership is harnessed. The mention of gender, identification of vulnerable groups, and identification of gender targeted mitigation and adaptation actions in INDCs (except Tanzania), and climate change policies and strategies represents a progressive trend in Africa. As African countries are in the process of transitioning their INDCs to NDCs, and implementing their own national policies on climate change, this presents a critical time to make gender equitable mitigation and adaptation actions.

Policies that support gender equality in access, use and control over science and technology, formal and informal education and training will enhance the nation's capability in disaster reduction, mitigation and adaptation to climate change. Gender equality is important precondition for successful climate change adaptation and mitigation responses and equitable social development. Furthermore, participation of women and men is essential to policy making and implementation of response measures should be a high priority in Africa. Finally, countries should determine gender responsive financial and budgetary needs to effectively implement gender-responsive climate change policies and strategies.

## References


Adejuwon JO, Jegede MO (2011) The impacts of drought open school enrollment and attendance In a Rural-urban setting of Sokoto state. J Meteorol Clim Sci 9:1–7

Adger WN, Dessai S, Goulden M, Hulme M, Lorenzoni I, Nelson DR, Naess LO, Wolf J, Wreford A (2009) Are there social limits to adaptation to climate change? Clim Change 93:335–354

AGRA AfaGRiA (2014) Africa Agriculture Status Report 2014 CLIMATE CHANGE AND SMALLHOLDER AGRICULTURE IN SUB-SAHARAN AFRICA. In: https://www-cif.climateinvestmentfunds.org/sites/default/files/aasr-2014climate-change-and-smallholder-agriculture-in-ssa.pdf (ed)

Akresh R (2016) Climate change, conflict, and children. Future Child 51–71

Bradshaw S (2010) Women, poverty and disasters: exploring the links through Hurricane Mitch in Nicaragua. The international handbook of gender and poverty Northhampton, Edward Elgar Publishing

Buhaug H, Benaminsen TA, Sjaastad E, Theisen OM (2015) Climate variability, food production shocks, and violent conflict in Sub-Saharan Africa. Environ Res Lett 10:125015

Cleaver KM (1993) A strategy to develop agriculture in sub-Saharan Africa and a focus for the World Bank, vol 23. World Bank Publications

Deprez A, Colombier M, Spencer T (2015) Transparency and the Paris agreement: driving ambitious action in the new climate regime. In. Institut du développement durable et des relations internationales. Working paper

Essegbey GO, Totin E, Karbo N, Traoré PS, Zougmoré R (2016) Assessment of climate change policy and institutional context: the case of Ghana

Hampshire K, Porter G, Owusu SA, Mariwah S, Abane A, Robson E, Munthali A, DeLannoy A, Bango A, Gunguluza N (2015) Informal m-health: How are young people using mobile phones to bridge healthcare gaps in Sub-Saharan Africa? Soc Sci Med 142:90–99

Ho-Lem C, Zerriffi H, Kandlikar M (2011) Who participates in the Intergovernmental Panel on Climate Change and why: a quantitative assessment of the national representation of authors in the Intergovernmental Panel on Climate Change. Glob Environ Change 21:1308–1317

Huang J (2015) The 2015 climate agreement: key lessons learned and legal issues on the road to Paris. Available at SSRN 2724109

Iyun BF (1999) Impact of environmental change on women in a peripheral coastal area, Nigeria. GeoJournal 48:223–229

Kalame FB, Nkem J, Idinoba M, Kanninen M (2009) Matching national forest policies and management practices for climate change adaptation in Burkina Faso and Ghana. Mitig Adapt Strat Glob Change 14:135–151

Leal Filho W, Rao KP, Sridhar G (2015) Adapting African agriculture to climate change, vol. Springer

Madzwamuse M (2011) Climate governance in Africa-adaptation strategies and institutions

Mburia R (2015) Africa climate change policy: an adaptation and development challenge in a dangerous world. Climate Emergency Institute, UNEP, Kenya

McCarthy JJ (2001) Climate change 2001: impacts, adaptation, and vulnerability: contribution of Working Group II to the third assessment report of the Intergovernmental Panel on Climate Change, vol. Cambridge University Press

Midgley GF, Bond WJ (2015) Future of African terrestrial biodiversity and ecosystems under anthropogenic climate change. Nat Clim Change 5:823–829

Mora C, Caldwell IR, Caldwell JM, Fisher MR, Genco BM, Running SW (2015) Suitable days for plant growth disappear under projected climate change: potential human and biotic vulnerability. PLoS Biol 13:e1002167

Moser SC (2010) Communicating climate change: history, challenges, process and future directions. Wiley Interdisc Rev Clim Change 1:31–53

Neumayer E, Plümper T (2007) The gendered nature of natural disasters: the impact of catastrophic events on the gender gap in life expectancy, 1981–2002. Ann Assoc Am Geogr 97:551–566
Roger C, Belliethathan S (2016) Africa in the global climate change negotiations. Int Environ Agreements Polit Law Econ 16:91–108
Röhr U (2009) Gender into climate policy: Toolkits for climate experts and decision makers. GenderCC–Women for Climate Justice, Wiesbaden
Rose RM (2015) The impact of climate change on human security in the Sahel Region of Africa. Donnish J Afr Stud Dev 1:009–014
Sarpong DB, Anyidoho NA (2012) Climate change and agricultural policy processes in Ghana
Shackleton S, Ziervogel G, Sallu S, Gill T, Tschakert P (2015) Why is socially-just climate change adaptation in sub-Saharan Africa so challenging? A review of barriers identified from empirical cases. Wiley Interdisc Rev Clim Change 6:321–344
Sonwa DJ, Dieye A, El Mzouri E-H, Majule A, Mugabe FT, Omolo N, Wouapi H, Obando J, Brooks N (2016) Drivers of climate risk in African agriculture. Clim Dev 1–16
Souza K, Kituyi E, Harvey B, Leone M, Murali KS, Ford JD (2015) Vulnerability to climate change in three hot spots in Africa and Asia: key issues for policy-relevant adaptation and resilience-building research. Reg Environ Change 15:747–753
Stavins RN (2015) 20 Linkage of regional, national, and sub-national policies in a future international climate agreement. In: Towards a workable and effective climate regime, p 283
Tall A, Hansen J, Jay A, Campbell B, Kinyangi J, Aggarwal PK, Zougmoré R (2014) Scaling up climate services for farmers: mission possible. Learning from good practice in Africa and South Asia
Vincent K, Cull T, Archer ER (2010) Gendered vulnerability to climate change in Limpopo province, South Africa. In: Gender and climate change: an introduction, pp 160–167
Yamin F, Depledge J (2004) The international climate change regime: a guide to rules, institutions and procedures, vol. Cambridge University Press
Zhang W, Pan X (2016) Study on the demand of climate finance for developing countries based on submitted INDC. Adv Clim Change Res

## Other Materials

Alliance for a Green Revolution in Africa (AGRA) (2014) https://www-cif.climateinvestmentfunds.org/sites/default/files/aasr-2014climate-change-and-smallholder-agriculture-in-ssa.pdf. Accessed Sept 2016
Ghana National Climate Change Adaptation Strategy (NCCAS) document
Ghana National Climate Change Policy (NCCP) document (2013) Ministry of Science and Environment
National Adaptation Strategy and Plan Of Action On Climate Change For Nigeria (NASPA-CCN) 2011, Available: http://citeseerx.ist.psu.edu/viewdoc/download;jsessionid=27C7F6A7FD5FCAAD495C69457F2C5F7D?doi=10.1.1.367.6707&rep=rep1&type=pdf. Accessed Sept 2016
Kenya National Climate Change Response Strategy (NCCRS) document (2010) Government of Kenya
National Climate Change Strategy (NCCS) document (2012) United Republic of Tanzania
Uganda National Climate Change Policy (NCCP) document (2012) Final Version for Approval

## Author Biographies

**Mary Nyasimi** is a Science Officer—Gender and Social Inclusion, the CGIAR Research Program on Climate Change, Agriculture and Food Security (CCAFS)—International Livestock Research Institute (ILRI), Nairobi, Kenya.

**Ayansina Ayanlade** is Environmental Scientist, Researcher and Lecturer—Department of Geography, Obafemi Awolowo University, Ile-Ife, Nigeria. He is also a CIRCLE Postdoc Fellow under CGIAR Research Program on Climate Change, Agriculture and Food Security (CCAFS), International Livestock Research Institute, Nairobi, Kenya.

**Catherine Mungai** is partnerships and policy engagement specialist—the CGIAR Research Program on Climate Change, Agriculture and Food Security (CCAFS) based at the International Livestock Research Institute (ILRI), Nairobi, Kenya.

**Mercy Derkyi** is a lecturer—Department of Forest Science, School of Natural Resources, University of Energy and Natural Resources (UENR), Sunyani, Ghana.

**Margaret Olusolape Jegede** is a researcher and lecturer—African Institute for Science Policy and Innovation (AISPI), Obafemi Awolowo University, Ile-Ife, Nigeria.

# Balancing Paradigms in Climate Change Communication Research to Support Climate Services

**Inez Ponce de Leon and Charlotte Kendra Gotangco**

## 1 Introduction

Crafting effective solutions to challenges posed by climate change requires evidence-based actions. Climate services is the arm of climate change science that deals precisely with using the results of science to inform practice. Climate services draws from state-of-the-art of climate science to customize knowledge products to meet the needs of specific stakeholders towards the goal of initiating and sustaining appropriate action versus climate change (AMS 2014; CSP 2014; WMO n.d.). Climate services are not limited to repackaging and popularizing science. Rather, climate services aim for increased capacity in interpreting and using scientific information in sectors or groups outside of climate scientists (Changnon and Kunkel 1999; Munang et al. 2010; Scott et al. 2011). Thus, climate services need to employ increasingly inter- and trans-disciplinary approaches to address these varied, changing audiences (Cooney 2010; Weaver et al. 2014). These approaches require iterative dialogue to give academic researchers and other stakeholders the opportunity to share expertise, identify solutions together, and adapt and adjust as needed through time (Dilling and Lemos 2011).

Active dialogue, however, requires a structure that is beyond the capacity and original intent of the Intergovernmental Panel on Climate Change (IPCC) assessments (Visbeck 2008). The IPCC has been criticized for its focus on the bench sciences, rather than consultation with a wide variety of stakeholders—a practice

I. Ponce de Leon (✉)
Department of Communication, Ateneo de Manila University, 3rd Floor, Social Sciences Building, Katipunan Avenue, Loyola Heights, 1108 Quezon City, Philippines
e-mail: iponcedeleon@ateneo.edu

C. K. Gotangco
Department of Environmental Science, Ateneo de Manila University, 3rd Floor, Manila Observatory, Katipunan Avenue, Loyola Heights, 1108 Quezon City, Philippines
e-mail: kgotangco@ateneo.edu



W. Leal Filho et al. (eds.), *Handbook of Climate Change Communication: Vol. 1*,
Climate Change Management, https://doi.org/10.1007/978-3-319-69838-0_12

that contradicts the complexity of the climate change issue (Glover 2006). Global initiatives have since been established that help address this gap. For instance, the WMO established the Global Framework for Climate Services (GFCS) in 2009 with a framework that explicitly calls for continued interaction amongst physical and social science researchers, practitioners, and decision-makers (Asrar et al. 2012; WMO 2014). Future Earth: Research for Global Sustainability Initiative, established by the International Council for Science in 2012, recognizes that sustainability is possible only through wide systemic and behavioral changes (Reid et al. 2010), and espouses co-design and co-production towards that end (Future Earth 2014). It builds on the work of various international sustainability programs, including the Earth System Sciences Partnership (ESSP), which early on recognized the importance of continuous stakeholder dialogue (Leemans et al. 2009).

Both GFCS and Future Earth are relatively new endeavors, and we only starting to grapple with what the large scale operationalization of “dialogue” might entail in the long term. Multi-stakeholder dialogue, co-design, and co-production need to be based on sound climate science, but also defined and planned with a sound social sciences foundation (Weaver et al. 2014). Therefore, these concepts must be supported by a firm research base. Given these needs, communication research emerges as a crucial field of inquiry, with academics and researchers needing to confront the challenge of transdisciplinary work. How has communication research been conducted? Does the literature provide enough grounding to support activities related to co-production and co-design? How should researchers be trained or oriented to address the needs of climate services through communication research design? This paper reflects on the underlying paradigms employed in recent literature and proposes ways by which different paradigms can help diversify and balance climate change communication research, for current and future researchers and higher education institutions to consider. With a stronger theoretical base undergirding climate change communication research, climate change communication practice can be more focused, systematic, and organized.

## 2 Articulating Paradigms Used by Communication Research

Communication research, and the social sciences in general, need to be supported by a paradigm to inform and drive research into various aspects of climate services (Cooney 2010; Hackmann et al. 2014; Weaver et al. 2014). A research paradigm or worldview contains assumptions about the nature of reality and knowledge. Paradigms guide how research should be conducted, and how results can be interpreted (Guba and Lincoln 1994). To some extent, paradigms can also inform how research results can be applied (Trench 2008); in this case, paradigms can drive how climate change research is used in climate change communication activities.

Some basic worldviews are summarized below and explained in detail in the literature (including Guba and Lincoln 1994; Hatch 2002; Ponce de Leon and Gotangco 2013; Trench 2008). Note, however, that this is not an exhaustive list as different disciplines may have their specific worldviews.

Post-positivism assumes that the world is made up of objective, measurable patterns that can be approximated by careful research. Under such a worldview, research can take a quantitative, qualitative, or mixed methods approach. Post-positivism aims for generalizations that apply to large groups or cut across different audiences. For example, researchers could use large-scale surveys, aggregate findings, and use the findings to create educational materials. Post-positivism guides dissemination models of communication, which have been used in various climate change information campaigns in different countries (Chambliss and Lewenstein 2012; Krantz et al. 2013; Lyytimaki et al. 2013). Under a post-positivist worldview, researchers can use one-on-one interviews or group discussions, but they would still be in charge of determining the content to be disseminated.

A critical paradigm assumes that the world comprises measurable patterns, but while there are objective truths, researchers cannot measure them directly because centuries of history, culture, and socialization prevent researchers from being completely objective. Critical research is concerned with power issues and structures, which might exist among different stakeholders, or arise from the nature of media itself. Any critically-oriented research cannot aim for generalization, because each group, culture, and society is different. Critical research is characterized by depth and location-specific work and must be pursued using qualitative methods. The critical worldview can be used to create contextualized, community-based climate change communication materials and initiatives. This method is closest to the participatory approach. Questions that can be explored include "How does this group perceive its problems?" and "How can this group formulate its own solutions?"

Constructivism assumes that humans can never measure or report reality, and can only represent it through symbols, such as numbers, words, or images. No human being, therefore, will ever have a full view of the truth, approximate or otherwise. The truth is constructed in the moment of research. Like critical work, constructivist work is unique to a location and point in time and must be pursued using qualitative methods. Researchers have used this worldview to study how communities make meaning of climate change (Becken et al. 2013; Rudiak-Gould 2012). Both critical and constructivist studies allow climate change communication initiatives to see how climate change is perceived and understood at local levels, and how these perceptions and understandings can drive (or hinder) climate change mitigation and adaptation programs. Research under this paradigm can be built on questions like "What does climate change mean to this group?" and "How do these meanings lead to solutions to the problem?"

Post-modernism assumes that the truth is viewed differently by every single individual, and there are as many truths as there are people who view it, with all truths equally valid. Research in post-modernism does not seek generalizations, but a reporting of all these truths. Post-modernist work is always qualitative, and stakeholders contribute equally to a final project. This is considered co-production, but can be difficult to carry out in culturally heterogeneous nations. Interesting questions can arise, such as "How do these different views reflect the cultural milieu and stakeholder group characteristics?"

No worldview or paradigm is better or worse than the other. Each worldview can be used to address different facets of climate services research (Ponce de Leon 2011; Trench 2008). The success of climate services efforts will rest on a good match among goals, audiences, resources, and context (Johnson 2012) as guided by a balanced application of research paradigms. Research paradigms should be articulated and applied from when a climate change communication problem is formulated, to the time the research is published and its findings are interpreted and utilized. This will ensure that findings are valid, and any conclusions made about the findings do not overreach the assumptions of the paradigm under which the research operates. Research in support of climate services should be cognizant of different worldviews and their advantages and limitations, especially since the field involves diverse stakeholders who must be heard and addressed (Carlton and Jacobson 2013). Each paradigm can lead to more interesting questions about various aspects of climate change communication. However, are all worldviews currently being used in a balanced way to structure climate change communication research, or are there dominant paradigms in recent publications? What have we learned from research guided by specific paradigms?

## 3 Paradigms in Recent Climate Change Communication Literature

To identify the paradigms employed in literature that have emerged since initiatives like the GFCS and Future Earth were launched, a quick scoping of peer-reviewed articles was performed. EBSCOHOST was used to search for research articles (including empirical articles, review articles, and commentaries) published between January 2010 and August 2014, using the key phrase "climate change communication". However, research articles from the fields of policy and politics, and education were excluded. The assessments and reflections in this paper are based on 145 unique articles, representing different fields of study. Major categories of studies focus on media use and content, context analysis, sociological studies, and psychological studies.

Studies on media use and content comprised articles that examined how climate change could be framed, how audiences understood these frames, and how these frames influenced their perceptions and behavior (for example: DiFrancesco and Young 2010; Dirikx and Gelders 2010; Eskjaer 2013; Nerlich et al. 2010; Pasquare and Oppizzi 2012). Other researchers examined the effects of mass media exposure over time on taking action versus climate change (Howell 2014). Research within the time frame, for the greater part, was post-positivist. Most research used experimental set ups, followed by questionnaires, workshops, and/or interviews to evaluate how people perceived and acted on mass media messages regarding climate change (for example: Gifford and Comeau 2011; Jang 2013; Lieske et al. 2014; Maibach et al. 2010; Morton et al. 2011; Myers et al. 2012; Nisbet et al. 2013). A good number of articles carried out large-scale surveys on mass media use (for example: Feldman et al. 2014; Stoutenborough and Vedlitz 2014; Tam and McDaniels 2013; Williams 2011; Zhao et al. 2011). In some studies on the effects of mass media on climate change attitudes, researchers assumed that they can predict behavior and perception by looking at media messages alone. In general, researchers called for more studies on whether message framing can lead to actual behavior change, and more participatory programs to engage audiences above and beyond media exposure.

Research on interactions between context and communication comprised articles that examined climate change from the audience's perspective, and then cross-checked these perspectives against both environmental and social records. Research published under this classification from 2010 to 2014 was also generally post-positivist, as it sought out large scale patterns to examine how environmental and economic contexts could affect people's perceptions about climate change (for example: Akerlof et al. 2012; Brulle et al. 2012; Howe and Leiserowitz 2013; Safi et al. 2012; Scruggs and Benegal 2012; Ungar 2014). However, some researchers did use in-depth, qualitative studies of specific locations to examine different environmental contexts and their effects on climate change, and even engaged participants in bottom-up approaches to design community-based, localized materials to help communities deal with climate change (Gentle and Maraseni 2012; Khan et al. 2012; Leonard et al. 2013; Schweizer et al. 2013). Recommendations included conducting more research in indigenous knowledge, as well as multidisciplinary studies (Scott et al. 2011), framing climate change as a local problem, and using more specific, concrete information to enable adaptation at local levels.

Sociology-based articles comprised research focused on group dynamics, behavior, and perceptions regarding climate change. Studies published from 2010 to 2014 were mostly post-positivist and based on surveys. Surveys were conducted at the country level (for example: Barnes et al. 2013; Hamilton et al. 2012; Hine et al. 2013; Marshall et al. 2013; Milfont 2012; Zia and Todd 2010) and across country boundaries (Bostrom et al. 2012; Crona et al. 2013). Some researchers chose qualitative methods instead of surveys to gather data on cultural understandings and beliefs. These included interviews and focus group discussions to examine various aspects of climate change perceptions from a sociological standpoint (for example: Becken et al. 2013; Buys et al. 2012; Nursey-Bray et al. 2012; Poortinga et al. 2011; Rudiak-Gould 2014; Sjogersten et al. 2013; Wibeck 2014a). Researchers found that various cultural communities varied in their understanding of climate change due to deeply-ingrained values and worldviews.

Psychology-based studies comprised articles on the cognitive aspects of climate change communication. Research into the psychological aspects of climate change communication focused on the public perceptions of climate change, risk, and the mass media. Most research into the psychology of climate change communication was based on surveys, which provided a broad-based, post-positivist look at how people from varying backgrounds perceived climate change (for example: Beard and Thompson 2012; Binder 2010; Carlton and Jacobson 2013; Hedlund-De Witt et al. 2014; Spence et al. 2012; Van der Linden 2014; Wendling et al. 2013). In general, research found that people interpret risk in different ways, but awareness of an issue does not necessarily lead to action. In general, researchers in psychology-based studies in climate services recommended more research on local worldviews and understandings of climate change. In terms of practice, these researchers recommended creating communication messages that are localized, and engaging in participatory projects to foster engagement among small communities.

All these research articles support the needs of climate services initiatives; however, based on this initial review, there seems to be over-reliance on post-positivism as a guiding research paradigm. Research in climate change communication published from 2010 to 2014 was largely post-positivist, seeking large-scale patterns to make generalizations. Researchers also called for more localized work, and some studies took on the challenge. These localized studies were critical or constructivist in nature, and documented local understanding and meaning-making regarding climate change.

Some post-positivist research, especially those that dealt with content analysis and media use, assumed that simply changing the message can lead to attitude and behavior change. However, this assumption is based on one-way dissemination models of science communication, where science alone holds the knowledge, and where knowledge flows to a lay public that will willingly change its behavior (Burns et al. 2003; Gregory and Miller 1998). However, research shows that awareness is not enough to take action, and that climate change knowledge does not translate immediately to environmentally-sustainable behavior (Jang 2013; Lieske et al. 2014; Morton et al. 2011; Whitmarsh et al. 2013). Dialogue cannot be gained through a one-way model of disseminating communication. The one-way model might not engender trust that would otherwise be gained through engagement with science and scientists (Cooper 2011; Whitmarsh et al. 2013; Wibeck 2014b). In the same manner, scientists might be able to catch a deeper glimpse of the complexity of social, economic, institutional, political, and cultural contexts of non-science stakeholders through dialogue, which can also help inform future research in climate change (Burns et al. 2003; Hackmann et al. 2014; Sharma and Gosain 2010).

Thus, dialogue entails more than "bringing down the science"; it should encourage contributions from all sides of research and practice to transform cross-disciplinal and multi-sectoral work into a coherent whole (Podesta et al. 2013). Research shows that the public has its own valid form of lay knowledge, which might be based on personal experience or beliefs (Weiler et al. 2012), but which is not necessarily wrong or subordinate to scientific knowledge (Brace and Geoghegan 2010; Etkin and Ho 2007; Howe and Leiserowitz 2013; Moser 2010; Naustdalslid 2011; Weber 2010; Weber and Stern 2011). For instance, indigenous knowledge can inform how people understand climate change, and researchers might need to understand how indigenous knowledge can help scientists understand climate at local levels (Brace and Geoghegan 2010; Leonard et al. 2013; Whitmarsh et al. 2013).

Dialogue should therefore engage non-scientists, with a focus on critical thinking (Cooper 2011) instead of simply filling people with the type of knowledge that scientists believe they should have. This is where more critical and constructivist research paradigms can fill in the gap: meaningful dialogue begins with understanding of local contexts (Cooper 2011). Granted, such research takes time to implement, and even longer to publish, which may be a disincentive to the research and academic world. But such research is valuable to climate services because it meets the insistent call for localizing messages and framing them in terms of values and issues pertinent to specific populations and groups. Qualitative, in-depth studies can unearth these values and issues, and future research should address this need.

## 4 Concluding Remarks: Reflections on Balancing Communication Research for Climate Services

Climate services require support from the field of climate change communication to effectively promote the dialogue and collaboration necessary for co-production and co-design. This paper reviewed climate change communication research published from 2010 to 2014, and found that most research used a post-positivist worldview, with very few research articles guided by critical or constructivist paradigms. This limits the dialogue needed by climate services. Climate change communication research must take a step forward by considering what approaches would value and use stakeholders' differing perspectives of the issue toward co-designing and implementing long-term solutions.

The drive toward more dialogue-driven climate services needs to be supported by collaboration among the many different stakeholders involved, which calls for researchers to adapt a more transdisciplinary approach. Climate change communication research is not meant to problematize and then seek to harmonize diverse perspectives, but rather to channel those voices toward developing more robust dialogue and co-production strategies. However, climate change communication must not only change on its own, but must be supported by infrastructure that lends importance to dialogue, from the moment that research problems are framed, to the time that research results are used to benefit a public that contributed to that research. To this end, research and higher education institutions can play an important role in directing and forming current and future change communication work, with the aim to:

- Examine how various cultures are unique in their understanding of climate change, and how these understandings might either hamper future action or actually encourage action versus climate change, even if the understandings do not completely match those of scientists. Research can involve close contact and iterative interactions with various groups.

- Engage the public in dialogue at all stages of the research, from inception to long-term implementation, including monitoring and evaluation. The public will need assistance in understanding the science of climate change; but scientists, too, need assistance in understanding how non-scientists view the phenomenon.
- Keep using research findings to inform further work. There is no end to the many understandings and perceptions about climate change. As the audience keeps changing, so will their communication needs. Researchers are dealing with moving targets, and they must be grounded in good research philosophy before they can proceed.
- Consider using constructivist and critical worldviews to push research forward. Interesting questions might include: How might indigenous knowledge enrich bench-sciences-derived knowledge? How can this knowledge be used in climate services? What meaning do people make of the climate change issue?

These recommendations are based on an initial review of the recent academic literature; however, the current review is limited. Future reviews should include books, use other databases, expand the time frame, and assess constructivist or critical research in detail. Furthermore, projects and initiatives utilizing the constructivist or critical or post-modern worldviews may currently be conducted in other arenas (e.g. by practitioners, civil groups, non-government or aid organizations) but are not sufficiently reflected in peer-reviewed academic literature. This assessment focuses specifically on work implemented by the research and academic sector.

Inter- and transdisciplinary initiatives such as the GFCS and Future Earth require more research to support the need for salient climate services and meaningful co-production and co-design. By exploring other worldviews, climate change communication research can truly support climate services, which, in turn, can better engage the diverse publics that must be served. The researchers hope that this paper can help institutions involved in climate services and/or in training current and future generations of researchers to understand the information needs and unique contexts of various groups and cultures; and, in so doing, address the climate change issue by facilitating the development of useful and actionable climate services.

**Acknowledgements** This paper was partially supported by funding from the Philippine Higher Education Research Network (PHERNet) of the Commission of Higher Education (CHED) to the Ateneo de Manila University.

## References

Akerlof K, Maibach EM, Fitzgerald D, Cedeno AY, Neuman A (2012) Do people 'personally experience' global warming, and if so, how does it matter? Glob Environ Change 23:81–91

American Meteorological Society (AMS) (2014) Climate services: a policy statement of the American Meteorological Society. Retrieved from http://www.ametsoc.org/policy/2012statement_climate_services.html

Asrar G, Ryabinin V, Detemmerman V (2012) Climate science and services: providing climate information for adaptation, sustainable development and risk management. Curr Opin Environ Sustain 4:88–100

Barnes AP, Islam M, Toma I (2013) Heterogeneity in climate change risk perception amongst dairy farmers: a latent class clustering analysis. Appl Geogr 41:105–115

Beard CA, Thompson JL (2012) Engaging visitors in climate change communication: a case study of Southern Florida's national parks and wildlife refuges. Appl Environ Educ Commun 11(1):25–34

Becken S, Lama AK, Espiner S (2013) The cultural context of climate change impacts: perceptions among community members in the Annapurna Conservation Area, Nepal. Environ Dev 8:22–37

Binder AR (2010) Routes to attention or shortcuts to apathy? Exploring domain-specific communication pathways and their implications for public perception of controversial science. Sci Commun 32(3):383–411

Bostrom A, O'Connor RE, Bohm G, Hauss D, Bodi O, Ekstrom F, Halder P, Jeschke S, Mack B, Qu M, Rosentrater L, Sgelensminde I (2012) Causal thinking and support for climate change policies: international survey findings. Glob Environ Change 22:210–222

Brace C, Geoghegan H (2010) Human geographies of climate change: landscape, temporality, and lay knowledges. Prog Hum Geogr 35(3):284–302

Brulle RJ, Carmichael J, Jenkins JC (2012) Shifting public opinion on climate change: an empirical assessment of factors influencing concern over climate change in the U.S. 2002–2010. Clim Change 114:169–188

Burns TW, O'Connor DJ, Stocklmayer SM (2003) Science communication: a contemporary definition. Public Underst Sci 12:183–202

Buys L, Miller E, Van Megen K (2012) Conceptualising climate change in rural Australia: community perceptions, attitudes, and (in)actions. Reg Environ Change 12:237–248

Carlton S, Jacobson SK (2013) Climate change and coastal environmental risk perception in Florida. J Environ Manage 130:32–39

Chambliss EL, Lewenstein BV (2012) Establishing a climate change information source addressing local aspects of a global issue. J Sci Commun 11(3):1–8

Changnon SA, Kunkel KE (1999) Rapidly expanding uses of climate data and information in agriculture and water resources: causes and characteristics of new applications. B Am Meteorol Soc 80:821–830

Climate Services Partnership (CSP) (2014) What are climate services? Retrieved from http://www.climate-services.org/content/what-are-climate-services

Cooney CM (2010) The perception factor: climate change gets personal. Environ Health Persp 118 (11):484–489

Cooper CC (2011) Media literacy as a key strategy toward improving public acceptance of climate change science. Bioscience 61:231–237

Crona B, Wutich A, Brewis A, Gartin M (2013) Perceptions of climate change: linking local and global perceptions through a cultural knowledge approach. Clim Change 119:519–531

DiFrancesco DA, Young N (2010) Seeing climate change: the visual construction of global warming in Canadian national print media. Cult Geogr 18(4):517–536

Dilling L, Lemos MC (2011) Creating usable science: opportunities and constraints for climate knowledge use and their implications for science policy. Glob Environ Change 21:680–689

Dirikx A, Gelders D (2010) Ideologies overruled? An explorative study of the link between ideology and climate change reporting in Dutch and French newspapers. Environ Commun 4 (2):190–205

Eskjaer M (2013) The regional dimension: how regional media systems condition global climate change communication. J Int Intercult Commun 6(1):61–81
Etkin D, Ho E (2007) Climate change: perceptions and discourses of risk. J Risk Res 10:623–641
Feldman L, Myers TA, Hmielowski JD, Leiserowitz A (2014) The mutual reinforcements of media selectivity and effects: testing the reinforcing spirals framework in the context of global warming. J Commun 64:590–611
Future Earth (2014) Future Earth initial design report. Retrieved from http://www.futureearth.info/sites/default/files/Future-Earth-Design-Report_web.pdf
Gentle P, Maraseni TN (2012) Climate change, poverty and livelihoods: adaptation practices by rural mountain communities in Nepal. Environ Sci Policy 21:24–34
Gifford R, Comeau LA (2011) Message framing influences perceived climate change competence, engagement, and behavioral intention. Global Environ Change 21:1301–1307
Gregory J, Miller S (1998) Science in public: communication, culture, and credibility. Plenumtrade, New York
Glover L (2006) Post modern climate change. Routledge, London, UK
Guba G, Lincoln YS (1994) Competing paradigms in qualitative research. In: Denzin NK, Lincoln YS (eds) Handbook of qualitative research. SAGE Publications, Thousand Oaks, CA
Hackmann H, Moser SC, St. Clair AL (2014) The social at the heart of global environmental change. Nat Clim Change 4:653–655
Hamilton LC, Cutler MJ, Schaefer A (2012) Public knowledge and concern about polar region warming. Polar Geogr 35:155–168
Hatch J (2002) Doing qualitative research in education settings. State University of New York Press, NY
Hedlund-De Witt A, de Boer J, Boersema JJ (2014) Exploring inner and outer worlds: a quantitative study of worldviews, environmental attitudes, and sustainable lifestyles. J Environ Psychol 37:40–54
Hine DW, Reser JP, Phillips WJ, Cooksey R, Marks ADG, Nunn P, Watt SE, Bradley GL, Glendon AI (2013) Identifying climate change interpretive communities in a large Australian sample. J Environ Psychol 36:229–239
Howe PD, Leiserowitz A (2013) Who remembers a hot summer or a cold winter? The asymmetric effect of beliefs about global warming on perceptions of local climate conditions in the U.S. Glob Environ Change 23:1488–1500
Howell RA (2014) Investigating the long-term impacts of climate change communications on individuals' attitudes and behaviour. Environ Behav 46(1):70–101
Jang SM (2013) Framing responsibility in climate change discourse: ethnocentric attribution bias, perceived causes, and policy attitudes. J Environ Psychol 36:27–36
Johnson BB (2012) Climate change communication: a provocative inquiry into motives, meanings, and means. Risk Anal 32(6):973–992
Khan AS, Ramachandran A, Usha N, Aram IA, Selvam V (2012) Rising sea and threatened mangroves: a case study on stakeholders, engagement in climate change communication, and non-formal education. Int J Sustain Dev World 19(4):330–338
Krantz S, Monroe M, Bartels W (2013) Creating extension programs for change: forest landowners and climate change communication. Appl Environ Educ Commun 12(4):272–279
Leemans R, Asrar G, Busalacchi A, Canadell J, Ingram J, Larigauderie A, Mooney H, Nobre C, Patwardhan A, Rice M, Schmidt F, Young O (2009) Developing a common strategy for integrative global environmental change research and outreach: the Earth System Science Partnership (ESSP). Curr Opin Environ Sustain 1:4–13
Leonard S, Parsons M, Olawsky K, Kofod F (2013) The role of culture and traditional knowledge in climate change adaptation: Insights from East Kimberley, Australia. Glob Environ Change 23:623–632
Lieske DJ, Wade T, Roness LA (2014) Climate change awareness and strategies for communicating the risk of coastal flooding: a Canadian maritime case example. Estuar Coastal Shelf Sci 140:83–94

Lyytmaki J, Nygren NA, Alla-Ketola U, Pellinen P, Ruohomaki V, Inkinen A (2013) Climate change communication by a research institute: experiences, success, and challenges. Appl Environ Educ Commun 12(2):118–129
Maibach EW, Nisbet M, Baldwin P, Akerlof K, Diao G (2010) Reframing climate change as a public health issue: an exploratory study of public reactions. BMC Public Health 10:299–309
Marshall NA, Park S, Howden SM, Dowd AB, Jakku ES (2013) Climate change awareness is associated with enhanced adaptive capacity. Agr Syst 117:30–34
Milfont TL (2012) The interplay between knowledge, perceived efficacy, and concern about global warming and climate change: a one-year longitudinal study. Risk Anal 32(6):1003–1021
Morton TA, Rabinovich A, Marshall D, Bretschneider P (2011) The future that may (or may not) come: how framing changes responses to uncertainty in climate change communications. Glob Environ Change 21:103–109
Moser SC (2010) Communicating climate change: history, challenges, process and future directions. Wiley Interdisc Rev Clim Change 1:31–53
Munang R, Rivington M, Takle ES, Mackey B, Thiaw I, Liu J (2010) Climate information and capacity needs for ecosystem management under a changing climate. Proc Environ Sci 1:206–227
Myers TA, Nisbet MC, Maibach EW, Leiserowitz AA (2012) A public health frame arouses hopeful emotions about climate change. Clim Change 113:1105–1112
Naustdalslid J (2011) Climate change—the challenge of translating scientific knowledge into action. Int J Sustain Dev World 18(3):243–252
Nerlich B, Koteyko N, Brown B (2010) Theory and language of climate change communication. Wiley Interdisc Rev Clim Change 1:97–110
Nisbet EC, Hart PS, Myer T, Ellithorpe M (2013) Attitude change in competitive framing environments? Open/closed-mindedness, framing effects, and climate change. J Commun 63:766–785
Nursey-Bray M, Peel GT, Frusher S, Gardner C, Haward M, Hobday AJ, Jennings S, Punt AE, Revill H, van Putten I (2012) Communicating climate change: climate change risk perceptions and rock lobster fishers, Tasmania. Mar Policy 36:753–759
Pasquare FA, Oppizzi P (2012) How do the media affect public perception of climate change and geohazards? An Italian case study. Glob Planet Change 90–91:152–157
Podesta GP, Natenzon CE, Hidalgo C, Toranzo FR (2013) Interdisciplinary production of knowledge with participation of stakeholders: a case study of a collaborative project on climate variability, human decisions and agricultural ecosystems in the Argentine Pampas. Environ Sci Policy 26:40–48
Ponce de Leon I (2011) Science communication beliefs of researchers based in the Philippines and the United States: a qualitative analysis of research cultures and worldviews. PhD dissertation. Department of Youth Development and Agricultural Education, Purdue University, Indiana, USA
Ponce de Leon I, Gotangco CK (2013) Reconciling post-positivist and post-modern worldviews in climate research and services. WMO Spec Bull 62:45–48
Poortinga W, Spence A, Whitmarsh L, Capstick S, Pidgeon NF (2011) Uncertain climate: an investigation into public skepticism about anthropogenic climate change. Glob Environ Change 21:1015–1024
Reid WV, Chen D, Goldfarb L, Hackmann H, Lee YT, Mokhele K, Ostrom E, Raivio K, Rockström J, Schellnhuber HJ, Whyte A (2010) Earth system science for global sustainability: grand challenges. Science 330:916–917
Rudiak-Gould P (2014) Progress, decline, and the public uptake of climate science. Public Underst Sci 23(2):142–156
Rudiak-Gould P (2012) Promiscuous corroboration and climate change translation: a case study from the Marshall Islands. Glob Environ Change 22:46–54
Safi AS, Smith WJ Jr, Liu Z (2012) Rural Nevada and climate change: vulnerability, beliefs, and risk perception. Risk Anal 32(6):1041–1060
Schweizer S, Davis S, Thompson JL (2013) Changing the conversation about climate change: a theoretical framework for place-based climate change engagement. Environ Commun 7(1): 42–62

Scott DJ, Lemieux CJ, Malone L (2011) Climate services to support sustainable tourism and adaptation to climate change. Clim Res 47:111–122
Scruggs L, Benegal S (2012) Declining public concern about climate change: can we blame the Great Recession? Glob Environ Change 22:505–515
Sharma KD, Gosain AK (2010) Application of climate information and predictions in water sector: capabilities. Proc Environ Sci 1:120–129
Sjogersten S, Atkin C, Clarke MI, Mooney SJ, Wu B, West HM (2013) Responses to climate change and farming policies by rural communities in Northern China: a report on field observation and farmers' perception in dryland North Shaanxi and Ningxia. Land Use Policy 32:125–133
Spence A, Poortinga W, Pidgeon N (2012) The psychological distance of climate change. Risk Anal 32(6):957–973
Stoutenborough JW, Vedlitz A (2014) The effect of perceived and assessed knowledge of climate change on public policy concerns and empirical comparison. Environ Sci Policy 37:23–33
Tam J, McDaniels TL (2013) Understanding individual risk perceptions and preferences for climate change adaptations in biological conservation. Environ Sci Policy 27:114–123
Trench B (2008) Towards an analytical framework of science communication models. In: Cheng D, Claessens T, Gascoigne J, Metcalfe B, Schiele S, Shi S (eds) Communicating science in social contexts: new models, new practices. Springer, Netherlands
Ungar S (2014) Media context and reporting opportunities on climate change: 2012 versus 1998. Environ Commun 8(2):233–248
Van der Linden S (2014) On the relationship between personal experience, affect, and risk perception: the case of climate change. Eur J Soc Psychol 44:430–440
Visbeck M (2008) From climate assessment to climate services. Nat Geosci 1:2–3
Weaver CP, Mooney S, Allen D, Beller-Sims N, Fish T, Grambsch AE, Hohenstein W, Jacobs K, Kenney MA, Lane MA, Langner L, Winthrop R (2014) From global change science to action with social sciences. Nat Clim Change 4:656–659
Weber EU (2010) What shapes perceptions of climate change? Wiley Interdisc Rev Clim Change 1:332–342
Weber EU, Stern PC (2011) Public understanding of climate change in the United States. Am Psychol 66(4):315–328
Weiler CS, Keller JK, Olex C (2012) Personality type differences between PhD climate researchers and the general public: implications for effective communication. Clim Change 112:233–242
Wendling ZA, Attari SZ, Carley SR, Krause RM, Warren DC, Rupp JA, Graham JD (2013) On the importance of strengthening moderate beliefs in climate science to foster support for immediate action. Sustainability 5:5153–5170
Whitmarsh L, O'Neill S, Lorenzoni I (2013) Public engagement with climate change: what do we know, and where do we go from here? Int J Media Cult Polit 9(1):7–27
Wibeck V (2014a) Social representations of climate change in Swedish lay focus groups: local or distinct, gradual or catastrophic? Public Underst Sci 23(2):204–219
Wibeck V (2014b) Enhancing learning, communication, and public engagement about climate change – some lessons from recent literature. Environ Educ Res 20(3):387–411
Williams AE (2011) Media evolution and the public understanding of climate science. Polit Life Sci 30(2):20–30
World Meteorological Organization (WMO) (n.d.) Climate services introduction. Retrieved from http://www.wmo.int/pages/themes/climate/climate_services.php
World Meteorological Organization (WMO) (2014) Implementation plan of the Global Framework for Climate Services. Retrieved from http://www.gfcs-climate.org/implementation-plan

Zhao X, Leiserowitz A, Maibach EW, Roser-Renouf C (2011) Attention to science/environment news positively predicts and attention to political news negatively predicts global warming risk perceptions and policy support. J Commun 61:713–731

Zia A, Todd AM (2010) Evaluating the effects of ideology on public understanding of climate change science: how to improve communication across ideological divides? Public Underst Sci 19(6):743–761

## Author Biographies

**Inez Ponce de Leon** has undergraduate and graduate degrees in molecular biology and biotechnology, as well as a Ph.D. in Science Communication from Purdue University. Her research interests include science communication, social sciences communication, genetics, molecular biology, and audience studies. Her research projects include studies of various publics' understanding of climate change and extreme flooding events, stakeholders' climate change communication needs, and, in 2017, vulnerable communities' understanding of flooding and storm surge risks.

**Charlotte Kendra Gotangco** completed a Ph.D. in Earth and Atmospheric Science from Purdue University under a Fulbright Fellowship. She holds a Master in Environmental Management from the University of San Francisco and the Ateneo de Manila University, and a Bachelor of Science degree in Physics, Minor in Philosophy. She is interested in human-environment dynamics, particularly, in interdisciplinary approaches. Recently, she led a project on exploring indirect vulnerabilities of Metro Manila applying a supply chain approach, and the system dynamics modeling efforts for Metro Manila resilience.

# Communicating Climate Change Through Narratives: A Cross Pollination of Science and Theology

**Anna Huxley**

## 1 Introduction

Scientific outputs are often used as a method for communicating climate change, with the Intergovernmental Panel on Climate Change (IPCC) being the scientific body that appraises the latest 'scientific, technical and socioeconomic' data worldwide, which is 'relevant to the understanding of climate change' (UNFCCC 2014). The IPCCs multi-disciplinary approach assists the United Nations Framework Convention on Climate Change (UNFCCC) as political negotiations need to be informed by expert scientific knowledge (Hallegatte et al. 2016). Thus, the output from the IPCC is used by the UNFCCC 'as a baseline on the state of knowledge on climate change in making science based decisions' (UNFCCC 2014).

Many organisations, including faith-based, also use the output from the IPCC to frame climate change campaigns. The complexities of climate science and the uncertainties (those linked to the limitations of our knowledge and to human choices) make communication problematic, regardless of whether it is being communicated to policy makers or the public (Hallegatte et al. 2016, pp. 665–666). Using scientific terminology, such as 'tipping points' and 'feed-back loops', is complex and can be difficult to grasp. When climate change is presented in the form of predictions and graphs it can also appear inaccessible, as too big, or disengaging, when not paired with solutions. Hulme (2009, p. 74) identifies that many have high expectations of climate science and expect too much of scientific predictions. However, Hulme notes that using polar bears as the symbol of climate change may have an emotional appeal, regarding the loss of the Arctic, but asks if can they encapsulate all of the challenges of climate change? (Hulme 2009, p. 242). Klein asks in the documentary *This Changes Everything*, 'What is it about those melting

A. Huxley (✉)
University of Manchester, Oxford Road, M13 9PL Manchester, UK
e-mail: anna.huxley@manchester.ac.uk



W. Leal Filho et al. (eds.), *Handbook of Climate Change Communication: Vol. 1*,
Climate Change Management, https://doi.org/10.1007/978-3-319-69838-0_13

glaciers and desperate polar bears that make us want to look away? Is it really possible to be bored by the end of the world?' (*This Changes Everything* 2015).

So if polar bears signal that climate change is distant and unconnected or even irrelevant, how can the science of climate change be communicated in an accessible way? Climate Outreach have identified that it is not the failure of the IPCCs presentation of science but rather it 'reflects an outdated model of how science is incorporated into society, and how social change occurs' (Corner and Eck 2014, p. 4). This raises questions of whether science, and more specifically the outputs of the IPCC, are the most effective method of communicating climate change to the public? The aim of this paper is to examine if scientific data on climate change can be presented in accessible ways that incorporate messages of hope and inspire action. Moreover, this paper aims to discuss the role faith organisations can play in communicating these messages.

According to Gardner (2006, p. 49) around 85% of the world's population adhere to a faith tradition. The world's faiths have numerous texts, traditions and rituals regarding stewardship and care for the Earth which infuse worship with hope, celebration and praise. 'Theologies have been created which stress the spiritual value of nature, our kinship with the non-human, and our ethical responsibilities to the Earth. New concepts of the divine, holiness, spiritual life, and sin are being forged [and] …[i]nnovative liturgies and rituals are being practiced' (Gottlieb 2006, p. 7). Faith traditions have, historically, provided inspiration and guidance to their followers with regard to human interaction with nature. It is therefore natural, that many faith traditions have become proactive in their engagement with the implications of climate change and attempt to communicate it in ways which lead to action. Accordingly, faith is an important lens with which to examine climate change communication.

When faith traditions communicate climate change it is clear they understand the psychology and importance of hope. This is something Pope Francis presented in the 2015 encyclical, *Laudato Si': On Care for our Common Home*. The encyclical upholds that change is possible and all humanity need to look honestly within their hearts to harness goodness and peace to override 'mental and social conditioning' to respond to climate change with dignity and 'embark on new paths to authentic freedom' (Francis 2015, p. 98). Through education and motivation humanity can change from being masters of the Earth and from superficially consuming Earth's resources to create new lifestyles based on an 'integral ecology' in 'serene harmony with creation' (Francis 2015, pp. 106–107). The encyclical acknowledges the anthropogenic nature of climate change with scientific research and whilst recognising the distinctiveness of science and theology, it states that dialogue is beneficial to both (Francis 2015, p. 33). Science and religion, however, are not straightforward companions (Russell 1997, pp. 7–18) which makes climate change an interesting case because the faiths are consciously engaging with science. There is therefore a question mark over how the faiths use science to communicate climate change.

## 2 FBOs and Climate Change

The data for this paper was generated for a Ph.D project aimed at understanding how and why FBOs are responding to the challenges of climate change and what those responses are. The author decided to focus attention on FBOs in the UK as there are over thirty UK FBOs engaged in climate action. Three were selected for the research sample due to their varying levels of engagement in action on climate change and differing organisational structure (see Table 1). The FBOs in the sample are the Alliance of Religions and Conservation, CAFOD, and Hope for the Future. Interviews and observations took place from January to March 2015 and were accompanied by document analysis to form the data set for this research. NVivo was used to organise and find insights into the data and the cross pollination of science and theology was an interesting theme which demanded further analysis. For the purpose of this paper the organisations shall be categorised according to their level of engagement in action on climate change, being referred to as High Level Organisation (HLO), Medium Level Organisation (MLO) and Low Level Organisation (LLO) (see Table 1).

**Table 1** Details of the sample FBOs

| | High Level Organisation HLO | Medium Level Organisation MLO | Low Level Organisation LLO |
|---|---|---|---|
| Faith based organisations | Alliance of religions and conservation (ARC) | Catholic agency for overseas development (CAFOD) | Hope for the future |
| Established | – 1986 as part of WWF UK<br>– 1995 independently (ARC n.d-d) | – 1960 (Mba n.d.) | – 2013 (Hope for the Future 2014a) |
| Financial benefactors | – Examples include: the World Bank, UNDP, WWF, the Government of Norway (ARC n.d.-a) | – UK supporters<br>– UK government (CAFOD n.d.-d, p.6) | – Individual supporters<br>– MLOs including; Tear Fund, Christian Aid and A Roche (Hope for the Future 2016) |
| Organisational aim | – To create partnerships between the faiths and secular organisations for environmental projects (ARC n.d.-c) | – To work with their partners overseas to tackle injustice and poverty regardless of faith<br>– To eradicate poverty (CAFOD n.d.-e) | – To motivate and resource congregational political responses to climate change in the UK |

(continued)

**Table 1** (continued)

| | High Level Organisation HLO | Medium Level Organisation MLO | Low Level Organisation LLO |
|---|---|---|---|
| Faith based organisations | Alliance of religions and conservation (ARC) | Catholic agency for overseas development (CAFOD) | Hope for the future |
| Climate change campaign aims | – Advises MLOs and LLOs providing a framework for long-term climate change campaigns<br>– To imbed environmental protection into the faith's rituals, practices, education programs | – To encourage supporters to lobby the UK government before the 2015 General Election and the UNFCCC talks in Paris<br>– Support transition to renewable energy for all (CAFOD n.d.-b) | – To get climate change on the political agenda before the 2015 General Election<br>– To equip people with the skills to communicate climate change with local politicians (Hope for the Future 2014b) |
| Scale of political engagement | – Works in partnership with UN organisations<br>– Advisory role in national governments (not UK) | – Lobby's the UK government<br>– Encourages supporters to engage with UK MPs and MEPs | – Encourages supporters to engage with MPs and MEPs at local and national scales |
| Role in overseas projects | – Instigates environmental projects rather than monitoring and guiding them | – Creates environmental projects overseas and is directly involved in all aspects of the projects | – Supports and raises awareness of climate change through the overseas work of other FBOs |
| Cross over engagement | – Supports and advises MLOs and LLOs on creating long-term environmental plans using the HLOs framework<br>– Promotes MLOs and LLOs | – Created a 7 year plan for the HLOs climate campaign launched in 2009<br>– Provided a speaker for the LLO supporter training day | – Uses campaign materials from other HLOs and MLOs<br>– Worked in partnership with another MLO on training supporters on political engagement (Hope for the Future 2015a) |

*Source* Author (2016)

## 3 Limitations

The research is limited by the number of participants included in the research sample and this was restricted by the confines of the Ph.D project from which this data sits. It is also limited by looking specifically at organisations which are inspired primarily by Christianity. The three month time frame for the completion of the interviews and observations could also be considered to be a limitation. The research was conducted with representatives of the FBOs rather than their supporters which is limiting in understanding the effectiveness of the campaigns. However, this provides further scope for research with the organisations supporters to understand the effectiveness of the methods of communication employed by the FBOs.

## 4 FBOs' View of Science

A cross pollination of science and theology can be seen in the perspectives of climate and environmental change presented by the FBOs which are used to communicate with supporters in two distinct ways. Firstly, they are used to raise awareness of the local and global impacts regarding the effects of climate change, and secondly to teach responsibility and inspire action at individual, local, national and global scales.

For the HLO the world is seen as 'data obsessed' and the documents produced by the IPCC are considered 'utterly incomprehensible' (ARC 2015). The HLOs founder stated in 1986 that, if it was data that was required to save the planet, then the planet would not be in this predicament as the data exists and the problems are clearly stated (ARC 2015). The HLO regularly uses the quote that 'nobody was ever converted by a pie chart' and the interviewee noted that 'nobody reads [the IPCC reports] and nobody understands them' (ARC 2015). However, it is not the value of the science that the interviewee is questioning but rather science being used as the method of communication[1] (ARC 2015). From their 30 years of experience the HLO maintains that 'nobody pays attention to statistics' and it is narratives which effectively transmit information (ARC 2015). The importance of using narratives to communicate climate change is a point echoed by Climate Outreach in, *Science & Stories: Bringing the IPCC to Life* (Corner and Eck 2014). The HLO believe that communication through narratives has a greater potential to create lasting change (ARC 2015). The HLO focus on narratives as they are not only a crucial element of faith traditions but they resonate with people's hearts and minds (ARC 2015).

The HLO launched the Climate Change Partnership Initiative (CCPI) at the UNFCCC Conference of Parties (COP) in 2000. All nations were invited to work with the world's faiths and lead by example working with environmental agencies to set energy reduction targets and reduce emissions (ARC n.d.-b). Part of the CCPI programme is myth, metaphor and memory and the HLO stresses their importance to inspire deeper engagement in environmental awareness and protection (ARC 2007). Rather than focusing on 'consumption, economics or policy', which according to the HLO have very few 'real roots', they harness solutions encapsulated in faith based narratives (ARC 2007). The HLO encourages the faiths to revive or rediscover narratives or 'half-forgotten traditions' which emphasize how

[1]The HLO have completed scientific testing on environments prior to projects being set up (ARC 2015).

the faith teaches care for creation and the natural environment (ARC n.d.-f). Old wisdom from faith narratives is retranslated, preserving 'ancient ecological knowledge' which can then be communicated (ARC n.d.-f). Contained throughout the HLOs guidelines for creating long-term plans are narratives from faith projects established with secular partners which have been brokered by the HLO (Colwell et al. 2009).

In 2009, the HLO, in partnership with the United Nations Development Program (UNDP), launched *Many Heavens, One Earth: Faith Commitments to Protect the Living Planet*. Through this, the HLO has guided eleven of the world's great faiths to work in partnership on projects with secular environmental organisations with funding from international bodies like the World Bank to develop long-term environmental commitments. Based on the HLOs guidelines the campaign output was the creation of over fifty, five to nine year long-term environmental action commitments, by the world's great faiths and others by secular organisations for worldwide environmental action (ARC n.d.-g).

The MLO features science in its campaign and references the IPCC in its materials, under the heading of 'climate change facts' (CAFOD n.d.-a). Presenting information from the IPCC as 'fact' signals to supporters that climate change is a real and serious issue. The IPCC output is harnessed by the MLO to make a clear connection between climate change and poverty and to stress the effects on the world's most vulnerable people. The use of the output of the IPCC, presented as facts on climate change, provides legitimacy to the MLOs supporters for the purpose of the campaign. It is also used to stress the anthropocentric nature of climate change and the MLO connects this to the impacts it is having on the world's poorest people. However, the real driver for the campaign is action out of solidarity and love of the world's people, who the MLO refers to as, 'our brothers and sisters' (CAFOD n.d.-a).

The LLO trains 'climate ambassadors' to visit congregations in the UK to hold 'climate write-ins' with the aim of getting climate change on the UK's political agenda (Hope for the Future 2015a). According to a video launched by the LLO in 2016 they have facilitated over 20 meetings with MPs and councillors and trained over 1000 people to effectively engage with MPs (Hope for the Future 2016). While they do not cite the IPCC materials in the 'raising awareness resources' on their website, they do cite the Royal Society, the Geology Society, NASA, the UK's Independent Committee on Climate Change, and the British Science Museum, as well as reports produced by WWF and Operation Noah (Hope for the Future 2014c).

According to the LLO, the most effective way to communicate climate change is in the context of worship (Hope for the Future 2015a). This point is demonstrated in how they engage supporters. At an LLO climate ambassador training day the LLOs patron stated that there 'was a growing scientific consensus on climate change' (Hope for the Future 2015b) yet in his online reflection of the day there is no mention of science (Hope for the Future 2014d). The focus of the patron's talk was on understanding climate change from a Christian perspective rather than a scientific one. Instead of referencing science or scientists, the LLOs patron cited Rowan Williams, the president of the MLO Christian Aid; 'climate change is a present reality, not a distant threat of the future' (Hope for the Future 2015b). The data shows that for the LLO, faith rather than science, is the main driver when communicating climate change. From examining the FBOs views of science in the context of communicating climate change, it is important to turn to how the FBOs practically adopt (or not) scientific language.

## 5 The Language of Science and Theology

There are differing opinions on the use of science when communicating climate by the FBOs. For the HLO science might not be seen as an effective way to communicate care for the Earth, environmental protection or teach responsibility to take action. However, it is seen as crucial to many of the MLOs and LLOs and faiths who created long-term plans for the HLOs *Many Heavens One Earth* campaign.

The MLO created a 7 year plan for the HLOs *Many Heavens One Earth* campaign. In it, the MLO details their role in responding to climate change, as for them, '[t]he environment and climate change is a cross-cutting corporate priority' (CAFOD 2009). For this reason, the MLO says that they will be 'increasingly called upon by partners to help reduce people's vulnerability to climate variability through a range of adaptation and disaster risk reduction (DRR) strategies' (CAFOD 2009). The MLO created partnerships with University College, London and Liverpool University looking at short term impacts of climate change which are used to inform their 'policy of climate change mitigation and adaptation' (ARC n.d.-e). However, this is for high level purposes in understanding potential impacts on overseas partners, rather than in developing strategies to communicate climate change to supporters (Humanitarian Future Programme 2009).

The science used in the MLOs own campaign materials is taken from the IPCC's 2014 *Fifth Assessment Report—Impacts, Adaption, and Vulnerability* which is used in conjunction with the effects on poverty. The MLO report *Climate Change and Vulnerability: Pushing people over the edge* examines the links between climate change and poverty, and focuses heavily on the IPCCs *Fifth Assessment Report* (CAFOD 2014). However, it is clear from data analysis that the MLO recognises that science cannot be the sole way to communicate climate change to their supporters.

For the MLOs campaign, science and theology are brought together in narrative form, with experiential narratives from their partners worldwide. The MLOs campaign guide opens with a quote from Pope Francis from his 2013 inauguration. Moving on to present narratives from the MLOs partners, detailing individual experiences of climate change from Kenya, the UK, Guatemala and Myanmar and presenting solutions, including the benefits of solar power (CAFOD n.d.-b). These narratives highlight the effects of climate change but also stress how it is not just an environmental problem. Their narratives highlight the importance of action, which the MLO states is a call from faith, for people in the UK to act for 'our brothers and sisters' worldwide, 'for creation and for future generations' (CAFOD n.d.-b).

According to the MLO, campaign planning took two years and research was conducted to establish what messages resonate with people when communicating climate change (CAFOD 2015a). Thus, the campaign was framed so as not to present messages of 'doom and gloom' (CAFOD 2015a). This was noted by the MLO as being disengaging (CAFOD 2015a) a point echoed, especially after the failure of COP15 in Copenhagen in 2009 (Feinberger et al. 2011). The science was selected to ensure that climate change is not presented in a way which makes it 'too big for people to grasp' with no 'real entry point' (CAFOD 2015a). The importance to the MLO of the data from the IPCC focuses on the impacts on the world's poorest people and how it will affect tackling poverty (CAFOD 2014). The MLOs campaign is driven by the urgent goal of 'the transition from polluting fossil fuels to sustainable energy for all' which they highlight, requires long-term solutions (CAFOD n.d.-b).

The LLO did not reference science as a driver for the organisation being created or for action on climate change. Rather they were driven by the idea of what the world would be like when their children and grandchildren are adults (Hope for the Future 2015a). The patron of Hope for the Future reiterated this point at the Sheffield Diocesan Synod. He used it as a driver to raise awareness of climate change and encourage others members of the clergy to do the same with their congregations:

> We need to be very clear. Left unchecked, global warming will wreak havoc in the earth [sic]. If we take action together, climate change can be reduced and, God willing, reversed for the sake of future generations. Hope for the Future is part of that desire to transform God's world so that it remains good and safe and beautiful for future generations (Croft 2014).

In this instance, science is not used as the driver in communicating a call to action rather it is Christian principles (Croft 2014). The 'invisible' and 'stealthy' nature of climate change are mentioned by the LLOs patron and he also presents science from the IPCC's 2013 report plus information from *The Rough Guide to Climate Change* (Croft 2014). The LLOs patron speaks of how the science is 'clear and accepted' and how predictions by scientists vary, due not to the uncertainty of science, but due to the uncertainty of the different scenarios regarding the choices people make (Croft 2014). 'Christians have a responsibility to speak out and take action on climate change along with everyone else on the planet [yet] Christians have a unique contribution to make because of… faith' (Croft 2014). According to the LLO, environmental action was not the Church's whole mission but was part of it and the Church is failing its responsibility if it does not take action on climate change (Hope for the Future 2015a). This is encapsulated in their initial organisational slogan of prayer, practice, and politics (Hope for the Future 2015a).

Scientists reporting on the changes to Earth's climate are not necessarily versed in the nuances of communication, the psychology of how change is achieved, or the specifics of implementing risk assessments (Anderegg 2010). The data shows that science may only form a part, if any, of the organisation's campaigns so how and why are these organisations surpassing the science or cross pollinating it with theology?

## 6 Moving Beyond Science

When communicating climate change to the public it involves highlighting how changes to the climate will affect people, the planet in local and global contexts, and the systems and life that the Earth supports. It also involves considering the required responses to this information. How can climate change be communicated in ways that promote and inspire action? And how should people respond if they are driven to take action?

The HLO does not understand human nature to be intrinsically bad, which can often be a conclusion reached by those wishing to persuade people to make changes due to the anthropogenic nature of climate change (ARC 2015). For the HLO creating long-term change means to imbed teachings, regarding environmental protection and relationships with the Earth, into rituals and traditions (ARC 2015). The faiths are taking action and trying to bring change as they are motivated by their own beliefs and communicate through the medium of narratives (ARC 2015). The HLO asks the faiths to share stories of the environmental actions and projects they have established which the HLO publicises through global media and news agencies (ARC 2015). These environmental success stories allow the HLO to move beyond the science, to not be burdened by scientific predictions and language. Instead they publicise and praise environmental success stories of projects they brokered through listening, engaging and celebrating narratives of action. The actions from projects developed through partnerships forged by the HLO, between faith traditions and secular organisations, emphasise hope, which in turn inspires further action (ARC 2015).

The approach of utilising narratives by the HLO has, since 2007, gained their Secretary General the position as of one of the special advisors to the UN Secretary General on climate change, the environment and the faiths (Weldon 2007). In 2015 for the UNFCCC Paris COPs the special advisors guided the French Government to ask all COP delegates two crucial questions: why do you care about the environment? And, when things are bad what poem, prayer, piece of music, or art provides light in the darkness? (King 2015). This demonstrates the importance of the HLOs approach in how they ask questions which prompt people to think in narratives. It is these narrative responses which the HLO believes can shift worldviews and create lasting change.

The MLO provides climate reflection postcards for supporters to contemplate climate change. Interestingly, the word science and the IPCC are not referenced on the postcards instead they are purely theological (CAFOD n.d.-c). The postcards raise many questions to inspire reflection: why is it important to take responsibility and act for the climate? Why is it a Christian duty to care for and protect the Earth? How do you treat the Earth in way God would want? How can Catholic faith and commitment to protect the Earth be reflected in action? Who is suffering the effects of climate change? How can political leaders be encouraged to act? Can we see the Earth as a neighbour? (CAFOD n.d.-c). The postcards are accompanied by online videos supporting deeper theological reflection on 'Faith and Climate' and '[t]he earth [sic] as neighbour' (CAFOD n.d.-c). These materials demonstrate that, although science is used to frame the campaign and to emphasise the links with climate and poverty, the heart of the campaign is Catholic teachings and responses aligned to the faith. The MLO surpasses science to move into a spiritual philosophic realm inspiring supporters to engage with ideas of deep ecology and biocentrism.

The MLO transcends the confines of the IPCC reports to speak of loving the Earth as one loves God and treating all life on Earth, human and other than human beings, as 'brothers and sisters' (Zampini Davies 2014). These perspectives are supported by biblical narratives including Genesis and the Good Samaritan and also narratives of the experiences of their overseas partners (Zampini Davies 2014) thus stressing Catholic responsibility for climate action. By presenting human stories of the direct experiences of climate change by the MLOs partners overseas, it stresses the human element which builds connections to the effects of climate change and promotes action. Communicating climate change in this way the MLO believes will lead to the following actions; convincing policy makers to commit to action, praying and fundraising for the suffering, making individual and congregational changes, and rethinking relationships with the Earth and all life (CAFOD 2015b).

For the LLO science is a feature in the examples provided in their online 'worship and service' resources. Yet a key point is reiterated in materials; warnings regarding burning fossil fuels have not just come from scientists, 'the earth [sic] itself has also been warning us' (Jowett 2015). This is legitimised with reference to the IPCC, yet entwined with the sayings of Jesus from Luke's Gospel (Jowett 2015). It is Christian teachings that are the basis for the LLOs campaign which aims to educate people about climate change and to encourage congregations to engage with MPs to ensure that climate change is on the UKs political agenda (Hope for the Future 2015a). Although grounded in science the campaign message is one of hope. Accordingly, it frames climate change in the context of worship, highlighting the optimism of the Christian faith accentuating teachings of resilience, the importance of action, and the place of prayer and celebration in challenging times.

## 7 Conclusion

One of the many challenges in communicating climate change for the FBOs is establishing the roles that science and theology have in campaigns. This is challenging as the FBOs aim for their campaigns to be driven by hope and unity, rather than fear and division. The FBOs also need to ensure compaigns are accessible and appealing to their target audiences. The IPCCs assessments of the future impacts of climate change can appear bleak with little hope. The FBOs understand these disengaging effects and move beyond the science, or even choose not to make it a contributing feature, when they communicate climate change. The sample FBOs all harness narratives when commutating climate change to their supporters. These are not only audience specific, but the use of narratives as a communication tool remains true to the faith traditions that these organisations are influenced and driven by. The narratives not only inform supporters about the current and future impacts of climate change. They are also connected to the actions they wish for supporters to make, whether individual and congregational, or at local, national or global scales.

When comparatively examining the methods of communicating climate change by the three organisations, the variances can be attributed to the different roles of the FBOs in action on climate change. The data shows that the HLO sees less value in utilising science into their campaign message and materials. The framework the HLO have developed to be used by the faiths to create their own long-term plans for environmental protection does not promote information presented in the form of projections and statistics. Rather the HLO promotes the use of narratives as they believe that they connect with people in ways which will lead them to positive action and that promote hope in creating long-term change. They use their 'no science' narrative approach in communicating climate change to influence individuals indirectly, through working with the faiths and organisations themselves, which is a contributing factor as to why their approach differs.

For the MLO science is seen as important when communicating climate change to provide authority and to emphasise the links between climate change and poverty. The science is used specifically to target the MLOs audience to demonstrate the effects climate change has, and will have, on the world's poorest most vulnerable people and the impacts this is having on the MLOs aim of eradicating poverty. For the MLO, the data shows science to be crucial in planning campaigns and in ascertaining the potential impacts for people and places considered at risk from climate change. However, it is clear from the data that the MLO does not see science as the best tool with which to communicate climate change. The data shows that the substance of their campaign is in surpassing science to inspire deeper philosophical spiritual reflections regarding relationships with all life and the Earth in line with Catholic teachings.

The LLO sees worship as an effective way to communicate climate change and although they use science in some of their resources the focus is more about listening to the narrative that the Earth is providing. The narrative comes in the form of storms, floods, drought, rising global temperatures and their global impacts. These are accompanied by biblical narratives and stories of Jesus which move beyond the scientific data with aims to inspire individual, political and congregational action on climate change and to promote hope for the future.

The data shows that there is a certain harmony between the different levels of engagement by the organisations and the different ways the organisations approach communicating climate change. This paper demonstrates the significance of a movement of movements when communicating climate change due to the importance of all humanity understanding the vital need for action and the variety of ways with which to engage people. A multiplicity of ways must be used to communicate climate change and these approaches to communication need to be intrinsically bound to Earth centered responses. This paper highlights that the communication of climate change by the sample FBOs is not confined to raising awareness but is connected to greater attempts to change people's lifestyles, shift their world views and create long-term changes.

This paper emphasises that further research is required to examine the effects of the campaigns with the organisations supporters. This is especially important for the MLO and LLO whose supporters are members of the public rather than organisations as is the case for the HLO. This proposed research would allow for campaign effectiveness to be measured through qualitative and quantitative research.

Importantly this research emphasises that scientific data, although crucial with regard to forming the knowledge base and informing political decisions, may not be the best tool to communicate climate change to the public. It is the use of narratives, which are considered more operative by the FBOs when communicating, not only the potential effects of climate change, but also when presenting attainable solutions. The psychology of hope is utilised by the FBOs according to the specific audience they are communicating too. By not including, or moving beyond the science of climate change, the FBOs distance themselves from the hopelessness of the IPCC assessments and the bleak appraisal of the negative actions of humanity. The research findings show that by using narratives, discussions on climate change move beyond the reliance on scientific language, to speak in terms applicable to the target audiences with hope. This narrative approach alters how climate change is presented and rather than the focus being a series of great threats, it is approached as positive opportunities to change. The narratives attempt to unify humanity globally and to inspire the celebration and worship of all life and the Earth. They aim to motivate individual and congregational political engagement to bring wider change and overall they use narratives to assist supporters in imagining a more hopeful future.

## References

Anderegg WRL (2010) The ivory lighthouse: communicating climate change more effectively. An editorial review essay. Clim Change 101:655–662. Available from: http://doi.org/10.1007/s10584-010-9929-z

ARC (2007) UN and ARC launch programme with faiths on climate change December 7th 2007. Available from: http://www.arcworld.org/news.asp?pageID=207. Accessed 10th Jan 2015

ARC (2015) Interview 1 Interviewed by: Huxley, Anna. (21st Jan 2015)

ARC (n.d.-a) Available from: http://www.arcworld.org/about.asp?pageID=7. Accessed 14th Jan 2015

ARC (n.d.-b) Available from: http://www.arcworld.org/projects.asp?projectID=137. Accessed 15th Jan 2015

ARC (n.d.-c) Available from: http://www.arcworld.org/about.asp?pageID=4. Accessed 10th Feb 2015

ARC (n.d.-d) Available from: http://www.arcworld.org/about.asp?pageID=2. Accessed 6th Jan 2015

ARC (n.d.-e) Available from: http://www.arcworld.org/downloads/Long-Term-Plans-Eco-Twinning.pdf. Accessed 15th Aug 2015

ARC (n.d.-f) Available from: http://www.arcworld.org/projects.asp?projectID=362. Accessed 10th Jan 2015

ARC (n.d.-g) Available from: http://www.arcworld.org/projects.asp?projectID=497. Accessed 10th Jan 2015

CAFOD (2009) CAFOD's seven year plan for presentation to the alliance of religions and conservation. Available from: http://www.arcworld.org/downloads/CAFOD-7YP.pdf. Accessed 25th Jan 2015
CAFOD (2014) Climate change and vulnerability: pushing people over the edge. Available from: http://cafod.org.uk/content/download/20298/142968/version/3/file/PUSHING%20PEOPLE%20OVER%20THE%20EDGE%20high%20res.pdf. Accessed 12th May 2015
CAFOD (2015a) Interview 1 Interviewed by: Huxley, Anna. (13th Jan 2015)
CAFOD (2015b) Interview 2 Interviewed by: Huxley, Anna. (13th Jan 2015)
CAFOD (n.d.-a) Available from: http://www.cafod.org.uk/Campaign/One-Climate-One-World/About-the-campaign. Accessed 2nd Mar 2015
CAFOD (n.d.-b) Available from: http://www.cafod.org.uk/Campaign/One-Climate-One-World/About-the-campaign. Accessed 2nd Mar 2015
CAFOD (n.d.-c) Available from: http://cafod.org.uk/content/download/22668/158320/version/3/file/Prayer_climate_postcard-reflections-2.pdf. Accessed 20th Oct 2015
CAFOD (n.d.-d) Trustees Report 2014/2015. Available from: http://cafod.org.uk/content/download/27347/269712/version/1/file/Report%20of%20the%20Trustees%20and%20Financial%20Statements%202015.pdf. Accessed 30th Aug 2015
CAFOD (n.d.-e) Available from: http://cafod.org.uk/. Accessed 14th Jan 2015
Colwell M et al (2009) Many heavens, one earth: faith commitments to protect the living planet: guidebook on the seven key areas for creating a long-term faith plan to protect the living planet. United Nations and ARC (UNARC). Available from: http://www.arcworld.org/downloads/UNARC%20-%20guide%20to%20creating%20your%20seven%20year%20plan.pdf. Accessed 5th Jan 2015
Corner A, van Eck C (2014) Science & stories: bringing the IPCC to life. Climate Outreach. Available from: http://www.comunicarseweb.com.ar/sites/default/files/biblioteca/pdf/1400631585_COIN-Science-and-Stories-Bringing-the-IPCC-to-life.pdf. Accessed 14th Nov 2016
Croft, The Rt. Revd. Dr. Steven (2014) Hope for the future: fighting the giant of climate change. Presidential address to the Sheffield Diocesan Synod 8th March 2014. Available from: http://blogs.oxford.anglican.org/hope-for-future-fighting-giant-of/. Accessed 12th Nov 2016
Feinberger M, Willer R (2011) Apocalypse soon? Dire messages reduce belief in global warming by contradicting just-world beliefs. Psychol Sci 22 (1):22–34
Francis P (2015) Laudato Si': on care for our common home. Encyclical letter. The Incorporated Catholic Truth Society: Publishers to the Holy See, London
Gardner GT (2006) Inspiring progress. Religions' contributions to sustainable development. Worldwatch Institute, Washington, DC
Gottlieb RS (ed) (2006) Introduction: religion and ecology—what is the connection and why does it matter?' In: The oxford handbook of religion and ecology. Oxford University Press, Oxford, pp 3–24
Hallegatte S et al (2016) Mapping the climate change challenge. Nat Clim Change 6:663–668
Hope for the Future (2014a) Available from: http://www.hftf.org.uk/who-we-are/. Accessed 20th Feb 2015
Hope for the Future (2014b) Available from: http://www.hftf.org.uk/ourvision/. Accessed 29th Nov 2016
Hope for the Future (2014c) Available from: http://www.hftf.org.uk/other-sources-of-information/. Accessed 15th Nov 2016
Hope for the Future (2014d) Available from: http://www.hftf.org.uk/training-day-outcomes/. Accessed: 21st May 2015
Hope for the Future (2015a) Interview 1 Interviewed by: Huxley, Anna. (12th Feb 2015)
Hope for the Future (2015b) Observation Climate Ambassador Training Day, St. Mark's Church, Broomhill, Sheffield. Observer: Huxley, Anna. (31st Jan 2015)
Hope for the Future (2016) What have we been up to this year? [Twitter] Available from: https://twitter.com/HopeFTFuture/status/799578879748927488. Accessed 18th Nov 2016
Hulme Mike (2009) Why we disagree about climate change: understanding controversy, inaction and opportunities. Cambridge University Press, Cambridge

Humanitarian Futures Programme (2009) Climate exchange: Liverpool Uni/ CAFOD April 21st 2009 Available from: https://humanitarianfutures.wordpress.com/tag/hfp/. Accessed 21st Nov 2016

Jowett, Rev. Canon NPA (2015) Available from http://www.hftf.org.uk/suggested-sermon/. Accessed 12th Mar 2015

King, Ed (2015) Climate Envoys to receive letter, asking: Why do you care? 22nd July 2015 Available from: http://www.climatechangenews.com/2015/07/22/climate-envoys-to-receive-letter-asking-why-do-you-care/. Accessed 21st Nov 2016

Mba Claud (n.d.) Available from: http://cafod.org.uk/About-us/Our-history. Accessed 6th Jan 2015

Russell B (ed) (1997) Grounds of conflict. In: Religion and science. Oxford University Press, Oxford, pp 7–18

This Changes Everything (2015) Directed by Avi Lewis. [Documentary Film] Canada: Klein Lewis & Luverture Films Production

UNFCCC (2014) United Nations Framework Convention on Climate Change 'Intergovernmental Panel on Climate Change (IPCC)' Available from: http://unfccc.int/bodies/body/6444.php. Accessed 9th Nov 2016

Weldon S (2007) Put your faith in a Greener Future. Western Daily Press, 13th Aug 2007 Available from: http://www.arcworld.org/downloads/Western%20DP%20Aug%2019%202007.pdf. Accessed 7th Jan 2015

Zampini Davies A (2014) Loving God's creation: biblical stories for CAFOD's one climate one world. Available from: http://cafod.org.uk/content/download/22392/156708/version/2/file/Prayer_climate_Genesis-and-Good-Samaritan.pdf. Accessed 15th Oct 2015

## Author Biography

**Anna Huxley** is a Ph.D student at the University of Manchester. She completed her Undergraduate and Master's Degrees at the University of Wales, Lampeter in Religious Studies and Religion and International Politics. Her Ph.D research is a qualitative investigation into Faith-Based Organisations in the UK and action on climate change. She is examining the FBOs motivations; methods of communication and their responses to climate change. The thesis examines the FBOs individually and comparatively to provide a greater understanding, in practice, of the role of faith in responding to climate change.

# Framing Climate Change: A Multi-level Model

**Renée Moernaut, Jelle Mast and Luc Pauwels**

## 1 Introduction

Climate change is *the* threat of the 21st century (IPCC 2014). However, the ways in which media depict the issue may influence which concerns are foregrounded, or whether the audience is encouraged to engage. Indeed, newspapers or TV broadcasts can be very influential sources, especially when it comes to abstract issues such as climate change, which most people have little experience with: Media may affect the boundaries of public and political debates and, as such, whether and which mitigative and/or adaptive action is taken (Graber 1988). Therefore, it is of crucial importance to gain better insights into the main representations which circulate in the media. Accordingly, it is no surprise that the number of studies on media representations of climate change have been expanding. Framing studies, in particular, are popular, especially for looking at verbal (e.g., Brand and Brunnengräber 2012; Fletcher 2009; Kenix 2008; Nisbet and Scheufele 2009; Shehata and Hopmann 2012; Van Gorp and van der Goot 2009; Vihersalo 2008; Weathers and Kendall 2015), but also visual (Borah 2009; Grittmann 2014; O'Neill 2013) or multimodal (e.g., Wessler et al. 2016) representations. However, many of these studies do not exploit the full potential of the framing concept/method. As such, they largely prevent more profound, and much-needed, insights into the breadth and depth of the media framing of the issue.

Firstly, most studies do not look beyond the mainstream media. However, only by taking a broader range of sources into account, one may draw a more

R. Moernaut (✉) · J. Mast · L. Pauwels
Department of Linguistics and Literary Studies (Applied Linguistics),
Vrije Universiteit Brussel, Pleinlaan 5 (Room 308), 1050 Brussels (Elsene), Belgium
e-mail: renee.moernaut@vub.be

R. Moernaut · J. Mast · L. Pauwels
Department of Communication Studies, University of Antwerp,
Sint-Jacobsstraat 2 (Room S.M.479), 2000 Antwerp, Belgium



W. Leal Filho et al. (eds.), *Handbook of Climate Change Communication: Vol. 1*,
Climate Change Management, https://doi.org/10.1007/978-3-319-69838-0_14

comprehensive picture of available frames. Alternative media are, for instance, more likely to promote non-hegemonic (i.e., non-dominant) frames than mainstream media (Brand and Brunnengräber 2012), which may engage a broader audience (Brulle 2010). Besides, by exposing ourselves to different perspectives, we have a better chance to identify the dominant assumptions we live with—and which determine our framing (analyses)—and to take a more critical distance (Hertog and McLeod 2001).

Secondly, many frames are rather thematic labels, which help to capture large amounts of communication, but "neglect the ideological nature and consequences of the framing process as well as the power relationships that influence that process" (Carragee and Roefs 2004, p. 219). Yet, as Dryzek (1997) illustrates, for instance, climate change and environment are, in particular, subjects of intense ideological debates and stake-holders' interests. Besides, most frames do not encompass an underlying argumentative structure (Entman 2004). Accordingly, many researchers fail to explicate exactly how they have detected frames, or how they come up with a particular name. Thus, there is a need for more clarity, and for more reliable and replicable coding measures (Tankard 2001). Frame matrices may be useful tools for summarizing the underlying patterns of frames, while also allowing for easy comparison and testing (e.g., Van Gorp and van der Goot 2009).

Finally, although the communicative power of visuals is broadly acknowledged (e.g., Messaris and Abraham 2001), most framing analyses are purely verbal.

Hence, the authors have tried to take the existing, and useful, research to a next level. Based on a framing analysis of 1256 climate change articles published in five Flemish (i.e., northern Belgian) outlets (period: 28/02/2012–28/02/2014), this conceptual study will attempt to answer these research questions: (1) Which set of multimodal climate change frames can be found drawing on a corpus of mainstream *and* alternative media articles?; (2) How do these climate change frames account for ideological diversity?; (3) How can comprehensive frame matrices summarize the main characteristics of those (climate change) frames? In short, the aim of this study is to contribute to the enhancing and broadening of the research on climate change framing, and, accordingly, to framing research in general.

## 2 Framing

A frame is an immanent structuring idea which gives coherence and meaning to a verbal-visual text. Framing, then, is applying a particular frame to structure an area of life: It involves selecting, omitting, expanding and giving salience to certain aspects of a perceived reality in a text, providing context and suggesting a particular problem definition, causal responsibility, moral evaluation, and/or treatment recommendation to the audience (Entman 1991, 2004; Gamson and Modigliani 1989; Van Gorp 2006).

'Frame packages' comprise a—latent—central organizing idea (often a generally shared motif, key event, narrative), and are made immanent by (manifest/latent) reasoning devices (problem definition, causal responsibility, moral evaluation, and/or treatment recommendation) and verbal/visual framing devices (Gamson and Modigliani 1989; Pan and Kosicki 1993; Van Gorp 2006).

Frames are in essence multimodal: They emerge in the verbal *and* visual, and in the interactions/relationships between the modes. Yet, the visual mode is highly salient, and particularly apt to framing (see e.g., Messaris and Abraham 2001).

## 2.1 Ideology

Framing and ideology are tightly interwoven. The authors define ideology, drawing on Carvalho (2007), as "(…) a system of values, norms and political preferences, linked to a program of action vis-à-vis a given social and political order" (p. 225). Frames are used to naturalize and justify ideologies. However, one ideology never equals one frame, or vice versa (Gamson and Modigliani 1989; Van Gorp 2006). Thus, one frame can be used by various ideological groups. Incorporating the frameworks of authors like Benford and Snow (2000), Baden (2010) or Brulle (2010) into this widely accepted view, the authors distinguish between three interacting and mutually dependent framing levels:

(1) Masterframes are broad collectives of language, images and ideas about how the world is and should be (see ideology). Masterframes may evolve through the interaction with other masterframes or through the adaptations by various interest groups, who use masterframes to structure their differing concerns, in various contexts (see below).
(2) ('Generic substance') frames (see Baden 2010) help to structure particular problems or concerns (e.g., economics, justice). Being part and parcel of human culture (Van Gorp 2006), they constitute a rather stable, 'universal' group, which may reappear across various issues (e.g., climate change, asylum debate). Yet, the 'culturally available frames' (Gamson and Modigliani 1989) in any given space-time may differ.
(3) Subframes are shaped by the available masterframes. Thus, while various interest groups may use the same frames, the realization of their subframes depends on the masterframe they abide by. Accordingly, subframes may contribute to the development of the masterframes (see above). Clearly, then, subframes are less stable than frames.

While mainstream media rather have a 'top-down' character in terms of context (e.g., sponsors), content (e.g., sources) and production process (e.g., journalists), alternative media tend to be 'bottom-up'. For instance, mainstream media are likely to abide by the 'hierarchy of access', giving precedence to elite sources, like politicians. Alternative media more often give a voice to 'alternative' sources, like grassroots groups, trying to open up the debate for other perspectives (Downing 2001). Hence, we may expect that mainstream and alternative media draw our attention to similar frames. Yet, alternative media are likely to use more non-hegemonic subframes than their mainstream counterparts, and vice versa (e.g., Brand and Brunnengräber 2012).

## 3 Environmental Frames in the Literature

The existing empirical research on the (media) framing of climate change and related issues is extensive. Table 1 provides a—non-exhaustive—overview, focusing on comprehensive, original and/or broadly cited contributions.

Classifying this large number of frames, the authors end up with seven categories, which provide a good overview of the general tendencies in the current debate (i.e., they may inform the current research): 'Uncertainty/Conflict', 'Economics and Technology', 'Alarmism/Disaster', 'Responsibility (Human-Induced Climate Change)', 'Morality and Ethics', 'Health and Safety' and 'Transformation'. Yet, the current frame sets show some limitations.

The current field of climate change frames is quite messy, with many researchers identifying their own—often non-exhaustive and/or not mutually exclusive (Tankard 2001)—sets of frames along slightly different lines. This prevents comparison among studies and issues, thus inhibiting the further development of the framing concept and methodology (Hertog and McLeod 2001).

The problems are largely due to the fact that many of the 'frames' work on different levels: Themes or topics—e.g., 'political' (Borah 2009)—attributes—e.g., risk and benefit frames—journalistic formats or routines—e.g., 'conflict' (Weathers and Kendall 2015)—or myths or mythological archetypes—e.g., 'Pandora's Box' (Nisbet and Scheufele 2009). The former two categories are described as issue-specific, conflicting with the idea of 'universal frames'. The latter are generic, but lack substance (see above). Others—e.g., 'polluters and causes' (Grittmann 2014)—focus on only one part of the argumentative structure (Entman 1991, 2004). As such, many empirically identified 'frames' do not represent a 'central organizing idea' or underlying narrative, despite of the fact that the conceptual parts of most studies draw on the definition of Entman (2004) and/or Gamson and Modigliani (1989).

Accordingly, the existing frames often lack a clear argumentative structure: Most studies only describe frames in generic terms, failing to explicate clear sets of reasoning and framing devices (Reese 2007; Tankard 2001). Exceptions are, among others, Van Gorp and van der Goot (2009) who provide frame matrices. These are useful tools, which allow for clarity and reliability, at least when combined with a comprehensive set of replicable multimodal coding measures (Tankard 2001) (see below). Most studies, however, only use a very limited and largely monomodal (verbal) set of devices.

As such, the majority of the existing frame(s) (sets) have one or multiple limitations which prevent them from fully informing us about the communicative effects of (climate change) frames, their underlying ideological interests and,

**Table 1** Climate change and environmental frames in the literature

| | *Uncertainty / conflict* | *Economics / technology* | *Alarmism / disaster* | *Human-induced climate change / responsible actors* | *Morality and ethics* | *Health and safety* | *Transformation* |
|---|---|---|---|---|---|---|---|
| ***Shehata & Hopmann* (2012)** | Scientific uncertainty | Economic consequences | | Climate change | | | |
| ***Nisbet & Scheufele* (2009)** | Scientific / technical uncertainty | Economic development | Pandora's box | Public accountability and governance | Morality / ethics | Public health | |
| ***Kenix* (2008)** | Conflict | Economics | Sensationalism | New evidence or research<br>Scientific background<br>Domestic politics<br>International relations<br>Current weather<br>Consequences | Morality<br>Consequences | Consequences | |
| ***Weathers & Kendall* (2015)** | Conflict<br>Uncertainty | Consequence | | New evidence<br>Action | Ethics | Consequence | |
| ***Fletcher* (2009)** | Scientific scepticism | Economic opportunity | | | | Security threat | |
| ***Semetko & Valkenburg* (2000)** | Conflict | Economic consequences | | Responsibility | Morality | Human interest | |
| ***Vihersalo* (2008)** | | Climate change as a problem of greenhouse gas emissions<br>Climate change as a problem of vulnerability | | Climate change as a problem of greenhouse gas emissions | Climate change as a problem of vulnerability | | |
| ***Brand & Brunnengräber* (2012)** | | Maneagability<br>Compatability<br>Adaptation<br>Technological fix / behaviour | | | | | Transformation<br>Anti-adaptation |
| ***Van Gorp & van der Goot* (2009)** | | Progress | Frankenstein<br>Undermining of fundaments | Progress | Responsibility<br>Mother Earth | | |
| ***Grittmann* (2014)**<br>*Visual frames* | | Problem definition<br>Polluters and causes<br>Solutions | | Problem definition<br>Polluters and causes<br>Solutions | Moral evaluation | Problem definition | |
| ***O'Neill* (2013)**<br>*Visual frames* | Contested | | | Distancing | | | |
| ***Borah* (2009)**<br>*Visual frames* | | | | Political | | Lives saved / lost<br>Human interest<br>Pragmatic<br>Global warming victims | |
| ***Wessler et al.* (2016)**<br>*Multimodal frames* | Political negotiations | Global warming victims<br>Civil society demands<br>Political negotiations<br>Sustainable energy | | Global warming victims<br>Civil society demands<br>Political negotiations<br>Sustainable energy | | | |

The grey colours mark those 'frames' which fit multiple categories notably, potential hegemonic struggles within the context of frames (Carragee and Roefs 2004).

Taking these considerations into account, the authors will attempt to provide a limited, but comprehensive, set of climate change frame matrices, based on a systematic analysis method, which allows for reliability and replicability (Tankard 2001).

## 4 Research Design

The sample comprises articles of three Flemish mainstream newspapers, 'De Standaard' (DS) (broadsheet, centrist), 'De Morgen' (DM) (broadsheet, leftist) and 'Het Laatste Nieuws' (HLN) (popular, liberal background), and two—openly left-wing—alternative outlets, 'DeWereldMorgen' (DWM) and Mo* Magazine (MO). The selection was based on maximum variation sampling: The three mainstream outlets arguably represent the various ideological viewpoints found in the Flemish media landscape. Besides, the two broadsheet newspapers are the most influential and widely circulated dailies, while HLN is the most widely read (popular) newspaper among all strata of the Flemish population. DWM and MO are the major alternative outlets in Flanders, reaching respectively 15.000 and about 3.500 readers a day (cim.be; DeWereldMorgen.be). Employing the search tools of the Belgian press archive Gopress and/or the archives of the outlets the authors conducted the following key word searches: 'climate change', 'global warming' and 'greenhouse effect'. Additionally, a manual search was done in the paper archives of the national and university libraries. These searches yielded a total of 1256 articles, for the period February 28, 2012 to February 28, 2014.

An inductive-deductive qualitative framing analysis was applied. Firstly, a random sample of about 25% of the articles was analyzed in a qualitative-inductive way (this percentage was guided by data saturation). This approach does not limit the potential frames to those previously identified in the literature, but allows a researcher to detect new frame categories. Through repeated processes of open coding, axial coding and selective coding, the researchers constructed frame matrices for each (sub)frame identified in the corpus. These allow for transparency and replication. Drawing on the matrices, the remainder of the corpus was deductively analyzed. If (repeated) deviations from the initial frames were found, the frame matrices were updated. Later on, the overarching masterframes were constructed, based on the grouping of the subframes. The authors added one additional subframe, theorizing on the literature review and analogical patterns within other frames. This was not found in the corpus, but may well appear in other contexts (see below). Moreover, during the entire analysis process, the authors kept the literature review at the back of their mind (Glaser and Strauss 2008; Van Gorp 2006). Based on Entman (1991) the most salient multimodal frame for each article was determined. After all, ordinary readers are unlikely to discern, and be affected by, all frames in a text.

## 4.1 Analytic Method

The multimodal framing analysis toolkit (Table 2), which was used to identify the (sub/master)frames, is informed by the existing framing literature (Entman 1991; Grittmann 2014; Pan and Kosicki 1993; Rodriguez and Dimitrova 2011; Van Gorp 2006). Yet, the authors maintain that the analytical frameworks provided do not fully allow for the systematic and convincing excavation of multimodal frames,

**Table 2** Multimodal Framing Analysis Toolkit

| Verbal | Multimodal Framing Analysis Toolkit | Visual |
|---|---|---|
| | *CENTRAL ORGANIZING IDEA*<br>– *Problem definition*<br>– *Causal responsibility*<br>– *Treatment recommendation*<br>– *Moral evaluation* | |
| – Naming and predicational strategies | ***(In)animate participants and attributes***<br>– Individualization/humanization. (e,g., 'the 30 year old Davy')—collectivization/generalization (e.g., 'the citizens')<br>– Identification (what one is, e.g., 'vulnerable')—functionalization (what one does, e.g., 'the teacher')<br>– Narrative roles (hero, victim, villain) | – Visual choices (e.g., facial expression, gesture, clothing, conceptual patterns ((dis)connection), focus on generic characteristics…) |
| – Choice for particular verbs<br>– Active/passive verb constructions, adjuncts, auxiliary verbs<br>– Nominalizations (e.g., 'the globalizing world') | ***(Inter)actions***<br>– Transitivity (a property of verbs that highlights the doings and relations between participants: material (e.g., 'we emit GHG'), behavioural (e.g., 'they laugh'), verbal (e.g., 'they discuss'), relational (e.g., 'he is the leader'), existential (e.g., 'it is raining') and mental (e.g., 'we fear') processes<br>– Active/passive voice ('we save them' versus, 'they are saved')<br>– Tense (indication of time) | – Vectors (e.g., gazes, arm lines)<br>– (Lack of) gaze (offer or demand)<br>– Ordering of images (e.g., causality) |
| – Personal pronouns, e.g., 'inclusive we' (includes speaker and addressee), 'exclusive we' (includes speaker, but not the addressee), they | ***Positionality***<br>– (Moral) alignment between participants inside/outside the picture frame | – Point of view (horizontally/ vertically)<br>– Distance (intimate, personal, social, public) |
| – Focus on time, causality, sequences<br>– Adverbs, prepositional constructions | ***Context (time, space, causality…)***<br>– Absence of presence<br>– Character (e.g., generic or specific)<br>– Definition of 'the other world' | – Focus on space, spatial relations and characteristics<br>– Distance/angle (focus on context or subject/scene)<br>–Sublimation (decontextualization) |

(continued)

**Table 2** (continued)

| Verbal | Multimodal Framing Analysis Toolkit | Visual |
|---|---|---|
| – Grammatical mood, modal auxiliaries, adjectives, adverbs, particles | ***Modality***<br>– (Degrees of) necessity, possibility, certainty, likelihood | – Articulation of detail, background, colour saturation, modulation or differentiation, quality, articulation of depth … |
| | ***Rhetorical figures***<br>– Juxtapositions, contrasts, metaphors (e.g., 'the Earth is a machine'), metonyms (e.g., smokestack stands for human emissions), similes (e.g., 'brave as lions'), hyperboles (strong exaggerations), ellipses (omissions), presuppositions, euphemisms, personifications, semantic reversals (misleading terms which suggests inaccurate facts or circumstances, e.g., 'development' is said to hide destructive imperialism)… | |
| – Depiction of narrator: Naming and predicational strategies<br>– Depiction of quotes: modality and hedging, choice for particular quoting verbs | ***Narration***<br>– (De)legitimization of sources ('narrators' (see previous levels)) and/or their quotes | – Depiction of narrator: Choice for particular attributes, poses, facial expressions, settings…<br>– Narrator as focalizer: global, objective, subjective… point of view |
| | ***Intertextuality (direct/broader context)***<br>– Key events, icons and symbols, myths and mythological archetypes…<br>– Similar (multimodal) texts (e.g., same theme) | |

Note that narrative elements or styles (content) can be found on multiple levels. Yet, a distinct 'narration level' is discerned in order to account for the formal/meta-level of the narration

let alone their underlying ideologies. Therefore, the existing views were complemented with insights from (multimodal) Semiotics and Critical Discourse Analysis (e.g., Chouliaraki 2006; Machin and Mayr 2012; Kress and van Leeuwen 2006; van Leeuwen 2008), photography and film studies (e.g., Monaco 2000) and (eco)linguistics (e.g., Halliday 2000; Stibbe 2015).

On several of these levels salience enhancing devices can be discerned as well (Entman 1991, 2004; Van Gorp 2006), for instance, repetition, size, placement on a page, in an outlet or in the structure of the news story. These devices may help to define which devices have been "magnified or shrunk in order to make them more or less salient" (Entman 1991, p. 9).

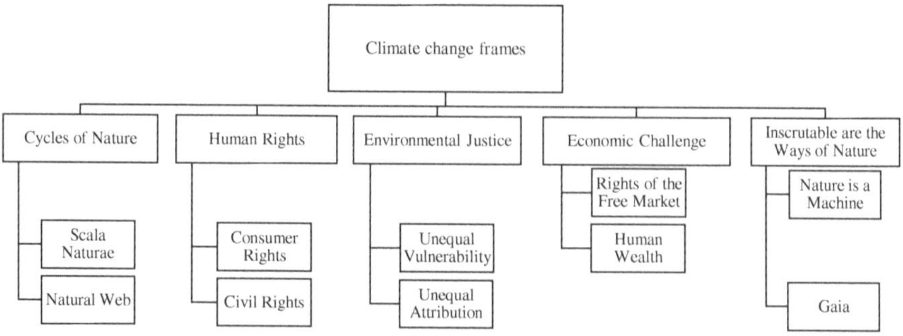


**Fig. 1** Overview of the climate change frames and subframes

# 5 Results

Based on the analyses, the authors introduce a frame set of five frames and ten subframes, which are visualized in Fig. 1.

## 5.1 *Five Frames*

*'Cycles of Nature'*

Cycles of Nature refers to the multiple cycles in nature which (non-)living beings are part of (e.g., life cycles). Accordingly, the underlying narrative reads: Due to the acting of human (e.g., contempt for nature), the natural system as we know it is under threat. Continuously interacting with nature, human is a secondary victim. Hence, humans need to rethink the ways in which they act and interact with (non-) living beings and processes. Typical devices are: repetition of terms like 'nature', 'the world'; juxtaposition of human and nature.

*'Environmental Justice'*

Environmental justice is a narrative used by various groups, in different contexts, to denounce the unequal distribution of environmental benefits and burdens. The frame contends that certain peoples with particular socio-economic, gender, ethnic and/or cultural characteristics (mainly (in) 'the South') will pay the bill of climate change as they are more sensitive to changes and/or lack coping abilities. The dominant groups (mainly (in) 'the West') carry most responsibility for the problems, due to their misguided (i.e., egocentric) priorities and beliefs. The problems can only be solved by certain—well-equipped/informed—groups. Recurring devices are the contrast between groups (e.g., 'West' and 'South') or the reference to disasters, like Typhoon Haiyan (2013), as key events.

*'Economic Challenge'*

Economic challenges are important concerns in today's society (see the prevalence of economic frames in the literature). 'Economic Challenge' argues that the wealth and development of human are under threat. This is due to climate change and, particularly, the response of (mostly) elite groups, who overlook fundamental economic interests. Indeed, an optimal reaction guarantees the status-quo situation, but ideally amplifies human prosperity. Framing devices encompass: a focus on economic terminology, numbers, comparatives and superlatives indicating economic loss or gain; intertextual references to stock market reports and the like; threatening present balanced with the promise of future economic development.

*'Human Rights'*

The modern conception of human rights (e.g., The UN Universal Declaration of Human Rights) encompasses a number of 'universal' principles all humans are entitled to. Accordingly, the 'Human Rights frame' exclusively highlights the fate of the human species, whose rights and well-being are being threatened. Lacking the (mental) ability to sufficiently protect themselves, however, ordinary people are affected most severely. Climate change constitutes a major threat. Yet, it is people —particularly, the elites—who are endangering their own kind, blinded by short-term (economic) interests. Therefore, all humans must take responsibility. Characteristic devices are the strong focus on the human world (e.g., repetition of 'human', 'society') or intertextual references to the Universal Declaration of Human Rights.

*'Inscrutable are the Ways of Nature'*

This—rather sceptic—frame evokes an idea which can be found across ideologies and religions: Human ought to respect nature, which he cannot entirely understand (e.g., Verhagen 2008). Climate change is depicted as a largely natural variance: Developments in the natural system are due to actors and processes in the system. If needed, the natural system will also mitigate the flaws. Hence, the intervention of human is unimportant. Typical devices are the exclusive focus on entirely natural environments and the backgrounding/deleting of human (responsibility).

The frame matrices (Table 3) provide a more detailed overview of the most characteristic reasoning and framing devices of each frame, as well as some typical examples from the corpus.

**Table 3** Frame matrices

| Problem definition | Causal responsibility | Solution | Moral basis | Framing devices | Examples |
|---|---|---|---|---|---|
| **Cycles of Nature Frame** | | | | | |
| The natural system as we know it, is seriously affected in—potentially—dangerous ways. Interacting with nature, (ordinary) human is also a—secondary—victim | The current climate change is not natural but caused by humans (i.e., mainly but not exclusively (top-down) elites). As such, human shows contempt for (the characteristics of) nature | Human will need to rethink and change the ways in which he acts and interacts with other (non-)living beings and systems. This will also benefit human. While all of us must act, we must be led by those who are best equipped (with knowledge, insights, means…) | – The current human acting is immoral and indefensible (i.e., overly selfish) towards the larger natural system<br>– Recognition of certain values of the natural system (vis-à-vis contempt), which are meaningful to human | **Participant and process depiction**<br>– Human, 'inclusive/ collective we' (especially particular groups), climate change: responsible agent (causes and solutions)<br>– Cause: human development, wrong (i.e., 'misguided') priorities and beliefs (e.g., short-term economic prosperity, contempt for nature), man-made climate disasters<br>– Object of help: green economy, sustainable development, altered human acting and interacting with the rest of the (natural) world + redefining of priorities with an eye on the survival and well-being of the natural system (and thus, human)<br>– The natural system, human (secondary): patient (problem)<br>– Unwanted object: vulnerability, loss of | – "Every year we dig up more from the Earth, emit more GHG and consume more energy." (DS, 04/25/ 2013, p. 10)<br>– "We cannot link climate change to one particular weather phenomenon, but we clearly see the fingerprint."(DWM, 01/ 16/2014)<br>– "(…) all energy is produced by wind turbines, hydraulic pumps and solar panels." (HLN, 10/27/2012, p. 85) |

(continued)

**Table 3** (continued)

| Problem definition | Causal responsibility | Solution | Moral basis | Framing devices | Examples |
|---|---|---|---|---|---|
| | | | | resilience, misrecognition of certain ways of living<br>– Context: generic ('the world'), often focused on human environments (general)<br>**Rhetorical devices**<br>– Juxtaposition human society and nature/climate change<br>– Repetition of terms like 'nature', 'the world': references to human-nature interaction | |
| **Environmental Justice Frame** | | | | | |
| Whereas some—socio-economic, ethnic, cultural gender…—groups (mainly, but not exclusively (in) the South) are less responsible for climate change, they (will) pay the bill: They are more sensitive to changes and/or lack coping abilities | The most powerful and dominant groups (mainly, but not exclusively (in) the West) carry a large responsibility, for the loss of lives, livelihoods, cultures… beyond their time-spaces | A combination of mitigation (GHG reduction) and (local) adaptation can solve the problems | – The current acting of the ('western') elites is immoral and indefensible (i.e., overly selfish) towards other groups in society, who live beyond the (cultural) time-space of the former<br>– Recognition (vis-à-vis contempt) of other lives, livelihoods, cultures | **Participant and process depiction**<br>– The elites (mainly, but not exclusively (in) the West), climate change: responsible agents (cause)<br>– Cause: elite ('western') development, wrong (i.e., 'misguided') priorities and beliefs (e.g., egocentric economic prosperity, contempt for 'others'), climate disasters<br>– Particular human(s) (groups) (with a certain socio-economic, ethnic, cultural, gender… | – "We pollute and they pay the price." (DWM, 09/20/2013)<br>– "As a result of the Kyoto protocol they [the industrialized countries] were obliged to reduce their GHG emissions during the past 15 years. They had taken on that responsibility because they bear a historical heritage for (…) gases which cause a gradual temperature rise on Earth." (DS, 11/19/2013, p. 14) |

(continued)

**Table 3** (continued)

| Problem definition | Causal responsibility | Solution | Moral basis | Framing devices | Examples |
|---|---|---|---|---|---|
| | | | | character): responsible agents (solution)<br>– Object of help: mitigation + adaptation (with a certain role for green economy/ sustainable development)<br>– Particular bottom-up groups (mainly, but not exclusively (in) the South): patients (problem)<br>– Unwanted object: vulnerability, e.g., socio-economic loss, misrecognition of certain ways of living, organizing one's life and economy…<br>– Context: continuous time, (mainly) human environments<br>**Rhetorical devices**<br>– Metaphors, hyperboles, numerals… emphasizing the extraordinary natural forces hitting certain groups<br>– Contrast/juxtaposition: 'us' versus 'them'<br>– Key events/intertextual references: Typhoon Haiyan (2013), Tsunami (2004) | |

(continued)

**Table 3** (continued)

| Problem definition | Causal responsibility | Solution | Moral basis | Framing devices | Examples |
|---|---|---|---|---|---|
| **Economic Challenge Frame** | | | | | |
| Human is, primarily, a(n economic) consumer and/or producer. However, his wealth and development are under threat | Climate change sets in motion major changes which have an impact on human (access to) wealth and development. The response of certain elites is, however, economically not beneficial, or even dangerous | Human response to climatic changes at least has to guarantee (status-quo) human wealth and development, but—ideally—will amplify these | – Economic decay and decline are unacceptable and must be counteracted at all times<br>– We must recognize climate change as 'a blessing in disguise' and reap the benefits of this change (as efficiently as possible) | **Participant and process depiction**<br>– Elites, climate change: responsible agent (cause)<br>– Cause: current (climate) acting, wrong (i.e., 'misguided') priorities and beliefs (e.g., overlooking crucial economic interests)<br>– Elites: responsible agents (solution)<br>– Object of help: human technology, development, ingenuity<br>– Man as economic agent: patient (problem)<br>– Unwanted object: (economic) loss and decay<br>– Deletion (decontextualization): non-economic agents, patients, objects—particularly strong utilitarian view on nature<br>– Context: the world, especially the West; mostly human environments—(one global) economic society<br>**Rhetorical devices**<br>– Metaphor: war | – "(…) Belgium might save 60 to 150 billion euros, because the country would no longer need to import oil from the Middle East or other fossil fuels." (DM, 09/19/2013, p. 16)<br>– "(…) the peripheral northern areas of Europe, with the message that the European Union needs and will have to dedicate more attention to this strategically important region." (DS, 03/10/2012, p. 38) |

(continued)

**Table 3** (continued)

| Problem definition | Causal responsibility | Solution | Moral basis | Framing devices | Examples |
|---|---|---|---|---|---|
| | | | | – Numbers: (economic) loss or gain (often combined with comparative/superlative)<br>– Juxtaposition economy/technology and climate<br>– Repetition of economic terminology, one-liners<br>– Key events/intertextual references: stock market reports, financial news (pages)<br>**Narration**<br>– Focus on threatening present/future, but balanced with ideal (i.e., promise) of (economic) development | |
| **Human Rights Frame** | | | | | |
| The human rights, freedom, well-being, ways of living of all humans on Earth are under threat. Bottom-up groups (i.e., ordinary people)—who (currently) cannot act for themselves—are among the ones who are affected most severely | Climate change is a threat. Yet, humans—particularly, but not exclusively, the elites—endanger their own kind, by holding on to wrong (misguided) priorities and beliefs | All actions that humans undertake must be inspired by the fundamental responsibility to defend their own rights and protect their well-being. While all of us must act, we must be led by those who are best equipped (with knowledge, insights, means…) | – Moral acting is acting that—first and foremost—takes into account, and respects, the greater (long-term) interests of humanity (i.e., human rights)<br>– We must have faith in the human capacity to lead ourselves to salvation | **Participant and process depiction**<br>– Man, 'inclusive/collective we' (especially particular groups), climate change: responsible agent (causes and solutions)<br>– Cause: human development, wrong (i.e., 'misguided') priorities and beliefs (e.g., short-term economic prosperity, overlooking human | "For years now, scientific and legal research has shown that climate change is caused by human, that it is life-threatening and that it will cause violations of human rights." (DM, 11/09/2013, p. 72) |

(continued)

**Table 3** (continued)

| Problem definition | Causal responsibility | Solution | Moral basis | Framing devices | Examples |
|---|---|---|---|---|---|
| | | | | rights), man-made climate disasters<br>– Object of help: green economy, sustainable development + redefining of priorities and beliefs with an eye on (long-term) human well-being<br>– Human ('inclusive/global we'), particularly certain (bottom-up) groups: patient (problem)<br>– Unwanted object: human suffering, lack of well-being, freedom, misrecognition of certain ways of living, or organizing one's life<br>– Context: human environments<br>**Rhetorical devices**<br>– Contrast/juxtaposition: top-down versus bottom-up<br>– Repetition of terms like 'human', 'society', 'human rights'<br>– Key events/intertextual references: The Universal Declaration of Human Rights | |

(continued)

**Table 3** (continued)

| Problem definition | Causal responsibility | Solution | Moral basis | Framing devices | Examples |
|---|---|---|---|---|---|
| **Inscrutable Are the Ways of Nature Frame** | | | | | |
| Developments can be witnessed in the natural system | The changes are internal to the system. They are—largely—due to specific agents | If needed, these flaws are mitigated by the natural system. Human's help is—largely—superfluous | – Human should acknowledge his limited impact on the natural system<br>– Demon-strating (human) self-knowledge is accepting our (rather) limited understanding of nature | **Participant and process depiction**<br>– The natural system: responsible agent (cause and solution), patient (problem)<br>– Cause: GHG<br>– Object of help: system-internal changes<br>– Unwanted object: end of the current state<br>– Nominalizations, personifications, passivizations, (visual) deletion: decontextualization of (causes and) solutions; responsibility of human is largely absent/ minimalized<br>– Context: (largely) natural environments; continuous time<br>**Rhetorical devices**<br>– Contrast/juxtaposition: (sublime) nature versus human<br>– Key events/intertextual references: 'climate change is a hoax' statements, for instance on social media (by Republican Americans) | – 'Nature will solve the problems' |

## 5.2 'The Anthropocentric Masterframe'

Six of the identified subframes—'Scala Naturae', 'Unequal Vulnerability, 'Human Wealth', 'Consumer Rights', 'Rights of the Free Market' and 'Natural Machine'—promote an anthropocentric ideology. Values are, among others, (economic) development, competition and human ingenuity. Also, there is a crisp human-nature dualism (Dryzek 1997; Stibbe 2015; Verhagen 2008). Accordingly, the central organizing idea of the Anthropocentric Masterframe can be summarized as follows: Human is the main victim, as his environment and/or natural services are threatened due to—largely—external pressures. The latter need to be removed for him to enjoy again what he deserves. Common framing devices are:

1. The 'nature is a machine' metaphor is pivotal: Nature is a distant object which mainly exists to fulfil human (economic, scientific, aesthetic…) needs. As such, it constitutes the background of an economic society, with (individual) consumers and producers. Nature encompasses (groups of) largely disconnected processes and (non-)living beings.
2. Climate change, its consequences (e.g., floods) and/or causes (mainly GHG) are often quite independent and largely external agents [see decontextualization (Iyengar 1990)]. If human responsibility for GHG emissions is acknowledged, it is often backgrounded or abstracted (e.g., nominalizations like 'human-induced climate change').
3. If agency is foregrounded, particular (generalized or identified) human or natural agents are often singled out, as villains and/or heroes. This backgrounds structural, collective responsibilities (decontextualization).
4. The main problems are economic/material maldistribution, for instance, the destruction of lives or livelihoods. Some references can also be found to cultural identity misrecognition, like the loss of long-standing—human or natural—traditions. Solutions need, accordingly, to focus on redistribution (and identity recognition) (Fraser 2005). More structural, interconnected problems are backgrounded or deleted (decontextualization).
5. The war metaphor (Kenis and Mathijs 2014) highlights the external threats which must be fought.
6. A threatening present and/or future is highlighted, with stories of suffering, loss and decay. Clear solutions or positive alternatives, which are more likely to engage (O'Neill 2013), are often absent or abstract (Brulle 2010).
7. Numbers, percentages and graphs—e.g., specifying GHG emissions—highlight the scientific, rational character of this masterframe.
8. The main claims-makers (i.e., sources) belong to ('western') elite groups, mostly scientists, politicians, economic players or NGOs. Journalists are usually (part of a small group of 'western') professionals.

There are two directions underlying this anthropocentric argumentation (see Fig. 2), 'Status Quo' and 'Reform' (Dryzek 1997; Hopwood, Mellor and O'Brien 2005):

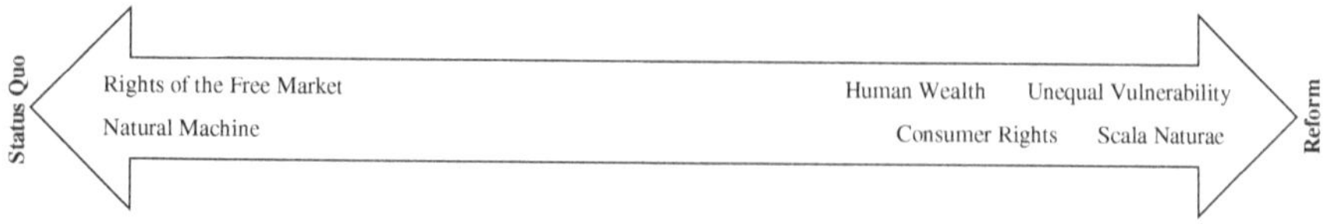


**Fig. 2** The Status Quo—Reform continuum underlying the Anthropocentric Masterframe

'Natural Machine' and 'Rights of the Free Market' can be situated within the 'Status Quo' view (see Table 4 for the frame matrix), which does not, or barely, question unlimited (economic) development. Accordingly, this perspective takes the described arguments and devices to an extreme, notably, the decontextualization strategies, the singling out of individual (non-)human agents, and the idea of nature serving human (as a forgiving, corrective Santa Claus).

The other subframes lean more closely towards 'Reform' (see Table 5), which highlights the need to limit growth, looking for solutions within the current society. Top-down groups are presented as the main responsible agents for causes (e.g., GHG emissions), while bottom-up groups are internally weak and the major victims of the changes (e.g., patients, conditional agents). Hence, the top-down groups must provide solutions (e.g., green economy/technology, transfer of means and knowledge). Yet, 'collective we' must also follow up instructions, on how to alter our consumption or production (Brulle 2010; Kenis and Mathijs 2014).

### 5.2.1 Six Anthropocentric Subframes

*'Scala Naturae Subframe'*

'Scala Naturae' (see Table 6 for the frame matrix) is a typical anthropocentric metaphor: Human is at the top of the natural ladder which represents a hierarchy of perfection. Every lower step is there to support the higher ones (Verhagen 2008). Accordingly, the 'Scala Naturae Subframe' argues that human-induced climate change threatens the natural system, some parts more than others. These changes also affect humans, at the top of the ladder. Although all humans are responsible for preferring short-term development (causing GHG) over the long-term services of nature, some countries, politicians or corporate organizations bear a larger responsibility. Therefore, (elite) human intervention is imperative, since nature lacks internal resilience or coping abilities.

*'Unequal Vulnerability Subframe'*

Central to this subframe (Table 7) is the acceptance that the inequality of humans—along socio-economic, ethnic, cultural, gender… lines—is given by nature (Chouliaraki 2006; Fraser 2005). The story underlying 'Unequal Vulnerability' can be summarized as follows: Some groups are characterized by internal weaknesses. They are economically, socially, culturally… underdeveloped. As such, they are disproportionally vulnerable to and less able to cope with climate

**Table 4** Anthropocentric Masterframe Matrix—Status Quo

| Problem definition | Causal responsibility | Solution | Moral basis | Framing devices | Examples |
|---|---|---|---|---|---|
| Nature caters to human needs. However, human's full access to these services is currently under pressure (maldistribution) | The acting of individual players threatens the services of nature | Nature—particularly climate change—is generous and forgiving. Humans can enjoy the fruits of nature if they recognize this (redistribution) | – Nature, as a perfectly designed machine, should be allowed to function in the ways, and for the purposes, it was meant | **Participant and process depiction**<br>– Individualized players: responsible agents (cause)<br>– Cause: pressures, changes—maldistribution<br>– Individualized players, climate change: responsible agents (solution)<br>– Object of help: taking away of pressures, changes—redistribution<br>– (Humans benefitting from) natural services: patients (problem)<br>– Unwanted object: maldistribution (vulnerability, loss)<br>– Nominalizations, personifications, passivizations, (visual) deletions, sublimation: decontextualization of broader responsibilities (human-induced climate change + hegemonic system)<br>– Environment: mainly resource, or abstract circumstance (background), deleted<br>– Context: (decontextualized) natural environments (for human usage)<br>**Rhetorical devices**<br>– Metaphor: nature as Santa Claus, nature is a machine<br>– Use of (stereo)typical climate change symbols (metonyms/metaphors) to associate discussion to idea of (natural) climate change | – "The European market is uncertain as Europe focuses on sustainable energy." (DS, 11/27/2012, p. 15)<br>– "Earthworms are the real culprits behind global warming." (HLN, 02/05/2013, p. 24)<br>– An (abstracted) globe (only) shows two new, more efficient shipping lanes (MO, 03/05/2013)<br>– "(…) over the last 65 million years, the mountain woods have been able to prevent that the temperature would rise to levels which would be catastrophic for living beings." (DS, 02/11/2014, p. 8)<br>– Comparative (before-after) satellite images, which are typically used to show the melting of (polar) ice sheets or glaciers, are used to demonstrate the growing of the ice, instead. (HLN, 09/10/2013, p. 24) |

(continued)

**Table 4** (continued)

| Problem definition | Causal responsibility | Solution | Moral basis | Framing devices | Examples |
|---|---|---|---|---|---|
| | | | | – Prevalence of numbers, percentages<br>– Contrast/juxtaposition: human versus nature (dualism)<br>– Deletion/concealing of natural web<br>– Presupposition: anthropocentric worldview as given—status-quo as the only possibility<br>**Narration**<br>Sources: mainly elites (top-down) (mainly scientists) | |

The highlighted devices are general (i.e., common) characteristics in the anthropocentric view

**Table 5** Anthropocentric Masterframe Matrix—Reform

| Problem definition | Causal responsibility | Solution | Moral basis | Framing devices | Examples |
|---|---|---|---|---|---|
| Human is *the* victim of climate change, as the environment—which caters to his needs—is changing. However, certain groups in society, as well as in nature, have **internal** characteristics which make them more vulnerable and less able to cope than others (maldistribution, cultural identity misrecognition) | Man-made climate change, its causes (GHG) and (direct) consequences constitute a **largely external** threat. While all of us humans carry some responsibility for this threat, some—mainly, but not exclusively elite (top-down) groups—are more responsible. As such, they show contempt for others (nature and/or humans) (maldistribution, cultural identity misrecognition) | Due to their responsibility, and/or their accomplishments, particular groups need to act for the sake of nature, and thus human. They need to mitigate their GHG emissions and adapt, by means of green economy/technology or sustainable development. Besides, they need to share their money, means or knowledge with others, and help them (to) protect (themselves) (adaptation and/or mitigation). (**external solution**) (redistribution, cultural identity recognition) | – The elites have the obligation (and potential) to protect the others, or to help them to protect themselves, by sharing their accomplishments (inequality as an accepted fact)<br>– The right to live, to develop, to prosper… as 'Universal (human) Justice' | **Participant and process depiction**<br>– Climate change (consequences), '(inclusive/collective) we', (abstracted, backgrounded) human, certain (top-down) groups (individualization)—mainly as consumers/producers: responsible agents (cause)<br>– Cause: GHG emissions, development, ignorance/unwillingness (i.e., preoccupation with short-term, egocentric interests), (man-made) climate disasters—maldistribution, cultural identity misrecognition<br>– Certain (top-down) groups, '(inclusive/collective) we' human (generalized, identified)—mainly as consumers/producers: responsible agents (solution)<br>– Object of help: mitigation and adaptation by means of sustainable development/green economy (production + consumption) (e.g., renewable energy, emission trading, techno-fix) + transfer of money, means, knowledge, 'development'—redistribution, cultural identity recognition<br>– Human, particular (bottom-up) groups in human society' and nature (generalized, identified): patients | – "The US did not ratify the Kyoto protocol out of fear that it would harm its economic interests." (DS, 11/12/2013, p. 6)<br>– "(…) the Arctic is *the* climate controller—let us say, *the* air conditioning - of this globe." (DS, 10/17/2013, p. 9)<br>– A drawing of a globe highlights (hyperbole) a number of causes (e.g., a smokestack, a car) and consequences (e.g., a southern woman and child, colour symbolism) (HLN, 06/21/2012, p. 6)<br>– "The shortage of water resources does not only have a huge impact on the economy (…) but also on the natural system itself." (DWM, 10/23/2013)<br>– "The citizens will have to move by foot or by bike." (HLN, 01/14/2014, p. 14)<br>– "(…) sharing green technology with the developing countries." (DM, 06/23/2012, p. 1) |

(continued)

**Table 5** (continued)

| Problem definition | Causal responsibility | Solution | Moral basis | Framing devices | Examples |
|---|---|---|---|---|---|
| | | | | (problem)/'vulnerable'—carriers or identifieds, patients, conditional agents<br>– Unwanted object: maldistribution (vulnerability, loss), cultural identity misrecognition (of cultural practices, characteristics)<br>– Nominalizations, personifications, passivizations, (visual) deletion, sublimation: decontextualization of broader responsibilities (mainly the hegemonic system)<br>– Environment: circumstance (background), abstract, source, deleted<br>– Context: one global (human) economic society<br>**Rhetorical devices**<br>– A limited number of quite stereotypical metonyms/symbols, like the polar bear or the suffering child<br>– Metaphors: nature is a machine, war<br>– Prevalence of numbers, percentages<br>– Contrast/juxtaposition: top-down versus bottom-up, human versus nature (dualism)<br>– Synecdoche/'collective we': We (human) are all responsible; we are all under threat; we all have to act<br>– Deletion/concealing of natural web<br>– Visual/verbal corporate speak | |

(continued)

**Table 5** (continued)

| Problem definition | Causal responsibility | Solution | Moral basis | Framing devices | Examples |
|---|---|---|---|---|---|
| | | | | – Presupposition: anthropocentric worldview as given—status-quo as danger; reform as solution<br>**Narration**<br>– Sources: mainly elites (top-down) (science, politics, economics, agriculture, industry…)<br>– Focus on threatening (climate) present/future | |

The highlighted devices are general (i.e., common) characteristics in the Anthropocentric view

**Table 6** Scala Naturae Subframe Matrix

| Problem definition | Causal responsibility | Solution/action perspective | Moral basis | Framing devices | Examples |
|---|---|---|---|---|---|
| The effects of climate change (and, secondarily, other human acting) can be witnessed on separate layers and in various parts (i.e., particular species, systems) of the environmental system. They are getting weaker, are undergoing change. Yet, they cannot protect themselves: They lack resilience and coping abilities. Nature caters to the needs of human, who is at the top of the system. Hence, human will also be affected | Man-made climate change (consequences/causes) is a serious threat. Although all of us are guilty of emitting GHG and/or refusing to change habits (i.e., prioritizing our own, (mainly) short-term interests), particular countries, politicians or corporate organizations are more guilty than others. As such, human disregards the (economic, aesthetic…) characteristics or services of the endangered species and systems | We—i.e., 'elite managers'', producers, consumers… —need to cut the GHG emissions (mitigation) and help vulnerable nature to adapt, mainly by means of a green economy, sustainable and technological development. By making sure that natural species and systems—and their particular workings and characteristics—are preserved or improved, we also protect our own interests | – Human has been assigned the (paternalistic) role of ruler *and* protector of nature. Hence, he is expected to live up to this role (human as a good housekeeper). Accordingly:<br>– Nature caters to human needs. Hence, acting for the sake of nature is acting for the sake of human, today and in the future | **Participant and process depiction**<br>– Climate change, 'inclusive/collective we', singled out countries, politicians, corporate organizations: responsible agents (cause)<br>– Cause: short-term growth, human (economic) interests over long-term natural services<br>– Singled out individuals, scientists, 'inclusive/collective we': responsible agents (solution)<br>– Object of help: emission reduction by means of sustainable development/green economy (mitigation) + aid, protection, development/modification (adaptation)<br>– Singled out species, parts of the ecosystem, 'sublime nature': patients (problem)<br>– Unwanted object: vulnerability, decay, loss of resilience, characteristics, practices (e.g., hibernation patterns)<br>– Context: (distant) natural environments, usually linked to human | – "Climate change threatens 80 percent of the Earth's ecosystems." (DWM, 10/10/2013)<br>– "The US are the only western economy which have (…) not conducted a single policy addressing climate change." (DWM, 06/28/2013)<br>– "Help, the polar bear is drowning." (DM, 12/07/2013, p. 49)<br>– "Invest-ments in energy in particular have to move rapidly from unsustainable sources to (…) renewable energy." (DS, 09/30/2013, p. 44)<br>– A drawing of a globe highlights a number of causes (e.g., a smokestack) and consequences (e.g., a southern woman and child, colour symbolism) (HLN, 06/21/2012, p. 6)<br>– "The Dutch are going to breed climate resistant great tits." (HLN, 11/21/2013, p. 10) |

(continued)

**Table 6** (continued)

| Problem definition | Causal responsibility | Solution/action perspective | Moral basis | Framing devices | Examples |
|---|---|---|---|---|---|
| | | | | **Rhetorical devices**<br>– Symbols/metonyms: the smokestack, the threatened 'polar bear' (or other animal), the melting ice/glacier, cracked ground—lack of visualized alternatives (e.g., wind turbine)<br>– Metaphors: sublime nature, survival of the fittest, colour symbolism, the globe<br>– Contrast/juxtaposition: individualized villains versus individualized heroes<br>– Key events/intertextual references: (decisions/protocols of) climate summits, IPCC, scientific journals like 'Nature', (scenes similar to) the painting 'Wanderer above the sea of fog' (C.D. Friedrich)<br>**Narration**<br>Sources: mainly scientists, politicians and NGOs | |

**Table 7** Unequal Vulnerability Subframe Matrix

| Problem definition | Causal responsibility | Solution | Moral basis | Framing devices | Examples |
|---|---|---|---|---|---|
| Particular ('bottom-up') groups in society are characterized by structural weaknesses: They are economically, socially, culturally, politically, physically and/or psychologically less developed. As such, they are more vulnerable to and less able to cope with climate change consequences | Man-made climate change (consequences/causes) are a serious threat. Due to their development the most powerful leading ('top-down') groups bear a major responsibility for emitting GHG. Besides, many of them refuse to profoundly change habits. As such, they do not (sufficiently) recognize the lives, economies, right to develop, cultural identities… of others | Their responsibility, and/or their power, knowledge and means entitle the ruling groups (mainly 'the West') as main agents. They need to mitigate their GHG emissions and aid the most vulnerable to adapt, mainly by means of a green economy and (the transfer of) technological development and other accomplishments. That is, they must save and improve—and thus acknowledge—the lives, economies and/or cultural identities of others | Charity | **Participant and process depiction**<br>– Climate change, particular groups of elites ('the West'), 'ex/inclusive we': responsible agents (cause + solution)<br>– Cause: mainly economic interests of a selected group of people over the (social, economic, environmental…) interests of others<br>– Object of help: emission reduction by means of green economy (mitigation) + (developmental) aid, charity, transfer (of money, means, knowledge) (adaptation)<br>– Particular bottom-up groups ('the South'): patients (problem)<br>– Unwanted object: vulnerability, socio-economic loss, loss of cultural traditions (e.g., hunting) | – "The US did not ratify the Kyoto protocol out of fear that it would harm its economic interests." (DS, 11/12/2013, p. 6)<br>– "(…) the combination of geography, poverty, unsafe constructions, and the extreme population growth." (DM, 11/13/2013, p. 4)<br>– Two starving women, holding their emaciated babies, look at the spectator, their gaze expressing a demand for help. (HLN, 08/09/2012, p. 12)<br>– Slogans on banners during a demonstration of locals: "Save us. We are drowning". (DWM, 11/12/2013)<br>– "(…) sharing green technology with the developing countries." (DM, 06/23/2012, p. 1) |

(continued)

**Table 7** (continued)

| Problem definition | Causal responsibility | Solution | Moral basis | Framing devices | Examples |
|---|---|---|---|---|---|
| | | | | – Context: continuous time (timelessness of passivity, vulnerability as an eternal state)/distant (mainly) human environments<br>**Rhetorical devices**<br>– Symbol/metonym: the helpless child<br>– Contrast/ juxtaposition: 'us' (the elites—'West') versus 'them' (the others—'South')<br>– Numbers: lives lost (and lives saved), economic loss<br>– Concatenation of verbal/visual images of passivity, helplessness, destruction<br>– Key events/ intertextual references: Kyoto Protocol (1997), Millennium Development Goals | |

(continued)

**Table 7** (continued)

| Problem definition | Causal responsibility | Solution | Moral basis | Framing devices | Examples |
|---|---|---|---|---|---|
| | | | | <u>**Narration**</u><br>– Prevalence of ('Western') elite voices (e.g., politicians), NGOs, or elite representatives of bottom-up groups<br>– Alarmist style (e.g., graphic images, narratives of despair) | |

change consequences. Elite groups bear a major responsibility: They give priority to their own interests and, thus, do not sufficiently recognize the lives, economies, cultural identities or right to develop of others. In order to save and improve the latter, the elites—who also possess the required knowledge or means—have the responsibility to compensate the victims (Vihersalo 2008).

*'Consumer Rights Subframe'*

Rather than as a collective group of citizens, the anthropocentric worldview approaches humans as individual ('passive') consumers, of policies or economic products (Dryzek 1997; Stibbe 2015). Hence, 'human rights' become 'consumer rights'. The 'Consumer Rights Subframe' (Table 8) contends that mankind's well-being, (mental/physical) health and safety are vulnerable to changes in (human) environments. Yet, some (bottom-up) individuals cannot guarantee their own safety: They are ignorant about the changes and/or lack the means to act. Apart from immensely powerful climate change, particular (mainly political elite) groups are the main villains: They do not use their knowledge and means to counter-act these threats, or to bolster the resilience of the victims ('their clients'). Therefore, politics, public services and science are urged to relieve human misery (mostly adaptation) and guide the mitigation by individual consumers.

*'Human Wealth Subframe'*

The Human Wealth Subframe (Table 9) highlights the negative economic, technological and/or cultural consequences of climate change for human, as producer and/or consumer. While we all run a risk, those groups who lack the incentive, as well as the necessary knowledge or means, to protect their own possessions, suffer the most severe (economic/material/cultural) losses. Meanwhile, those (mainly political, economic elites) who are well-aware of the risks and able to react, are too preoccupied with short-term gain to act for the sake of collective long-term benefits. Accordingly, it is those groups who need to take the responsibility to act, and encourage the acting of others. These investments may, preferably, stimulate further development (e.g., job creation).

*'Rights of the Free Market Subframe'*

This highly reactionary subframe (Table 10) highlights that there are no limits to human development as nature is forgiving and has corrective capacities. Due to overly rigorous environmental policies and/or unfair competition, however, some economic/political stakeholders are hampered in their struggle for the wealth, resources or opportunities nature, and particularly climate change, provides. As such, they are faced with revenue loss, inefficiency or a weak political/economic position. These pressures need to be taken away to allow those agents to prosper, drawing on technological development and ingenuity.

*'The Natural Machine Subframe'*

Natural Machine is a metaphor commonly found in the anthropocentric discourse and thinking. Nature is considered a perfectly designed machine, serving the needs of 'higher living beings', humans in particular (Verhagen 2008). The subframe (Table 11) argues that the changes in the human environment are mainly due

**Table 8** Consumer Rights Subframe Matrix

| Problem definition | Causal responsibility | Solution | Moral basis | Framing devices | Examples |
|---|---|---|---|---|---|
| Our well-being, mental and/or physical health and safety are vulnerable to the changes in our (human) environment. Some (bottom-up) individuals cannot guarantee their own safety: They are ignorant about the changes and/or lack the means to act | Man-made climate change (consequences) is an immensely powerful force, threatening human rights. However, particular (mainly political top-down) groups are also major villains. They do not sufficiently use their insights and means to counter-act these pressures or to bolster the resilience of the victims. As such, they run away from their responsibilities towards their 'clients' | The elites (politics, public services, science) need to protect human health, safety and well-being, largely by means of (green) economy/technology, (sustainable) development or emergency aid (focus on adaptation). (Bottom-up) individuals are instructed to change their (consumption) habits, mainly with an eye on mitigation (i.e., help them to save themselves) | – We cannot allow nature (i.e., climate change) to undo again the levels of health, safety, well-being… (development) we have managed to build up, and which separate us from other living beings<br>– Respect for the rights, and sensitivity to the needs, of the client (in a manager-client relationship) | **Participant and process depiction**<br>– Climate change, political elites, 'inclusive/collective we': responsible agents (cause)<br>– Cause: short-term, egocentric interests over long-term (general) human safety<br>– The political, scientific elites, social services, international (aid) organizations/ secondary: 'inclusive/collective we': responsible agents (solution)<br>– Object of help: emission reduction by means of sustainable development/ green economy (mitigation) + emergency aid, charity, transfer (of means, money, knowledge) (adaptation)<br>– 'Inclusive/collective we', mainly ordinary ('consuming') individuals: patients (problem)<br>– Unwanted object: loss of safe environments, natural resources = loss of human well-being, safety, health<br>– Context: recognizable places in human environments (the world, mainly the West) | – "Flash floods as 'water bombs' on Sardinia. Cyclone Cleopatra makes water level rise 3 meters: 18 dead." (DM, 11/20/2013, p. 5)<br>– A 'generic' western girl, suffering from hay fever (HLN, 04/18/2012, p. 22)<br>– World map with highlighted areas, identified as those (countries, groups, numbers of people) which suffer from droughts, water shortages, conflicts (DM, 02/10/2014, p. 5)<br>– "The citizens will have to move by foot or by bike." (HLN, 01/14/2014, p. 14)<br>– "Cameron promised to do all in his power to take sustainable measures." (DS, 02/14/2014, p. 18)<br>– "(…) the social relations in our neighbourhoods will change as well as the amount of time we allow our children to play outside (…)" (MO, 02/21/2014) |

(continued)

**Table 8** (continued)

| Problem definition | Causal responsibility | Solution | Moral basis | Framing devices | Examples |
|---|---|---|---|---|---|
| | | | | **Rhetorical devices**<br>– Metaphor: climate change is a monster<br>– Symbol/metonym: cracked earth/extreme weather + human face or reference to human: people in need/being saved by elite agents<br>– Numbers: lives threatened, lost, saved<br>– Repetition of 'lives', 'dead', 'man', 'save/protect'…<br>– Key events/intertextual references: immigration, social security… crises/debates, Biblical disasters (apocalypse, flood)<br>**Narration**<br>– Sources: Science, politics, public services, international organizations<br>– Alarmist style (e.g., dramatic word choices, hyperbole, accumulation/generalization…) | |

**Table 9** Human Wealth Subframe Matrix

| Problem definition | Causal responsibility | Solution | Moral basis | Framing devices | Examples |
|---|---|---|---|---|---|
| In their current state, our (collective) economic, technological and/or cultural wealth are highly vulnerable to climate change. Some (individual) consumers and/or producers, however, (will) suffer a larger part of this loss since they do not have yet the necessary (economic) incentives, nor the insights or means, to sufficiently protect their own possessions | Man-made climate change (causes/consequences) is a largely external force which hampers human (economic) growth and development. Some agents are aware of the threats and able to react. However, being too preoccupied with short-term gain, they fail to recognize the long-term benefits of climate action. They largely belong to elite (political/economic) top-down groups | The elites (politics, science, industry…), in particular, need to invest in green economy, technology and development (mitigation/adaptation). Besides, the (political) elites have to encourage the others to act and/or show them how (redistribution). These investments are likely to be repaid, by stirring (economic) development (e.g., profit, jobs) | – We cannot allow nature (i.e., climate change) to undo again the levels of economic, material and/or cultural development we have managed to build up, and which separate us from other living beings<br>– We need to manage well the costs and benefits of climate change, for the sake of human society | **Participant and process depiction**<br>– Climate change, political and industrial elites, 'inclusive/collective we': responsible agents (cause)<br>– Cause: prioritizing short-term interests over long-term gain<br>– Political, industrial, scientific elites/secondary: 'inclusive/collective we': responsible agents (solution)<br>– Object of help: (encouraging of) mitigation and adaptation by means of (the transfer of insights of) sustainable development/green economy; economic incentives<br>– 'Inclusive/collective we', certain (ordinary) consumers or producers: patients (problem)<br>– Unwanted object: environmental loss/loss of natural | – "But the climate twists also burden us with other costs." (DS, 05/28/2013, p. 40)<br>– Destruction caused by coffee rust on a Mexican plantation (MO, 05/30/2013)<br>– "We would carry an extra 'Greek debt' outside of the Union every year." (DWM, 06/25/2012)<br>– "Prospect of 65.000 green jobs" (DM, 09/19/2013, p. 16)<br>– Focus on a high-tech invention which might save the ski season (HLN, 10/27/2012, p. 85)<br>– "The Stern report demonstrates that the costs of controlling the climate are about 10 times lower than the economic damage the climate would cause if the warming is left unaddressed." (DWM, 07/25/2012) |

(continued)

**Table 9** (continued)

| Problem definition | Causal responsibility | Solution | Moral basis | Framing devices | Examples |
|---|---|---|---|---|---|
| | | | | resources = economic/material, cultural (e.g., skiing, drinking wine) loss, decay<br>– Secondary consequence of solution: (economic) development (efficiency, profit, job creation…)<br>**Rhetorical devices**<br>– Symbol/metonym: human technology, natural destruction + human face (or reference)<br>– Numbers: economic/technological/cultural loss or gain<br>– Key events/intertextual references: Stern Review on the Economics of Climate Change (2006), (late-2000s) economic crises<br>**Narration**<br>– Sources: science, economics, agriculture, industry… | |

**Table 10** Rights of the Free Market Matrix

| Problem definition | Causal responsibility | Solution | Moral basis | Framing devices | Examples |
|---|---|---|---|---|---|
| (Some) economic and/or political players are being hampered in the struggle for wealth and (new) resources/ opportunities nature (climate change) keeps providing. They are being faced with revenue loss, unfair competition, inefficiency… | Some governments take too far-reaching climate/environmental measures which—directly or indirectly—hamper the workings of the free market. Certain players—not/less restrained by such policies—further endanger the economic interests of others | Climate measures stop where the rights of the free market begin. With the political pressures lifted, economic and or political players can prosper thanks to climate change, drawing on technology and ingenuity. They enjoy, among others, economic profit and efficiency and a stronger economic/political position | – Liberalism must be protected, whatever the cost ('the end justifies the means')<br>– Economic development as supreme human goal<br>– Accep-tance of 'survival of the fittest' as inescapable 'natural' principle | **<u>Participant and process depiction</u>**<br>– Governments, countries: responsible agents (cause)<br>– Cause: climate action, policies, unfair competition<br>– Governments, countries, industry/ secondary: climate change: responsible agents (solution)<br>– Object of help/positive consequence: human ingenuity, technology, development/profit, efficiency, access to resources, strong economic/political position<br>– Governments, countries, industry: patients (problem)<br>– Unwanted object: economic loss, inefficiency, inaccessibility of resources, weak economic/political position<br>– Decontextualization: concealing of all values beyond the liberalist-economic<br>– Positionality: them, distanced | – "The new policy obligates all airlines flying into Europe to buy $CO_2$ emission rights." (DS, 03/13/2012, p. 30)<br>– "The European market is uncertain as Europe focuses on sustainable energy." (DS, 11/27/2012, p. 15)<br>– The lay-out highlights numbers, such as "500.000 barrels of oil" ("the revenue the Arctic can produce") (DM, 09/12/2012, p. 12)<br>– "The struggle for the black gold of Alaska has started" [DM, 09/12/2012, p. 12 (title)]<br>– An (abstracted) globe (only) shows two new, more efficient shipping lanes (MO, 03/05/2013)<br>– "(…) a worldwide march of oil companies to the Arctic, partly accommodated by the global warming" (DM, 07/30/2012, p. 5) |

(continued)

**Table 10** (continued)

| Problem definition | Causal responsibility | Solution | Moral basis | Framing devices | Examples |
|---|---|---|---|---|---|
| | | | | **Rhetorical devices**<br>– Use of typical climate change symbols (metonyms), particularly melting ice, to associate discussion to climate change debate; accepted fact of melting poles as starting point<br>– Contrast/juxtaposition: human versus human, country versus country<br>– Numbers: profit, efficiency, resources, (potential) loss<br>– Comparatives, superlatives: highlighting positive evolutions<br>– Key events/intertextual references: 'Arctic strategies' of various countries, American Gold Rush (19th century), Cold War (1945–1991)<br>**Narration**<br>– Journalists working for centrist, rightist outlets<br>– Sources: academics specialized in economics, political sciences…, politics | |

**Table 11** Natural Machine Subframe Matrix

| Problem definition | Causal responsibility | Solution | Moral basis | Framing devices | Examples |
|---|---|---|---|---|---|
| Changes can be witnessed in humans' natural environment | The increasing level of GHG—as major cause of climatic changes—is mainly due to individual natural processes, variances and species, and thus (largely) separate from human | (Other/separate) systems and species may evolve, adapting to their particular (new) environmental circumstances. As such, the perfectly designed natural machine can regulate and mitigate these changes, especially for the sake of 'higher' living beings (particularly, human) | – We must respect the natural state of the environment: sublime, pristine, free of human (intervention) | **Participant and process depiction**<br>– GHG, climate change, natural processes or species: responsible agents (cause)<br>– Cause: GHG emissions<br>– Natural processes or species, (human as a mechanic): responsible agents (solution)<br>– Object of help: mitigation of GHG<br>– The various parts/processes of the natural system, with a focus on human environments: patients (problem)<br>– Unwanted object: vulnerability, instability<br>– Context: (decontextualized) natural environments (for human usage)<br>**Rhetorical devices**<br>– Metaphor: sublime nature<br>– Use of typical climate change symbols (metonyms/ metaphors), such as smoke, taken out of the human context<br>– Natural 'villains' and 'heroes'<br>– Numbers: GHG | – "Earthworms are the real culprits behind global warming." (HLN, 02/05/2013, p. 24)<br>– "The researchers calculated that Hurricane Katrina killed or damaged millions of trees and—indirectly—released 100 tons of GHG into the atmosphere, as such." (MO, 11/19/2013)<br>– "Mountain woods are the thermostat of our climate" (title). "(…) over the last 65 million years, the mountain woods have been able to prevent that the temperature would rise to levels which would be catastrophic for living beings." (DS, 02/11/2014, p. 8)<br>– Comparative (before-after) satellite images, which are typically used to show the melting of (polar) ice sheets or glaciers, are used to demonstrate the growing of the ice instead. (HLN, 09/10/2013, p. 24) |

(continued)

**Table 11** (continued)

| Problem definition | Causal responsibility | Solution | Moral basis | Framing devices | Examples |
|---|---|---|---|---|---|
| | | | | – Repetition of terms referring to individualized natural agents (e.g., earthworms, sea grass, woods) and material (mechanical) processes as 'store', 'absorb', 'mitigate'<br>**Narration**<br>– Sources: scientists, balanced with counter-views which are less salient<br>– Stories of surprise, the unexpected (regarding nature repairing itself) | |

to the GHG emissions caused by particular natural mechanisms and species. However, the natural machine can regulate and mitigate these changes: (Other) species or processes are designed so that they can adapt to their particular (new) environments. Human intervention is (largely) unnecessary.

## 5.3 *'The Transformation Masterframe'*

The four remaining subframes—'Natural Web', 'Unequal Attribution', 'Civil Rights' and 'Gaia'—promote a biocentric worldview. Central values encompass harmony, diversity, mutual interconnectedness, equality and human moderation (i.e., excessive human growth is denounced). Humans are part of nature; nature is an important touchstone for human. The central organizing idea of the 'Transformation Masterframe', which structures all counter-hegemonic subframes except for Gaia (see below), is this: Nature *and* bottom-up groups 'close to nature' are victimized by the hegemonic socio-economic system in multiple, interconnected ways. As such, they become disproportionally vulnerable. Yet, drawing on alternative (biocentric) values, these groups are the only ones who can stimulate a (funda)mental (i.e., human-internal) transformation of society (Dryzek 1997; Hopwood, Mellor and O'Brien 2005; Stibbe 2015; Verhagen 2008). Central framing devices are (see Table 12 for the frame matrix):

1. Contextualization of causes (Iyengar 1990): Structural hegemonic processes are foregrounded (through personifications, nominalizations, passivizations) as main causes, rather than singled out agents. (Generalized) political/economic elite agents mainly act as metonyms for the ideological system they try to uphold. The focus lies on mental processes, particularly the inability to think outside the 'anthropocentric box'.
    a. Revealing and denouncing of anthropocentric processes like '(sustainable) development' or 'green economy' as deceptive 'semantic reversals'.
2. Nature *and* humans (the whole natural web) are victims. Yet, those bottom-up groups 'closer to nature' (from an anthropocentric perspective) are, in particular, foregrounded. Rather than being essentially weak, they are being victimized: While we find processes which highlight the lack of agency (e.g., passive voice, conditional agency) of these groups (see anthropocentrism), it is also made clear that this passivity is due to certain—foregrounded—responsible agents or causes. (Material/economic) multi-levelled maldistribution (e.g., economic loss due to climate change *and* unfair trade relations), cultural status misrecognition (e.g., ignoring the value of biocentric alternatives) and/or political misrepresentation (e.g., undemocratic processes) are described as major causes (Fraser 2005).
3. The bottom-up groups—'inclusive, exclusive or collective we'—are also the heroes who can facilitate redistribution. That is, if they are allowed to act as

**Table 12** Transformation Masterframe Matrix

| Problem definition | Causal responsibility | Solution | Moral basis | Framing devices | Examples |
|---|---|---|---|---|---|
| Nature, as well as those (bottom-up) human groups who stand close to nature, are being victimized by the hegemonic system. They are—among others—deprived of their agency, (access to) alternative (i.e., biocentric) values and/or ways of living, voice… This victimization makes them disproportionally vulnerable to (i.e., less able to protect themselves from) climate change (and other crises) (**external pressures**) | The hegemonic socio-economic system lies at the roots of multilevel injustices: (direct/indirect) maldistribution, misrecognition and misrepresentation. It is largely sustained by 'western(ized)' humans, particularly—but not exclusively—elite groups (**human-internal cause**) | The victimized groups—the only ones who are able to look beyond the hegemonic thinking—must become equal partners in the debate. Drawing on (biocentric) alternatives like harmony, mutual dependency or sufficiency, they can instigate (funda)mental change. This will lead to long-term mitigation and adaptation, and a more safe, equal coexistence for all (redistribution, (status) recognition, representation) (**human-internal solution**) | – The right to self-determination—Biocentric alternatives (like harmony, collaboration or sufficiency) need to be recognized as precondition for safe, satisfying and good living for all<br>– Recognition and redrawing of an unequal world—participatory parity as precondition for a just and equal world | **Participant and process depiction**<br>– Elites (mostly generalized), the hegemonic socio-economic system, human-induced climate change: responsible agents (cause)/often as sensers<br>– Cause: the hegemonic system, evident in 'green economy', 'sustainable development', 'unlimited consumption ideal'… —maldistribution, misrecognition, misrepresentation<br>– 'In/exclusive/collective we', mainly groups of (bottom-up) citizens: responsible agents (solution)/actors, sensers, sayers, sovereign agents<br>– Object of help: sustainable development/green economy + biocentric alternatives —redistribution, (status) recognition, representation<br>– The whole natural 'web', particularly bottom-up groups (generalized, identified): patients (problem)<br>– Unwanted object: maldistribution (multi-levelled vulnerability), misrecognition (of the value of (biocentric) alternatives), misrepresentation (no true democracy)<br>– Contextualization (foregrounding of agents or circumstances, nominalizations, personifications): highlights society-wide responsibilities (causes/solutions)<br>– Environment: 'sublime', circumstance (background), | – "The society we live in (…) does not recognize limits (…): The economy keeps developing and growing, investments are being done, new products are continually being launched." (DWM, 08/23/2013)<br>– "(…) a second wave will hit the Philippines soon: the wave of international emergency aid." (DM, 11/13/2013, p. 31)<br>– "The rainforest of the Amazon, also called 'the lung of the world', breathes carbon dioxide ($CO_2$) in and oxygen out (…)." (DS, 02/17/2014, p. 8)<br>– A cartoon of a man standing in a flooded area. The rain is pouring down (metonym for climate change consequences). His head is covered by an umbrella (metonym for current—insufficient—climate change action, within the system) full of dark clouds (metaphor for hegemonic system which blinds us for the true dangers to which we are exposed/more constructive alternatives) (MO, 03/11/2013).<br>– "(…) the Cuban delegation did not agree with the tight focus on 'green economy' (…)." (DWM, 06/15/2012) |

(continued)

**Table 12** (continued)

| Problem definition | Causal responsibility | Solution | Moral basis | Framing devices | Examples |
|---|---|---|---|---|---|
| | | | | source, deleted + complex and harmonious system, responsive subject, example for human (agent)<br>– Context: substitution of the global economic society with civil society/ies<br>**Rhetorical devices**<br>– Metaphors: Mother Earth + nature is a machine<br>– Contrast/juxtaposition: 'us' (bottom-up) versus 'them' (top-down), deconstruction (past/present) versus (reconstruction) future<br>– Hyperbole (usually bottom-up) versus ellipsis/litotes (usually top-down)<br>– The revealing and denouncing of 'sustainable development', 'green economy', 'emission trading'… as semantic reversals<br>– Presupposition: hegemonic system as sickening; transformation as healing<br>– Key events/intertextual references: the books, articles, speeches… of alternative thinkers like Harald Welzer or Andrew Simms<br>**Narration**<br>– Sources: mostly bottom-up<br>– Threatening present, (climate change + system) + inspiring alternatives | |

equal partners, or leaders (recognition, representation), they can set things right. They are usually presented as humanized agents, fully able to feel, reflect and act upon their own fate (i.e., sovereign agents, often in material or mental processes). They promote a (funda)mental transformation, combining sustainable development and biocentric alternatives (e.g., sharing, more traditional ways of living, bio-mimicry). Hence, the global economic society governed by elites ('them'), is substituted for (local) civil society/ies ('us') (Brulle 2010; Fraser 2005; Kenis and Mathijs 2014).

4. Nature, however, is not fully restored as equal partner. The 'Natural Web' (inclusive, complex, harmonious nature) and 'Mother Earth' (nature as active, responsive subject) metaphors (Lovelock 1988) are juxtaposed with the 'Natural Machine' metaphor (Verhagen 2008). The human-nature dualism is still largely there; humans often act as stewards for nature (Dryzek 1997).
5. Bottom-up voices are the dominant sources: Alternative think tanks, NGOs, grassroots groups, ordinary citizens… Citizen journalists—often members of these groups—work next to western/southern professional journalists.
6. A disillusioning image of today's reality (deconstruction)—the threat of climate change *and* hegemonic thinking—is balanced with positive, and potentially engaging (Brulle 2010), stories of a better ('the good') life (reconstruction).

### 5.3.1 Four Counter-Hegemonic Subframes

*'Natural Web Subframe'*

'Natural Web' is a central metaphor of the biocentric thinking, highlighting the interdependence and equality of all (non-)living beings, including (all) humans (Verhagen 2008). Accordingly, the underlying narrative of the subframe is this (Table 13): The anthropocentric worldview, responsible for several interconnected problems, harms the resilience of the whole natural web. Human stewards (mainly citizens, NGOs, activists), who are able to look beyond the ruling system, are willing and able to speak up for nature. Yet, they are largely silenced in mainstream debates, while the value of the biocentric alternatives (including the exemplary role of nature) is disregarded. Hence, only by recognizing the crucial role of these bottom-up groups, we may evolve towards a more harmonious, just and mutually dependent coexistence.

*'Unequal Attribution Subframe'*

Rather than the unequal vulnerability of some groups, the 'Unequal Attribution Subframe' (Table 14) highlights the unequal appropriation of resources, freedoms or rights as the main problem of environmental justice. Indeed, by enforcing 'development' upon those who do not meet their ideals, 'western(ized)' elites have built up considerable multi-levelled debts: These threaten the resilience and independence of those who are considered to be 'underdeveloped', rendering them disproportionally vulnerable to climate change or other hazards. Yet, it is exactly the internal resilience and knowledge of these bottom-up groups that allow them to

**Table 13** Natural Web Subframe Matrix

| Problem definition | Causal responsibility | Solution/action perspective | Moral basis | Framing devices | Examples |
|---|---|---|---|---|---|
| The natural web—all interconnected living beings and ecological systems—has (to a certain extent) internal resilience, but is largely deprived of it due to external human pressures, of which climate change is only one: Her agency and responsiveness, harmonious workings, exemplary function, 'voice'… are destructed and/or disregarded. Bottom-up groups can/are willing to speak up for this system, but are largely prevented from taking up this (steward) role | The (current acting, inspired by the) hegemonic socio-economic system does not help to solve, but mostly reproduces/ exacerbates the current situation: It lies at the roots of interacting injustice (environmental, economic, cultural, social, political…). Although being mainly sustained by a small group of elite (top-down) stakeholders, it is also a society-wide problem (i.e., '(western) human culture') | A universal ecological transformation is necessary. Participatory parity is a means and a goal. (Bottom-up) civil society—as representative of nature—must be an equal partner in the debate. They can engage 'us' and instigate (fundamental change towards a more just, inclusive, mutually dependent and modest (biocentric) coexistence. This will serve long-term mitigation and adaptation | – All living beings are equal. Yet, (ordinary) human —being a conscious creature —has a special duty to fulfil as a steward | **Participant and process depiction**<br>– Corporate, (international) political elites, the hegemonic system, 'inclusive/collective we' (mainly individual consumers), climate change: responsible agents (cause)<br>– Cause: the hegemonic system, evident in 'green economy', '(sustainable) development', 'emission trading', 'aid'…<br>– Natural 'web' + (bottom-up) ordinary citizens, 'inclusive/ collective we': patients (problem) + responsible agents (solution)<br>– Object of help; sustainable development/green economy + more traditional ways of living, modesty, circular economy, sharing, biomimicry, long-term thinking…<br>– Unwanted object: multi-levelled vulnerability, e.g., climate change consequences + detrimental consequences of emission trading, mass consumption…, | – "Meanwhile, however, human has become a 'super predator'. Master of the universe." (DM, 06/ 27/2012, p. 12)<br>– "The society we live in does not recognize limits (…)." (DWM, 08/23/2013)<br>– "The rainforest of the Amazon, also called 'the lung of the world', breathes carbon dioxide ($CO_2$) in and oxygen out (…)." (DS, 02/17/2014, p. 8)<br>– "While the civil society was kept quiet, the fossil fuel industry was able to seize a central seat at the summit." (DM, 11/23/2013, p. 58)<br>– Activists carry a banner saying "Bla Bla Bla – Act Now" (DWM, 11/21/2012)<br>– "'Lagom' is the Swedish term for just enough, adequate." (MO, 12/10/2012) |

(continued)

**Table 13** (continued)

| Problem definition | Causal responsibility | Solution/action perspective | Moral basis | Framing devices | Examples |
|---|---|---|---|---|---|
| | | | | disregarding/destructing of bottom-up alternatives and/or examples from nature, no equal voice (e.g., in international debates)<br>– Context: natural environments—but often focus on human environments<br>**Rhetorical devices**<br>– Symbols/metonyms: bottom-up protest, parts of the ecosystem (for the whole)<br>– Metaphor: human is a predator/god (denounced)<br>– Anthropomorphism/personification<br>– Contrast/juxtaposition: human masses/consumers versus nature<br>– Key events/intertextual references: 'Climate Train' during COPs<br>**Narration**<br>– Sources: alternative thinkers and think tanks, NGOs, scientists, ordinary people, (visual) first person perspective | |

**Table 14** Unequal Attribution Subframe Matrix

| Problem definition | Causal responsibility | Solution | Moral basis | Framing devices | Examples |
|---|---|---|---|---|---|
| Some ('bottom-up') groups —mainly those who do not meet the 'western' elite ideal (i.e., 'who are underdeveloped')—more than others are largely deprived of their livelihoods, cultural traditions, voice…, and thus resilience, due to external pressures. As such, they are disproportionally vulnerable. Climate change is the latest pressure exacerbating this victimized state | The ('top-down') 'western (ized)' elites—who sustain the hegemonic worldview —appropriate 'common goods' to themselves, in a process which they call 'development' (directed towards a western elite ideal). This (neo)-colonialism/capitalist imperialism results in a number of interconnected injustices: environmental, economic, social, cultural, political… | The victimized ('southern') groups need to be respected, and allowed to act, as ('sovereign') equals (i.e., participatory parity as means and goal). Drawing on their (local) experiences or cultures, they may provide (biocentric) alternatives, with an eye on local/diverse needs and long-term adaptation/ mitigation. As such, they can improve their own situation, but also inspire other groups to work (together) on a (funda)mental transformation | Multiple, locally defined justices, which are different but equally valuable as the western ideal | **Participant and process depiction**<br>– Political/corporate elites ('western (ized)'), NGOs, the hegemonic system, climate change: responsible agents (cause)<br>– Cause: the hegemonic system, evident in 'green economy', 'emission (trading)', 'development', 'aid', 'landgrabbing'…<br>– Particular bottom-up groups ('the South'), 'exclusive we'/'I' (generalized/ individualized, identified/ functionalized): patients (problem) + agents (solution)<br>– Particular (i.e., 'westerni(zed)') elites/ 'inclusive we': secondary responsible agents (solution)<br>– Unwanted object: multi-layered vulnerability, e.g., climate-induced economic loss + land grabbing, disregarding/destructing of alternative knowledge or cultures (e.g., climate-clever agriculture), no democratic voice (e.g., in international debates)<br>– Object of help: sustainable development/ green economy, scientific knowledge, means (mainly top-down) + biocentric alternatives (e.g., traditional knowledge)<br>– Reversal of hegemonic 'us-them' roles: others close to 'us' (humanization)<br>– Context: continuous time (timelessness of bottom-up resilience, harmony…)/ (clearly situated) human environments, but some recognition of the holistic nature of the world | – "(…) a second wave will hit the Philippines soon: the wave of international emergency aid." (DM, 11/13/2013, p. 31)<br>– A father and mother running from the devastation caused by typhoon Haiyan, carry/protect their child. (DM, 11/13/ 2013, p. 31)<br>– "The indigenous leaders show an enormous generosity, teaching us a different story." (DWM, 09/15/2013)<br>– "(…) the Cuban delegation did not agree with the tight focus on 'green economy' (…)." (DWM, 06/15/2012)<br>– "(…) we combine scientific weather forecasts with traditional knowledge." (MO, 06/29/2012)<br>– A Bangladesh farmer looks us proudly in the eyes, showing the produce of his combined farming project. (DWM, 12/21/ 2012) |

(continued)

**Table 14** (continued)

| Problem definition | Causal responsibility | Solution | Moral basis | Framing devices | Examples |
|---|---|---|---|---|---|
| | | | | **Rhetorical devices**<br>– Symbols/metonyms: the 'parent' looking after his/her children, the ('southern') hero/'militant'<br>– Metaphor: the elites as predator<br>– Concatenation of verbal/visual images of resilience, harmony, activity<br>– Key events/intertextual references: 'Principles of Environmental Justice' (1991)<br>**Narration**<br>– Sources: prevalence of locals (e.g., farmers), grassroots movements or officials, as well as ('western') NGOs speaking on their behalf | |

provide real alternatives. That is, if recognized by other groups as equal(ly developed), these people are able to help themselves, but also to encourage the rest of the world to work (together) on a (funda)mental transformation.

*'Civil Rights Subframe'*

From a biocentric perspective, human rights are the rights of the civil society. Rather than individual(istic) consumers, we are a community of citizens (e.g., Dryzek 1997). Accordingly, this subframe (Table 15) argues that the hegemonic system victimizes citizens, turning us into passive consumers. Preoccupied with our own surviving in the system, we are not able (nor allowed) to be critical, let alone to provide alternatives. However, this makes us disproportionally vulnerable to (climate) crises. Accepting this consumer role, we—unwittingly—aid the elites to sustain this system. Therefore, we need to reclaim our collective citizenship: Drawing on biocentric alternatives, we may work together towards a more meaningful, safe and sustainable life for all.

*'Gaia Subframe'*

Employing the biocentric view to argue for passive scepticism, the 'Gaia Subframe' (Table 16) clearly does not fit within the Transformation Masterframe. Besides, it is the only subframe which is based on theorizations (Gibbons 2015; Lovelock 1988; Verhagen 2008), as the authors were not able to identify it in the corpus. Gaia—'Mother Earth' in Greek mythology—is a well-known metaphor and theory: the Earth is considered an active, self-organizing and self-renewing subject, which constantly strives for internal balance. Accordingly, this subframe highlights the current pressures on Gaia's balanced state as problematic. The main cause are changes in the GHG balance, as part of the natural self-renewing processes. However, humans—as one species among others—also have been exerting multiple pressures, which largely prevent a flexible natural response. Yet, Gaia is not fundamentally endangered. If necessary, she will jump to a new stable state. This will, however, constitute a far less comfortable (and forgiving) living space for humans.

## 6 Discussion and Conclusions

In this study, an elaborate and nuanced set of frames, subframes and masterframes was presented. Now, these frames can be situated within the broader academic (and public/political) debate, demonstrating how they elaborate upon the existing discussions and respond to the formulated criticism. Figure 3 illustrates that the (sub) frames build upon—and expand—a valuable tradition. Indeed, the various continuums show that the 'frame categories' deduced from the literature can be found to a greater or lesser extent throughout all (sub)frames.

These continuums illustrate, for instance, that Consumer Rights is a non-sceptical subframe, which considers human as the main victim and employs the most alarmist style. Economics or technology are considered as the solution. Unequal Attribution is also a non-sceptical subframe, yet it is far more balanced:

**Table 15** Civil Rights Subframe Matrix

| Problem definition | Causal responsibility | Solution | Moral basis | Framing devices | Examples |
|---|---|---|---|---|---|
| All citizens are the victims of the hegemonic system, being reduced to passive/voiceless (individualistic) consumers. As such, preoccupied with our own surviving within (the 'here and now' of) this system, we are not able to consider more constructive (long-term) alternatives. This makes us disproportionally vulnerable to climate change, and other crises | The current socio-economic system constitutes a threat to civil society in multiple, interacting ways: environmental, economic, cultural, political… injustices. Sustaining the system, the (top-down) elites violate the rights of the (bottom-up) ones they govern/'control'. Yet, allowing ourselves to be deceived/passivized, all of us contribute to its reproduction | Reclaiming our role as citizens, we need to collectively work on (think/talk about) (funda)mental change. Participatory parity is a goal and a means. Drawing on values like sufficiency, sharing, harmony or (e) quality, we may improve our society, safety, well-being, environment and future (in the long run), in ways we consider most appropriate | Individualism, unlimited growth or consumption as stifling and blinding values, solely serving the interests of a minority (indoctrination) | **Participant and process depiction**<br>– Political and economic elites, the hegemonic system, climate change: responsible agents (cause)<br>– 'Inclusive/collective we', often as individual consumers: secondary responsible agents (cause)<br>– Cause: the hegemonic system, prescribing development, consumption, competition, human control... as preconditions for human well-being<br>– 'Inclusive /collective we', (bottom-up) citizens: patients (problem) + responsible agents (solution)<br>– Object of help: sustainable development/green economy + 'the good life', working less, returning to nature (e.g., gardening), collective endeavours based on mutual dependency or sharing (e.g., repair cafes), freedom to act well, for the sake of oneself and others<br>– Secondary consequence of solution: a better, more | – "It is clear: neoliberalism is a sickening system." (DWM, 11/21/2013)<br>– A cartoon of a man standing in a flooded area. The rain is pouring down (metonym for climate change consequences). His head is covered by an umbrella (metonym for current—insufficient—climate change action. within the system) full of dark clouds (metaphor for hegemonic system which blinds us for the true dangers to which we are exposed/more constructive alternatives) (MO, 03/11/2013)<br>– "(…) we must evolve from passive consumers to active producers who freely choose their destiny. Repairing everything, learning everything anew, doing it ourselves again." (DWM, 09/17/2012)<br>– "(…) the economy as servant of a better life" (DWM, 11/02/2012) |

(continued)

**Table 15** (continued)

| Problem definition | Causal responsibility | Solution | Moral basis | Framing devices | Examples |
|---|---|---|---|---|---|
| | | | | satisfying and meaningful life for human<br>– Unwanted object: multi-levelled vulnerability, e.g., climate change consequences + rebound effect of 'solutions', disregarding/destructing of (bottom-up) alternatives like 'the good life', moderation, collaboration<br>– Reversal of hegemonic 'us-them' roles: others close to 'us' (humanization)<br>– Context: generic (but rather western) human environments<br>**Rhetorical devices**<br>– Metaphors: human as a machine (denounced), suffocated/blind human<br>– Contrast/juxtaposition: individual consumers versus collective of citizens<br>– Repetition of terms like 'consumers', 'citizens', 'blindness', 'idea'<br>– Key events/intertextual references: 'I have a dream' (Martin Luther King, 1963), | |

(continued)

**Table 15** (continued)

| Problem definition | Causal responsibility | Solution | Moral basis | Framing devices | Examples |
|---|---|---|---|---|---|
| | | | | bottom-up initiatives such as city farming, repair cafes, community exchange networks…<br>**Narration**<br>– Sources: alternative think tanks/thinkers. NGOs, 'people like us' | |

**Table 16** Gaia Subframe Matrix

| Problem definition | Causal responsibility | Solution | Moral basis | Framing devices | Examples |
|---|---|---|---|---|---|
| The balanced state of Gaia is under pressure | The pressures are mainly due to increased levels of GHG. These are caused by natural developments in the system (i.e., the natural self-renewing process). Yet, humans—as one species among others—have also been exerting multiple, interacting pressures, which largely prevent a flexible natural response | Gaia is not fundamentally endangered. If necessary, she will respond by jumping to a new stable state. This may constitute a less favourable living space to humans than the one we are enjoying now | – Gaia is a living, responsive organism who owes our respect: We completely depend upon her for our survival<br>– A harmonious system of organisms, working together to sustain life on Earth | **Participant and process depiction**<br>– (Large-scale/ multi-layered) internal contradiction in the system, human as one species among others: responsible agents (cause)<br>– Cause: GHG + other destructive pressures<br>– Mother Earth, interconnected complex ecosystems: responsible agents (solution)<br>– Object of help: a new stable state, more favourable to some organisms than to others<br>– Mother Earth as a complex organism: patient (problem)<br>– Unwanted object: an unstable state, deconstruction of flexible natural response<br>– Context: multi-layered complex nature (natural web) | 'In the end. Gaia herself will make adjustments necessary to stabilize Earth's ecosystems.' (Based on Gibbons 2015) |

(continued)

**Table 16** (continued)

| Problem definition | Causal responsibility | Solution | Moral basis | Framing devices | Examples |
|---|---|---|---|---|---|
| | | | | **Rhetorical devices**<br>– Metaphor: Mother Earth<br>– Comparison with human body as self-regulating system; humanizations/ personifications<br>– Repetition of terms referring to complex ecosystems, 'Mother Earth'<br>– Presupposition: biocentric worldview as given<br>– Key events/intertextual references: the Gaia myth (and comparable narratives), 'The Ages of Gaia' (Lovelock 1988) | |

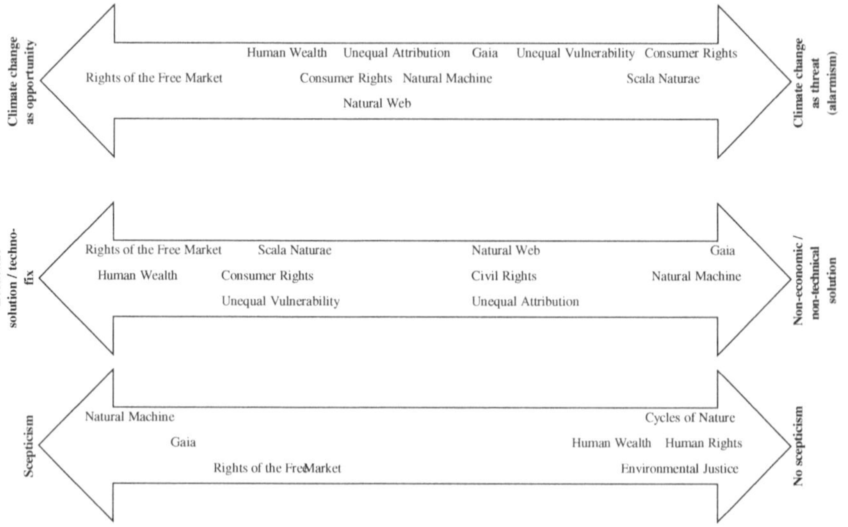


**Fig. 3** The position of the (sub)frames on four continuums, which draw on the 'framing categories' found in the literature

It considers climate change neither as a threat, nor (solely) as an opportunity. While economics and technology may play a role in the solutions, the focus lies on alternatives. The natural system at large is under threat, yet the human face is still particularly salient. Clearly, many of the existing frames constitute crucial arguments or structural elements of the detected comprehensive 'generic substance frames' (see Baden 2010). Particularly news routines—e.g., conflict, drama—and mythological archetypes—e.g., (certain) villains, victims—pave the way for particular (sub)frames, but do not constitute (sub)frames in their own right (Van Gorp 2006). This highlights that, indeed, most existing frame sets are not mutually-exclusive and not exhaustive (Tankard 2001). A contribution of this study, therefore, is that it takes the existing findings to a higher level.

Besides, this frame set also accounts for—and reflects—the ideological workings of the framing process, which do not always clearly emerge from other studies. Figure 4 illustrates the ideological diversity of the subframes, and thus the potential of the five frames to function as sites of ideological struggles:

Indeed, integrating framing and hegemony research (Carragee and Roefs 2004), this study attempts to further exploit the potentials of framing as 'a bridging model' (Reese 2007). More specifically, this framing analysis is one of the first to clearly illustrate the argumentation of Brulle (2010), and others. It is generally accepted that elite communicators should provide effective frames to engage the audience and encourage them to act in 'favorable ways' (e.g., Nisbet and Scheufele 2009).

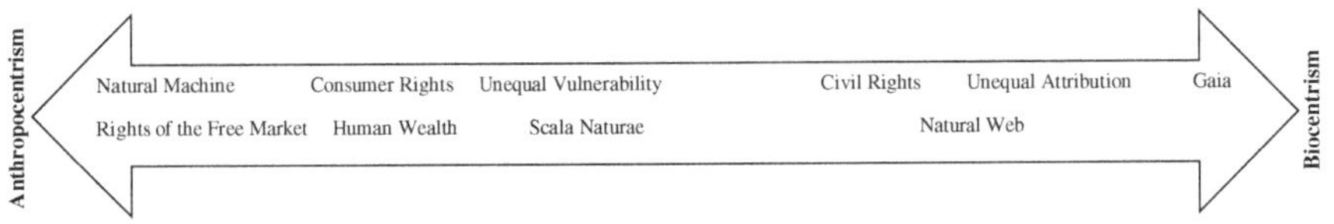


**Fig. 4** Position of the subframes on the ideological continuum anthropocentrism—biocentrism

Brulle argues, however, that such ‘top-down’ frames largely serve to manipulate passive individuals into supporting short-term pragmatic actions that fit within the hegemonic (anthropocentric) system (i.e., the interests of the dominant classes). Therefore, in order to constitute fundamental change, the involvement of (bottom-up) civil society is necessary. The latter is able to provide long-term (biocentric) alternatives in the interest of human *and* nature (Fraser 2005; Kenis and Mathijs 2014). We can, indeed, recognize this reversal of positionalities in the interplay of the Anthropocentric (reform) and Transformation Subframes/Masterframes. As such, this study provides tangible examples of the ‘deconstruction-reconstruction’ frames Brulle (2010) calls for: They face us with, and contextualize, the consequences of climate change and hegemonic thinking, (deconstruction), providing us—simultaneously—with constructive, positive alternatives.

Concluding, this study provides a comprehensive, yet detailed, set of frames and frame matrices which account more fully for the depth and breadth of the current field of climate change (media) frames. In particular, it demonstrates that alternative media, indeed, provide alternative (sub)frames, which are more likely to put the hegemonic ways of thinking into perspective. This might not only broaden our views, it may also affect the ways in which we interact with, and analyze, the mainstream framing. Clearly then, introducing a broader range of sources, as well as more elaborated analysis methods, also allows for the deepening of the (climate change) frames, adding ideological levels. Hence, the authors call for more (framing) research which takes alternative media into account. Also, they hope that their work may inspire other researchers to integrate framing and hegemony research (with the provided frames as starting point).

However, this study has some limitations: It lacks a comparison with other countries (beyond Western Europe). Besides, the frame set has not been tested on different issues, or beyond the context of media communication. The authors hope, thus, that the provided frames may inspire and benefit from (i.e., be further developed through their application) such endeavours. Finally, it would be highly valuable to incorporate this frame set into audience reception studies, to test how exactly the (sub)frames may influence public and political debates about climate change. After all, the current research design only allows for theorizations about ‘deconstructive and reconstructive’ framing effects (Brulle 2010).

## References

Baden C (2010) Communication, contextualization, & cognition. Patterns & processes of frames' influence on people's interpretations of the EU constitution. Eburon, Delft

Benford RD, Snow DA (2000) Framing processes and social movements: an overview and assessment. Ann Rev Sociol 26:611–639

Borah P (2009) Comparing visual framing in newspapers: Hurricane Katrina versus Tsunami. Newsp Research J 30(1):50–57

Brand A, Brunnengräber A (2012) Conflictive knowledge constructions on climate change through mainstream and alternative media? Transience 3(1):7–24

Brulle LJ (2010) From environmental campaigns to advancing the public dialog: environmental communication for civic engagement. Environ Commun 4:82–98

Carragee KM, Roefs W (2004) The neglect of power in recent framing research. J Commun 54:214–233

Carvalho A (2007) Ideological cultures and media discourses on scientific knowledge: Re-reading news on climate change. Public Underst Sci 16:223–243

Chouliaraki L (2006) The spectatorship of suffering. Sage, London

Downing JDH (2001) Radical media. Rebellious communication and social movements. Sage, Thousand Oaks, CA

Dryzek JS (1997) The politics of the earth: environmental discourses. Oxford University Press, Oxford

Entman RM (1991) Framing U.S. coverage of international news: contrasts in narratives of the KAL and Iran air incidents. J Commun 41(4):6–27

Entman RM (2004) Projections of power: framing news, public opinion, and U.S. foreign policy. The University of Chicago Press, Chicago, IL

Fletcher AL (2009) Clearing the air: the contribution of frame analysis to understanding climate policy in the United States. Environ Polit 18:800–816

Fraser N (2005) Reframing justice in a globalizing world. New Left Rev 36:69–88

Gamson WA, Modigliani A (1989) Media discourse and public opinion on nuclear power: a constructionist approach. Am J Sociol 95(1):1–37

Gibbons W (21 June 2015) Does climate change really matter to Mother Earth? The Savannah River Ecology Laboratory. University of Georgia. Published online: http://srel.uga.edu/outreach/ecoviews/ecoview150621.htm

Glaser BG, Strauss AL (2008) The discovery of grounded theory: strategies for qualitative research. AldineTransaction, New Brunswick, NJ

Graber DA (1988) Processing the news: how people tame the information tide, 2nd edn. Longman, New York, NY

Grittmann E (2014) Between risk, beauty and the sublime: the visualization of climate change in media coverage during COP 15 in Copenhagen 2009. In: Schneider B, Nocke T (eds) Image politics of climate change. visualizations, imaginations, documentations. Transcript Verlag, New York, NY, pp 127–151

Gopress (www.gopress.be)

Halliday MAK (2000) An introduction to functional grammar. Arnold, London

Hopwood B, Mellor M, O'Brien G (2005) Sustainable development: mapping different approaches. Sustain Dev 13:38–52

Hertog JK, McLeod DM (2001) A multiperspectival approach to framing analysis: a field guide. In: Reese SD, Gandy OH, Grant AE (eds) Framing public life: perspectives on media and our understanding of the social world. Lawrence Erlbaum, Mahwah, NJ, pp 139–161

IPCC (Intergovernmental Panel on Climate Change) (2014) Climate change 2014: synthesis report. Cambridge University Press, Cambridge

Iyengar S (1990) Framing responsibility for political issues: the case of poverty. Polit Behav 12(1):19–40

Kress GR, van Leeuwen T (2006) Reading images: the grammar of visual design. Routledge, London

Kenis A, Mathijs E (2014) Climate change and post-politics: Repoliticizing the present by imagining the future? Geoforum 52:148–156

Kenix LJ (2008) Framing science: climate change in the mainstream and alternative news of New Zealand. Political Science 60:117–132

Lovelock J (1988) The ages of Gaia: a biography of our living earth. Oxford University Press, Oxford

Machin D, Mayr A (2012) How to do critical discourse analysis: a multimodal approach. Sage, Thousand Oaks, CA

Messaris P, Abraham L (2001) The role of images in framing news stories. In: Reese SD, Gandy OH, Grant AE (eds) Framing public life: perspectives on media and our understanding of the social world. Lawrence Erlbaum, Mahwah, NJ, pp 215–226

Monaco J (2000) How to read a film. The world of movies, media, and multimedia: language, history, theory, 3rd edn. Oxford UP, New York, NY/Oxford

Nisbet MC, Scheufele DA (2009) What's next for science communication? Promising directions and lingering distractions. Am J Bot 96:1767–1778

O'Neill SJ (2013) Image matters: climate change imagery in US, UK and Australian newspapers. Geoforum 49:10–19

Pan ZP, Kosicki GM (1993) Framing analysis: an approach to news discourse. Polit Commun 10(1):55–75

Reese SD (2007) The framing project: a bridging model for media research revisited. J Commun 57:148–154

Rodriguez L, Dimitrova DV (2011) The levels of visual framing. J Vis Lit 30(1):48–65

Shehata A, Hopmann DN (2012) Framing climate change. A study of US and Swedish press coverage of global warming. Journal Stud 13:175–192

Stibbe A (2015) Ecolinguistics: language, ecology and the stories we live by. Routledge, London

Tankard JW (2001) The empirical approach to the study of media framing. In: Reese SD, Gandy OH, Grant AE (eds) Framing public life: perspectives on media and our understanding of the social world. Lawrence Erlbaum, Mahwah, NJ, pp 95–106

Universal Declaration of Human Rights (UDHR) http://www.un.org/en/universal-declaration-human-rights/

Van Gorp B (2006) Framing asiel. Indringers en slachtoffers in de pers [Framing asylum: intruders and victims in the press]. Acco, Leuven

Van Gorp B, van der Goot M (2009) Van Frankenstein tot de Goede Moeder: de inzet van frames in de strategische communicatie over duurzaamheid [From Frankenstein to Mother Earth: the use of frames in strategic communication about sustainability]. Tijdschr voor Commun 37(4):303–316

van Leeuwen T (2008) Discourse and practice: new tools for critical discourse analysis. Oxford University Press, New York, NY

Verhagen FC (2008) Worldviews and metaphors in the human-nature relationship: an ecolinguistic exploration through the ages. Lang Ecol 2(3):1–19

Vihersalo M (2008) Framing climate change in montreal 2005: an environmental justice perspective. In: Carvalho A (ed) Communicating climate change: discourses, mediations and perceptions. Centro de Estudos de Comunicação e Sociedade, Universidade do Minho, Braga, pp 34–52

Weathers MR, Kendall BE (2015) Developments in the framing of climate change as a public health issue in US newspapers. Environ Commun 10:593–611

Wessler H, Wozniak A, Hofer L, Lück J (2016) Global multimodal news frames on climate change: a comparison of five democracies around the world. Int J Press/Politics 21:423–445

## Author Biographies

**Renée Moernaut** is a Ph.D. Candidate in the Department of Languages and Linguistics (Applied Linguistics) of the Vrije Universiteit Brussel (Brussels, Belgium) and in the Department of Communication Studies of the University of Antwerp (Antwerp, Belgium). She is a member of the University of Gloucestershire Ecolinguistics Circle. Her research focuses on the multimodal framing of climate change in the Flemish (northern Belgian) media. The main aim of her work is to provide an overview, and qualitative discussion, of climate change 'frames' and their diverging ideologically coloured 'subframes'. Besides, she attempts to shed more light on synchronic differences across media outlets and types: The main fault lines are mainstream-alternative, popular-broadsheet and ideological backgrounds. As such, she wants, among others, to integrate framing with hegemony research. Further, she is interested in the production process of framing.

**Jelle Mast** received his Ph.D. in Communication Sciences from the University of Antwerp, Belgium in June 2011, with a doctoral dissertation entitled *Reality TV as Popular Hybrid Documentary: Questions of Genre Hybridization, Documentary Values and Ethics*. In November 2011, he joined the Department of Applied Linguistics at Vrije Universiteit Brussel, where he currently works as an Assistant Professor of Journalism Studies. As the coordinator of the Brussels Institute for Journalism Studies (BIJU), and an affiliate member of the Visual and Digital Cultures Research Center (ViDi) at the University of Antwerp, his main research interests are in the areas of visual communication, journalism practice, genre hybridization, and professional ethics.

**Luc Pauwels** is Professor of Communication Studies and Vice-Dean of the Faculty of Political and Social Sciences at the University of Antwerp, Belgium. He is currently chairing the program 'Master in Film Studies and Visual Culture' as well as the Research Group 'Visual Culture' at his university. Luc Pauwels is a board member of various academic associations and journals. He focuses on the study of the specific nature, the structure and the functioning of visual images. Most recently, he focuses on the complex role of visual representation in science, since issues of representation affect every aspect of scientific activity (encoding, display, analysis, data presentation, communication of concepts, etc.).

# Mass Media and Climate Change Induced Disaster Risk Reduction and Mitigation in Coastal Area of Bangladesh: A Sociological Study

**Joydeb Garai**

## 1 Introduction

Bangladesh is recently considered as one of the world's foremost disaster prone countries (Islam 2011). According to the Natural Disaster Risk Index 2010, Bangladesh is the top country in the "extreme risk" group of experiencing natural disaster. The situation is gradually aggravated, all the more by its densely population, high growth rate, severe poverty, poor institutional capacity and insufficient infrastructure in the world (Khan and Nahar 2014; Choudhury 2001). The name of Bangladesh is synonymous to natural hazards since its (natural hazards) frequency, severity and intensity (Khan and Nahar 2014).

Among the most climate change induced natural hazards, Bangladesh severely and frequently experiences tropical cyclones, storm surges, floods, soil salinity, droughts etc. (Nasreen 2004; Garai 2014). Over the last thirty years, the country has envisaged more than thirty eight severe cyclones of varying intensities. Among them the severe one is on 29 April 1991 when material damage was to the tune of about 2.4 billion US dollars and human casualty was about 140,000 lives. In addition to this, the flood of 1998 in Bangladesh caused massive economic loss that exceeds one billion US dollars (Choudhury 2001). The geographical location, land varieties, multiplicity of rivers, high spatial and temporal climatic variability and extreme weather events render Bangladesh highly vulnerable to natural calamities (Khan and Nahar 2014). Climate change induced natural hazards influence the costal Morphology of Bangladesh that affect the people especially in the south-western part and increase vulnerability and slow the process of social and economic development.

However, to overcome the vulnerability and mitigation of the disaster risks, well developed climate change communication system is needed and in this context

J. Garai (✉)
Department of Sociology, University of Chittagong, Chittagong 4331, Bangladesh
e-mail: garai.joydeb@gmail.com



W. Leal Filho et al. (eds.), *Handbook of Climate Change Communication: Vol. 1*, Climate Change Management, https://doi.org/10.1007/978-3-319-69838-0_15

Mass media can play a vital role in Bangladesh as well as in the world. Through the provision of better and effective information, awareness and planned communication system, it is possible to reduce the adverse impacts of hazards and avoid damage to property and infrastructure in the region and save many lives (Azad et al. 2013; Scanlon 2007; Duhe 2008). Timely and proper mass media response about impending disasters can lead to appropriate individual and community action, which is the key to implementing effective disaster prevention strategies, providing relief and rescue operations including evacuation of people to safe areas (Duhe 2008). Media can also contribute to educating public about disasters, warning of hazard, collecting and spreading news about affected areas and concerning government officials and relief organizations, and the general public to the specific needs about hazards preparedness, response and relief.

Fifteen years ago, the advanced warning system for tornados was only 45% but now the figure soar up to 89% and is growing (Cate 1994). In the United States of America, between the years of 1950 and 1959, approximately 1400 people lost their lives due to tornados, but the death toll dropped to 521 between the years of 1983 and 1992. According to the World Meteorological Organization, this achievement is only possible with the advancement in communication technology and media is doing this successfully as an effective instrument to climate change communication (Cate 1994).

Through rapid, widespread and timely dissemination of information, early warning and preparation can save thousands of lives from natural hazards. To illustrate the matter, an example can be given. In November 1970, a tropical cyclone along with high tidal surges struck southeastern Bangladesh, having lost 300,000 people's lives and making 1.3 million homeless. The same types of cyclone combined with storm surges hit the same locality in May 1985, but the loss of life was although still high (10,000) it was only 3% compare to the cyclone of 1970. The inherent cause of this dramatic difference was the introduction of new early warning system that allowed radio station to alert people in low-lying areas and media did this wonderful job (Cate 1994). This chapter sheds light on the role and responsibility of mass media to disaster risk reduction and mitigation in coastal area especially in the south-western part of Bangladesh. The chapter also explores the key issues in establishing good cooperation and efficient flow of information during emergency and measure the access of information to the rural masses and their preparedness for mitigation to the losses of upcoming hazards.

## 2 Mass Media and Disaster Management

In a study Azad et al. (2013) proposed a conceptual model for dealing the disaster and its management through mobile technology in Bangladesh. The paper claimed that in Bangladesh there are approximately more than 58.36 million mobile phone subscribers. Almost every single village in Bangladesh has been brought under the coverage of mobile network system even in disaster prone and remote areas of

Bangladesh. So by using mobile phone, effective disaster prevention strategies, relief and rescue operations including evacuation of people to safe areas can be done easily. Moreover, the paper claimed that this model can provide speedy alert of upcoming hazards that may save life and resources to the affected people (Azad et al. 2013). A seminar organized by Disaster Management Congress (2009), aiming to initiate a dialogue among various stakeholders on various issues related to the role of media during disaster events and to generate ideas on standard code of conduct for print and electronic media Journalists covering such events. The seminar further discussed the role of Government agencies, civil society and other organizations in informing the media and clarifying various issues that come to the center stage, before, during, and after natural hazards (Disaster Management Congress 2009).

The role of media in developing countries especially in the remote coastal area is much more important, because media can disseminate the knowledge about how to prevent and cope with the hazard. The opinions of different disaster experts about disaster response become vital component in this situation (Seid-Aliyeva 2006). Moreover, the detail loss of property or damage only can be displayed in media news not on peoples' opinions, because they are confused about estimating damage (Nazari et al. 2011). This can be an effective instrument to attend national and international agencies' concern to emergency relief/help to the affected people (Cowan et al. 2002). During the emergency periods Radio and TV are the most accessible means to get disaster news, because internet or other electronic means do not work properly, so people fail to get news from this means (Goldsborough 2002; Neuman and Guggenheim 2011).

In a study Kakonge (2012) highlighted that in Africa, as in other parts of the world, the media play a vital role in educating and informing the public about issues and events. In particular, during time of natural and man-made disasters the media often has an unintended dual role, both reporting the event itself and acting as a catalyst for evoking responses. In addition, he opinioned that the media is a vital actor for issuing warnings of disasters; for alerting government officials, relief organizations and the public to potential dangers; for gathering and transmitting information about affected areas; and for facilitating discussions about disaster preparedness and response (Kakonge 2012). Moreover, other study in 1990 Rattien, asserted that knowledge and technology can reduce the loss of natural hazards through a concerted cooperative international effort. According to him, communication is central to this endeavor for public education, early warning, evacuation and post-disaster relief. In addition, the role of training, education and knowledge dissemination will be given a high priority for disaster management through media. The article keeps focus on planning and preparedness for managing natural disaster rather than waiting passively for hazards to strike. He also emphasizes to prevent disaster before it strikes by employing different steps, dissemination knowledge/information among the masses through news media. The paper also refers to the benefit of facsimile which can help media disseminate scientific knowledge of disaster risk reduction after getting it from specialists (Rattien 1990).

In 2015 Soltani in a study revealed that media play a great role by involving society in disaster management especially raising awareness among people to engage in disaster management. According to this study, without mass media effective disaster management is not possible. The article indicates that mass media help in three phase to disaster management that are pre, during, and after natural hazard phase. In pre-disaster stage, Media's goal is to prevent disaster from occurring. During the hazard stage, media help stimulate the emotion and feeling of the populaces, inform people and other organization. At the post disaster stage, the strength and weakness of communication with the media and sharing of news and information can be determined. All of these stages are important in climate change communication for effective disaster management (Soltani 2015).

In a study Clerveaux et al. (2009) propose an effective model to disaster loss reduction. The key component of this disaster management model includes planning, communication and response during emergency. In the planning component, it focuses on identification of vulnerable population, location of their shelters and planning communication networks to disseminate emergency information. This paper highlights communication system, because effective communication system can reduce the loss of lives and assets by responding the hazards timely and properly. For communication purpose, they propose loudspeakers, use of patrol cars, cellular phones, mass media etc. as means to disseminate news among affected areas. In the last stage, the model focuses on response toward disaster which can save lives; reduce losses from the severity of natural hazards (Clerveaux et al. 2009).

## 3 The Methodology

This study is conducted by using both quantitative and qualitative methods. In order to obtain quantitative data, a semi-structured interview questionnaire has been used from six villages namely Burigoalini, Kalbari and Durgabati in Burigoalini union and Sawara, Dumurea and Cakbara in Gabura union under Shymnagar upazila (sub-district) of Satkhira district in Bangladesh (Fig. 1 show the location of the study area, village name is not displayed). Being located near to Bay of Bengal and adjacent to Mangrove forest (*Sundarban*), the study area is frequently envisaged with (emergency situation) different natural hazards such as cyclones, tidal surges, flash floods, soil salinity etc. All of these hardships trigger the researcher to choose this locality to examine the access of information got by mass media of the affected people during emergency periods. Moreover, for conducting this study in representative way, 200 samples were selected by using purposive sampling in the study area. Each male and each female (age 15–80) who were inhabitants of climate change induced natural hazards affected area were the unit of the study. In addition to this, to support the quantitative data 6 Focus Group Discussions or FGDs and 12 KIIs were adopted to get in-depth views and comprehensive panorama about mass media to disaster risk reduction and mitigation. For each of the FGD groups, approximately 10 respondents of heterogeneous occupations of a particular village were selected by using purposive sampling. Each of the selected respondents was informed formally to participate in the FGD. Prior to that, all of the invited respondents were informed about the objectives of study in detail. In addition, during the Focus Group Discussion, along with principle researcher, a research assistant was present there to verify the information that was come out from the discussion and to pick up the authentic data of the affected people of that particular locality. However, in case of KII, especially knowledgeable person, such as teachers, *union parishad* chairman, NGO workers, journalists, reporters, columnist and other affected people were given priority to collect information. Their deep knowledge, practical skills and long experiences help to provide authentic data to enrich the study. After conducting this study, data were processed and analyzed manually. In addition to this, for quantitative data, statistical analysis used SPSS Windows program (version 17.5) to process the data. Field work of this study was conducted between July and August of 2014. Secondary data were collected from several bangle and western literatures, research papers, dissertation, newspaper, magazines and so on.

The rationale behind the study is that there is hardly found any empirical study on climate change communication as well as mass media to disaster risk reduction in the existing literatures. Every year thousands of people face immeasurable

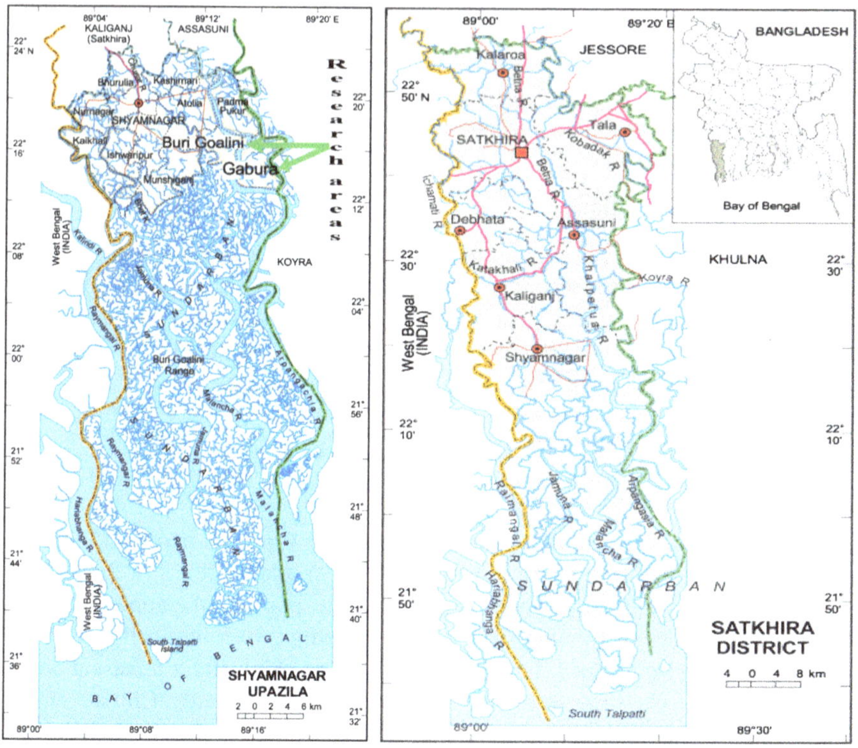


**Fig. 1** Shyamnagar upazila (sub-district) of Satkhira districts in Bangladesh

vulnerability due to the constraints to get timely and proper information/warning of imminent hazards. This chapter highlights that mass media can play a vital role by providing speedy alert of upcoming hazards and generating management strategies to remote area in disaster prone Bangladesh. This chapter focuses the attention of government and policy makers to manage disaster in new ways. Moreover, this chapter helps to reduce the paucity of empirical publication in the realm of climate change communication.

The study has some limitations as well. Firstly, the study was conducted with small sample size due to economic constraints and lack of sufficient manpower. However, researchers try to represent the real situation beyond the constraints. Secondly, it was really difficult to manage target informants (i.e. media person) to a place to collect information but researcher tackle that skillfully. Thirdly, the paucity of relevant documents related to this issue was another constraints that researcher face to write the chapter.

## 4 Mass Media and Disaster Risk Reduction and Mitigation in Coastal Area in the Context of Bangladesh

The major focus of this study is to examine the role of mass media in disaster risk reduction and mitigation in coastal area of Bangladesh. The chapter also delves into the access of information to the coastal people and their preparedness for the mitigation of losses of the upcoming natural hazards.

- **Educational Condition**

The study findings indicate that most of the respondents of the study area are either primary educated or secondary those are 27 and 41% respectably. Moreover, only 11% people able to get higher education like graduation or post-graduation in the coastal area. However, 12% respondents cannot get access to educate them in the locality. Frequent natural hazards, poor communication system, insufficient educational institutions and poverty are the major hindrances of this lower education rate in the coastal area (see Table 1).

- **Occupational Status and Income Level**

From the table, it is found that approximately 28% people involved in fishing activities such as catching fishes in the *Chunu* River and selling it in the adjacent market for maintaining livelihoods. Another large portion of people (36%) do Agro-labor activities or farming such as planting crops, weeding, cultivating, harvesting and some other activities but this ratio are gradually decreasing in the locality due to less yielding of crops by different natural hazards. Climate change induced natural hazard has become a great threat to this people to maintain their livelihoods because finding no other employment opportunity this remote coastal

**Table 1** Socio-economic status of the respondents. *Source* Field work

| Socio-economic variables | $N = 200$ | % |
|---|---|---|
| *Educational status* | | |
| Primary | 54 | 27.0 |
| Secondary | 81 | 40.5 |
| Higher secondary | 20 | 10.0 |
| Graduate | 12 | 6.0 |
| Post graduate | 9 | 4.5 |
| Illiterate | 24 | 12.0 |
| *Religious status* | | |
| Hindu | 49 | 24.5 |
| Muslim | 151 | 75.5 |
| *Occupational status* | | |
| Petty businessman | 17 | 8.5 |
| Agro-labor | 45 | 22.5 |
| Fisherman | 56 | 28.0 |
| Job-holder | 14 | 7.0 |
| Teacher | 8 | 4.0 |
| Farmer | 27 | 13.5 |
| Quack doctor | 2 | 1.0 |
| Wood/Pona collector | 17 | 8.5 |
| Unemployment | 14 | 7.0 |
| *Age of the respondents* | | |
| 15–25 | 44 | 22.0 |
| 25–35 | 77 | 38.5 |
| 35–45 | 43 | 21.5 |
| 45–55 | 12 | 6.0 |
| 55–65 | 12 | 6.0 |
| 65–75 | 7 | 3.5 |
| 75–above | 5 | 2.5 |
| *Monthly income of the respondents (in US dollar) ($1 = BDT 80 approximately)* | | |
| 0–63 | 118 | 59.0 |
| 63–100 | 42 | 21.0 |
| 100–150 | 24 | 12.0 |
| 150–above | 16 | 8.0 |
| *Monthly expenditure of the respondents (in US dollar) ($1 = BDT 80 approximately)* | | |
| 0–63 | 122 | 61.0 |
| 63–100 | 48 | 24.0 |
| 100–150 | 17 | 8.5 |
| 150–above | 13 | 6.5 |

people has largely been depend on natural resources (Mangrove forest). Approximately, 9% people especially women work as wood collector (small branches of dried trees from mangrove forest) or *pona*/larvae collector from *Chunu* River. In addition to this, another, 9% people work as small scale businessman and 7% people work in different government organizations or NGOs for maintaining their livelihoods. However, about 7% people are unemployed who either lose their working sources due to natural hazards or search for works for living (see Table 1).

Due to frequent natural hazards, the income rang of the respondents is very poor as majority of the respondents (59%) monthly income are below $63. Only a trifling portion of the respondents 8% earn $150 or more monthly. Moreover, due to low income, a large portion of the respondents (61%) maintain their life poorly by restricting their monthly cost within $63. Only 7% people have the ability to expend more than $150 for maintaining their family (see Table 1).

- **Age and Religious Status**

The observed data indicate that majority of the respondents 39% have age between 25 and 35 years, on the other hand, only 3% respondent's age range is above 75 year. The data also indicate that among the total 200 respondents, approximately 76% respondents are Muslim and rest of is Hindu (see Table 1).

- **Means of Getting Disaster News**

The study findings indicate that Radio and Television are the most news getting means among the respondents that are 50 and 30% respectively. Along with this, respondents also get disaster related news from relatives (7%), friends (6%) and Newspapers (4%) in their locality (see Table 2). About the type of news people get through news media in the locality, 36% of the respondents said that they get warning/alarming news and another same portion of people referred awareness type news of natural hazards. Reduction of optimum damage of upcoming hazard is the main objective for delivering this warning/awareness news through this media.

**Table 2** Role and responsibility of mass media in disaster risk reduction and mitigation. *Source* Field work

| Role of mass media to disaster management | $N = 200$ | % |
|---|---|---|
| *Means of getting disaster related news* | | |
| Television | 59 | 29.5 |
| Radio | 100 | 50.0 |
| Newspaper | 8 | 4 |
| Friends | 12 | 6.0 |
| Relatives | 13 | 6.5 |
| Mobile | 8 | 4.0 |

(continued)

**Table 2** (continued)

| Role of mass media to disaster management | *N* = 200 | % |
|---|---|---|
| *Types of news getting by news media* | | |
| Awareness type | 72 | 36.0 |
| Warning/Alarming | 72 | 36.0 |
| Disaster related | 38 | 19.0 |
| Good news | 6 | 3.0 |
| Security related | 12 | 6.0 |
| *Benefit of getting news* | | |
| Yes | 164 | 82.0 |
| No | 36 | 18 |
| *Types of benefit getting from media* | | |
| Saving life | 34 | 17.0 |
| Saving resources | 25 | 12.5 |
| Saving livestock | 18 | 9.0 |
| Saving family member | 43 | 21.5 |
| Preventing disaster | 22 | 11.0 |
| Being aware | 58 | 29.0 |
| *Evaluate the role of media in disaster risk reduction and mitigation* | | |
| Very well | 24 | 12.0 |
| Well | 67 | 33.5 |
| Moderate | 103 | 51.5 |
| Not well | 6 | 3.0 |
| *Type of medium plays more role in disaster management* | | |
| Television | 57 | 28.5 |
| Radio | 124 | 62.0 |
| News paper | 3 | 1.5 |
| Magazine | 5 | 2.5 |
| Postering | 6 | 3.0 |
| Mobile | 5 | 2.5 |
| *Feelings of disaster related news in media* | | |
| Real | 61 | 30.5 |
| Authentic | 53 | 26.5 |
| Correct | 64 | 32.0 |
| Circulate in proper time | 14 | 7.0 |
| Frequent circular | 5 | 2.5 |
| None of them | 3 | 1.5 |
| *More role of media in disaster management* | | |
| Develop communication system | 28 | 14.0 |
| Mass awareness news | 21 | 10.5 |
| Forming team | 37 | 18.5 |
| Strategy for saving property | 27 | 13.5 |

(continued)

**Table 2** (continued)

| Role of mass media to disaster management | *N* = 200 | % |
|---|---|---|
| Direct circular | 34 | 17.0 |
| News properly circular | 35 | 17.5 |
| Advice to go safe place | 18 | 9.0 |

Moreover, 82% interviewees claimed that this news help them to be prepared to face upcoming hazard. About the benefit of getting news, affected people asserted that they can save family members (22%), save resources (13%) and preventing disasters (11%). Moreover, 29% people said that they become aware about the hazards by getting the news (see Table 2).

- **Effective Medium in Disaster Management**

In order to evaluate the role of media to disaster risk reduction and mitigation, approximately 46% respondents claimed that it is well and 52% respondents asserted it as moderate. A trifling portion of people, 3% said that its role is not well (see Table 2). The observed data indicate that Radio (62%) and television (29%) (According to respondents) play more roles to disseminate disaster news among the respondents compare to other means due to its easy access and cheap cost. Other media means such as newspaper, magazine, etc. cannot play many roles due to poor communication system, illiteracy, cost effectiveness and so on. Inability to timely circulate disaster news in the locality is another impediment for not getting popularity of these means. The findings also indicate that the news people get from different means are authentic (27%), correct (32%) and real (31%). Moreover, a small portion of respondents (7%) said that this news is circulated in proper time (see Table 2). According to respondents, media can also play more roles in making strategy for saving property (14%), forming team (19%), making mass awareness news (11%) and circulating news properly (18%) (see Table 2).

- **Access of Information and Preparation of Coastal People to Upcoming Hazards**

From the study findings, it is found that a large portion of the respondents approximately 86% claimed that they can get news residing in the locality. On the other hand, 14% respondents cannot access of news in their family. The inherent cause of this is that their house is far remote from the locality to get the access of Radio frequency. Poor communication system and having less capability to purchase media means are some other causes of not accessing disaster news during natural hazards. Moreover, some other means, such as Daily newspaper, internet service, and magazine are not easily available in the locality. However, only Radio (48%) and Television (26%) are the most accessible means to get news. Only 19% interviewees said that they get news through mobile/Cell phone (see Table 3). Moreover, those who cannot get news from media means try to find it from their community (79%).

**Table 3** Measurement of the access of information to the rural masses and preparedness of upcoming hazards. *Source* Field work

| Access of information and preparation to upcoming hazards | *N* = 200 | % |
|---|---|---|
| *Facility of getting news in your family* | | |
| Yes | 172 | 86.0 |
| No | 28 | 14.0 |
| *Type of means to get news* | | |
| Television | 52 | 26.0 |
| Radio | 95 | 47.5 |
| Daily newspaper | 7 | 3.5 |
| Magazine | 2 | 1.0 |
| Internet | 6 | 3.0 |
| Mobile | 38 | 19 |
| *Sources of news in case of unavailability of media means* | | |
| Club | 5 | 2.5 |
| Community | 157 | 78.5 |
| Friends | 29 | 14.5 |
| Cannot get news | 9 | 4.5 |
| *Hearing of warning before cyclone* | | |
| Yes | 162 | 81.0 |
| No | 38 | 19.0 |
| *Steps taken after hearing warning* | | |
| Shifting to cyclone centre | 50 | 25.0 |
| Saving property | 24 | 12.0 |
| Saving livestock | 12 | 6.0 |
| Shifting baby and elderly people to safe place | 49 | 24.5 |
| Staying home | 65 | 32.5 |
| *Flow of news effectively during disaster* | | |
| Yes | 131 | 65.5 |
| No | 69 | 34.5 |
| *Causes of not effectiveness of flow of news* | | |
| For severity of disaster | 36 | 18.0 |
| Television, Radio become stopped | 38 | 19.0 |
| Communication system become disrupted | 60 | 30 |
| Having created dangerous situation | 50 | 25.0 |
| Other | 16 | 8.0 |
| *Quick medium to circulate news among people* | | |
| Television | 67 | 33.5 |
| Radio | 105 | 52.5 |
| Daily newspaper | 4 | 2.0 |
| Internet | 4 | 2.0 |
| Mobile | 15 | 7.5 |
| SMS/Message | 5 | 2.5 |

(continued)

**Table 3** (continued)

| Access of information and preparation to upcoming hazards | *N* = 200 | % |
|---|---|---|
| *Minimum time of having no electricity* | | |
| 1 day | 30 | 15.0 |
| 2/3 days | 26 | 13.0 |
| 7 days | 72 | 36.0 |
| No electricity | 72 | 36.0 |
| *Means for collection of news in absence of electricity* | | |
| Radio | 94 | 47.0 |
| Mobile | 36 | 18.0 |
| Relatives | 46 | 23.0 |
| Newspaper | 24 | 12.0 |
| *Types of news of local media after the hazards* | | |
| Informative | 92 | 46.0 |
| Realistic | 40 | 20.0 |
| Realistic and informative | 30 | 15.0 |
| Authentic | 17 | 8.5 |
| Like national news | 9 | 4.5 |
| None of them | 12 | 6.0 |
| *Comparison of roles of National or local media played after hazard* | | |
| National Television | 28 | 14.0 |
| Local Television | 55 | 27.5 |
| Radio | 105 | 52.5 |
| Newspaper | 7 | 3.5 |
| Internet | 5 | 2.5 |
| *Language structure circulating warning of natural hazards* | | |
| Very easy | 7 | 3.5 |
| Easy | 61 | 30.5 |
| Neutral | 104 | 52.0 |
| Tough | 18 | 9.0 |
| Very tough | 10 | 5 |
| *Opinions about getting disaster management knowledge from media* | | |
| Very satisfied | 16 | 8.0 |
| Satisfied | 104 | 52.0 |
| Neutral | 71 | 35.5 |
| Unsatisfied | 9 | 4.5 |

- **Steps Taken After Hearing Warning**

About the warning of cyclone, 81% respondents said that they hear it before the occurrence of natural hazards. After having heard the news, 25% respondents go to cyclone center, and another 25% respondents shift their family members i.e., baby and elderly people to safe place. For example, they shift their family member to adjacent high rise brick-built building or nearly school/college building to escape from the severity of the natural hazards. However, 33% respondents stay home to save their property from the hazards. For example, for saving livestock they keep it on high roadside or embankment. In the same way, to save valuable materials such as gold ornament, money, documents etc. they keep it in a locked iron box in the roof top or deep of soil in the floor of their room. Sometime they keep it neighboring brick-built house for saving it from the ruin of natural hazard.

- **Status of News Flowing During Natural Hazards**

From the study results, it is found that approximately 66% respondents claimed that disaster-news was effective during the natural hazards. On the other hand, 35% of the respondents claimed that circulation of disaster related news was not effective during natural hazards. The inherent causes of this ineffective flow of news are disrupted communication system, dysfunction of radio and television and severity of natural hazards as 30, 19 and 18% respondents claimed this respectively. Moreover, about 25% respondents mentioned that having created dangerous situation is the cause of poor circulation of news in hazards periods. According to the respondents, during natural hazard there has no electricity as 36% respondents asserted this and another 36% respondents claimed that sometime 7 days there has no electricity at a stretch due to hazards (see Table 3). The medium of news collection in this time are relatives (23%) and Radio (47%).

- **Nature of Local and National Media News in Hazards Periods**

Being located far distance from Satkhira district, modern communication facility i.e. satellite facility, internet service, high frequency radio etc. are not available in this particular area. For this why, people of this locality very much depend on local media for disaster news. Khulna Radio center and Bangladesh Television (BTV) are the main sources of getting latest disaster news to the affected people in the locality. In the same way, some local newspaper such as Daily Somoyer khobor, Daily Dristipat, Daily Kafala, Daily Potrodut, Daily Kaler Chitro etc. play good roles to circulate hazards related news among the affected people, even the national daily and international media depend on local media man to get authentic news of the affected area. Nolta Radio (local radio station) also plays a good role in this context. The observed data indicate that 46% of respondents said that the news of local media is informative and 15% claimed it as realistic and informative. In addition 5% respondents opinioned that it is like national news. According to the respondents, during natural hazards the role of local Television channel is much better (28%) then national one 14%. Moreover, the contribution of Radio in circulating and disseminating disaster news is higher (53%), compare to other means (see Table 3).

- **Views About Understanding Disaster News**

Since the people of coastal areas cannot get enough opportunity to educate themselves due to frequent envisaging of natural hazards, it is important keeping emphasize on circulating news in understandable language for the purpose of reaching it all level of people. The study result revealed that approximately 31% respondents opinioned that language structure of circulating hazards warning is easy to understand. Moreover, 52% respondents remain neutral to answer this question. On the other hand, 14% respondents mentioned their hardship to understand the news of natural hazards (see Table 3). In addition to this, about getting disaster management knowledge from the media, 60% respondents opinioned satisfactorily and 36% of the respondents remain neutral in their opinion. Only a trifling portion of respondents mentioned their dissatisfaction about this context. However, for gaining comprehensive disaster management goal, media should care for all level of people in the coastal area.

- **Key Issues for Effective Disaster News Circulation**

Being situated adjacent to the mangrove forest and lacking of employment opportunity, people of the study area always remain busy for maintaining their livelihoods. They barely get time to think about weather forecast or the news of their particular society. In this circumstance, this people should be encouraged to be careful to follow disaster related news along with showing sufficient income sources in their locality. The observed data indicate that for effective flow of news circulation awareness level should be enhanced as 42% respondents claimed that. Having increased the number of staffs in remote area and development of communication system are another two important factors for effective information dissemination in coastal area as 25 and 12% of the respondents asserted this respectively (see Table 4). Moreover, approximately 9% respondents keep emphasize on ensuring news circulation in remote area, for this why monitoring cell can be formed to monitor the situation. For example, in Gabura one of the study site of this project where daily news paper of a particular day reach the following day. This situation should be changed by raising awareness, developing communication system and increasing media staffs. In addition, about taking steps of circulating news in poor community, approximately 76% respondents answered positively. Raising awareness among people and circulating news by loudspeaker/mosque mike were the steps taken to circulate news among destitute community as 34 and 25% respondents mentioned it respectively (see Table 4).

- **Connection Between Media Persons and Affected People in News Circulation**

Due to poor frequency and bad weather situation, the radio wave many a time cannot cover the remote locality of coastal area. So this people remain in the realm of darkness without getting news from radio. Other means of news such as newspaper, magazine, internet service, satellite television are remained very far distance from their hand. The observed data indicate that 30% of the respondents have no access of high frequency radio system in their disaster prone area. However, to ensure authentic news circulation media persons keep connection to the affected people as well as with the affected area as 83% respondents claimed this (see Table 4).

**Table 4** Exploration of the key issues in establishing good co-operation and efficient flow of information during emergency. *Source* Field work

| Key issues for good cooperation and efficient flow of information | *N* = 200 | % |
|---|---|---|
| *Factors media should emphasize for effective news circulation* | | |
| Increase staffs in remote area | 50 | 25.0 |
| Reduce the cost of newspaper | 8 | 4.0 |
| Increase awareness | 84 | 42.0 |
| Agent-buyer good relation | 5 | 2.5 |
| Development of communication system | 24 | 12.0 |
| Accuracy in news circulation | 12 | 6.0 |
| Ensure news circulation in remote area | 17 | 8.5 |
| *Steps taken by authority to circulate news to destitute community and remote area* | | |
| Yes | 152 | 76.0 |
| No | 48 | 24 |
| *Types of step taken to circulate news to this community* | | |
| Raising awareness among people | 67 | 33.5 |
| Preventing disasters | 32 | 16.0 |
| Circulating news by loudspeaker/mosque mike | 49 | 24.5 |
| Unifying community members to combat hazards | 20 | 10.0 |
| Forming different teams | 21 | 10.5 |
| Organizing monitoring cells | 11 | 5.5 |
| *Connection between affected people and media persons in disaster related news circulation* | | |
| Yes | 165 | 82.5 |
| No | 35 | 17.5 |
| *High frequency radio system in disaster prone area* | | |
| Yes | 141 | 70.5 |
| No | 59 | 29.5 |

- **Economic Solvency and Mass Media**

From the table, it is found that 59% of respondents said that media guide them to be aware to mitigate the loss of natural hazards. Different daily newspapers as well as electronic media telecast different insightful and scientific column/lecture about the procedure of disaster risk reduction and mitigation along with coping strategy of hazards. The study data showed that, approximately 25 and 11% of respondents said that media help them through the guidelines of preventing disaster and training respectively (see Table 5). Media have a great influence to make attention of national and international agencies to support disaster affected people in terms of reliefs, medicines, cash money, foods, pure drinking water etc. From the study result, 84% respondents said that media suggest them to be solvent i.e. indicating new working opportunity (34%), showing the procedure of training (21%), and influencing to providing relief (17%) (see Table 5).

**Table 5** The role of media in public safety and generating public awareness about disaster risk reduction. *Source* Field work

| Role of media in public safety and awareness | *N* = 200 | % |
|---|---|---|
| *Guidelines to disaster risk reduction and mitigation* | | |
| Prevent disaster | 49 | 24.5 |
| Awareness | 117 | 58.5 |
| Training | 21 | 10.5 |
| Relief related | 13 | 6.5 |
| *Suggestion of media to be solvent after hazards* | | |
| Yes | 167 | 83.5 |
| No | 33 | 16.5 |
| *Types of suggestions of media to bring solvency in family after disaster* | | |
| Finding working sources | 38 | 19.0 |
| New working opportunity | 68 | 34.0 |
| Procedure of training | 41 | 20.5 |
| Influencing to provide relief | 33 | 16.5 |
| Lending money | 20 | 10.0 |
| *Security related role of media during and after disaster* | | |
| Yes | 188 | 94.0 |
| No | 12 | 6.0 |
| *Types of security related advice during and post hazard periods* | | |
| Providing forecast | 16 | 8.0 |
| Responsibility in disaster | 30 | 15.0 |
| Going cyclone center | 27 | 13.5 |
| Awareness type | 71 | 35.5 |
| Keeping safe place of female, baby and elderly people | 25 | 12.5 |
| Control unsocial activities | 6 | 3.0 |
| Protecting domestic resources | 6 | 3.0 |
| Preventive measures | 9 | 4.5 |
| Collecting foods | 10 | 5.0 |
| *Role of media in mass security and mass awareness* | | |
| Very satisfactory | 52 | 26.0 |
| Satisfactory | 130 | 65.0 |
| Unsatisfactory | 12 | 6.0 |
| Do not know | 6 | 3.0 |

- **Media's Role in Ensuring Security and Raising Awareness**

In the study area, media have a great contribution to increase security of the affected people in terms of circulating forecasting weather news, raising awareness, suggesting going cyclone center and preserving to domestic resources. People get comprehensive knowledge to ensure their safety from the natural hazards. The observed data indicate that most of the respondents (94%) answered positively that media circulate security related news during and post-disaster periods. According to the respondents, the news includes awareness type (36%), responsibility type (15%), going cyclone center (14%) and providing forecasting (8%). In addition to this, about the role of media for raising mass awareness and security 65% respondents expressed their opinions in satisfactory manner (see Table 5).

- **Strategies to Overcome Impediments of Circulating News in Hazard Periods**

The study results reveal that effective news circulation is disrupted in the study area due to several causes such as poor communication system, lack of having media means, unverified news, no scope of timely circulation of news and so on. About the impediments to disseminate news in the affected area, 42% respondents blame for poor communication system and approximately 19% respondents blame its inability to provide news in proper time. In addition, 12% respondents said that media have no scope to provide news due to remote area and poor communication system. However, to remove this impediments, 49% respondents keep emphasize on developing communication system. Moreover, 20% respondents highlight on circulating correct and faithful news. Furthermore, 11% respondents keep emphasize on circulating disaster related news with great care (see Table 6).

- **Media Persons and Disaster Management in Coastal Bangladesh**

From the findings of the study, it is found that media persons both electronic and print media in the local level try to collect news comprehensively. In collecting

**Table 6** Limitations of media for disaster risk reduction and mitigation. *Source* Field work

| Limitations of media for disaster risk reduction | $N = 200$ | % |
|---|---|---|
| *Major impediments of media to circulate news to prevent disaster* | | |
| No opportunity to get news | 24 | 12.0 |
| Cost effective | 15 | 7.5 |
| Unable to provide news in proper time | 37 | 18.5 |
| Uncertainty to get news | 13 | 6.5 |
| Cannot understand news | 13 | 6.5 |
| Lack of communication system | 84 | 42.0 |
| Insufficiency of accurate news | 14 | 7.0 |
| *Ways of removing impediments of media* | | |
| Getting news in cheap | 14 | 7.0 |
| Development of communication system | 97 | 48.5 |
| Correct and faithful news circulation | 39 | 19.5 |
| Keep emphasize on circulating disaster news | 21 | 10.5 |
| Not loss but how to reduce loss should be emphasized in news dissemination | 14 | 7.0 |
| Importance of newspaper should be highlighted | 8 | 4.0 |
| Ensure to circulate news in remote area | 7 | 3.5 |

disaster related news, they keep much emphasize then other news. Moreover, they focus much on preparedness and disaster management during the reporting of hazards related news. The study findings further indicate that media persons try to raise awareness among mass population about natural hazards in comprehensive way. In addition, most of the journalists mentioned that the role of media in disaster risk reduction and mitigation is very much effective. However, many of these reporters reported that the major limitations of news circulation are lack of expertise in collecting and reporting news in local level, economic problems and so on. The study findings reveal that rumor is a big problem in circulating disaster news. However, to avoid this situation most of the newscasters emphasize on collecting news by going to the root places. They also focus on neutrality and authenticity for reporting news. About the reporting of disaster news, most of the respondents (journalists) said that they try to focus the attention of government and international authority for relief and support to overcome the hazardous situation of affected people. In addition, they try to ensure the proper distribution of relief among affected people by reporting investigative news in media. The journalists of the affected area claimed that it is quite challenging to implement planning during emergency situation in coastal area. Disrupted environment due to natural hazards, along with interrupted electricity system and inexperienced press officials contribute to create challenging situation to gather and circulate timely and authentic news in the affected area. National level media take some initiatives about circulating the situation of hazards with the connection of weather forecasting organization. Some of the respondents claimed that local, national and international news agency work together to expose the situation with insightful and scientific guidelines and news with the help of respective scientists and experts. Government agencies provide update and create space for media to work during natural hazard. Almost all of the respondents, keep emphasize on recruiting expert broadcasters to collect news in affected area. They highlight on arranging training/workshops on how to collect news in disaster situation and overcome technical problems in hazard periods.

## 5 Conclusions

Bangladesh is one of the most vulnerable and climate change induced disaster stricken countries in the world (Azad et al. 2013). The frequent natural disasters impede the development process of Bangladesh (Nasreen and Hossain 2002). Moreover, in recent year the severity and intensity of hazards have increased to a great extend which contribute to create immeasurable destructions and losses in Bangladesh especially in south-western coastal belt (Garai 2014). However, mass media can play an important role to minimize this loss by properly disseminating news and preventive knowledge of disaster management among the affected people (Iqbal et al. 2014; Soltani 2015). In the study area media increase security of the affected people by circulating forecasting weather news/warning, raising awareness, suggesting to go secure place (cyclone center), indicating to save valuable resources before the imminent of natural hazards. It telecasts different insightful and scientific talking/writing about the process of disaster management along with different coping strategies of natural hazards. Media also try to make attention of different national and international agencies to come forward to support disaster affected people in terms of relief, foods, drinking water, medicine, house building materials, cash money and so on. Radio and television being easily accessible to the rural masses, play great role to disseminate disaster related news in the coastal area of Bangladesh. However, due to poor communication system, cost effectiveness, illiteracy, shortage of media workers and lack of awareness some other media means such as daily newspaper, magazines, journal, internet services etc. are hardly available in the locality. So government as well as policy maker should come forward to overcome these challenges by introducing effective use of information and knowledge dissemination throw mass media. Then the people of coastal area can utilize it more effectively to disaster risk reduction and mitigation in Bangladesh.

This study is conducted with the fund of 'Research Grants' of Chittagong University to investigate the role of mass media to disaster risk reduction and mitigation as a part of effective climate change communication. We hope in future different organizations will come forward to employ different projects to monitor coastal community to combat climate change induced natural hazards.

**Acknowledgements** I would like to acknowledge my cordial gratitude and indebtedness to Chittagong University (Research Grants) for funding to the research project. I would also like to express my gratitude to the department of Sociology of Chittagong University for providing me logistic supports to write the chapter. My special thanks go to the blind peer reviewers whose scholastic evaluation and priceless guidelines enrich the chapter.

## References

Azad SMY, Bahauddin KM, Himel TMS (2013) Disaster management through mobile technology: a conceptual model for Bangladesh. Bangladesh Res Publ J 8(1):34–40. Also available at http://www.bdresearchpublications.com/admin/upload/1308111/1308111.pdf

Cate FH (1994) The role of the media in disaster mitigation: roundtable on the media, scientific information and disasters. Int Disaster Commun Harnessing Power Commun Avert Disasters Save Live. Retrieved from http://www.annenberg.northwestern.edu

Choudhury AM (2001) Major disasters in Bangladesh and their impacts. Presented at the disaster management course held at PATC, Savar, Dhaka on 08 Nov 2001

Clerveaux V, Katada T, Hosoi K (2009) Information simulation model: effective communication and disaster management in a mixed-cultural society. J Nat Disaster Sci 30(1):1–11. doi:http://doi.org/102328/jnds.30.1

Cowan J, Mcclure J, Wilson M (2002) What a difference a year makes: how immediate and anniversary media reports influence judgments about earthquakes. Asian J Soc Psychol 5 (3):169–185. doi:10.1111/1467-839x.00102

Disaster Management Congress (2009) Thematic session on role of media in disaster management, organized by National Institute of Disaster Management (NIDM) held on 4–6 Nov 2009, New Delhi, India

Duhe SF (2008) Communicating Katrina: a resilient media. Int J Mass Emerg Disasters 26(2):112–127. Available at www.ijmed.org/articles/291/download/
Garai J (2014) The impact of climate change on the livelihoods of coastal people in Bangladesh: a sociological study. In: Leal FW, Alves F, Caeiro S, Azeiteiro UM (eds) International perspectives on climate change: Latin America and beyond. Springer International Publishing Switzerland, Cham, Switzerland, pp 151–163
Goldsborough (2002) New media and disasters (article). Teach Librarian 29(4):34
Iqbal MJ, Ali FM, Khursheed MB, Saleem S (2014) Analysis of role of media in disaster reporting in Pakistan. Eur Sci J 01(especial edition 2014)
Islam R (2011) Vulnerability and coping strategies of women in disaster: a study on coastal areas of Bangladesh. Arts Fac J (2010–2011):147–169, Dhaka University, Dhaka, Bangladesh. Available online at http://www.eldis.org/go/home&id=64076&type=Document#.VseewjZzpp4
Kakonge JO (2012) Improving the role of the media in disaster management for Africa. South News, New York, U.S.A
Khan MMH, Nahar N (2014) Natural disaster: socio-economic impacts in Bangladesh. Banglavision 13(1);58–67
Nasreen M (2004) Disaster research: exploring sociological approach to disaster in Bangladesh. Bangladesh e-J Soc 1(2) July 2004
Nasreen M, Hossain KM (2002) Sustainable development: Bangladesh perspective. Paper presented at the NGO session of world summit on sustainable development (WSSD), Johannesburg, South Africa, 26 Aug–4 Sept 2002
Nazari MR, Hassan MSBH, Parhizkar S, Osman MN (2011) The role of broadcasting in management of natural disasters. World Appl Sci 14(2):334–340
Neuman WR, Guggenheim L (2011) The evolution of media effects theory: a six-stage model of cumulative research. Commun Theor 21(2):169–196. doi:10.1111/j.1468-2885.2011.01381.x
Rattien S (1990) The role of the media in hazard mitigation and disaster management. Disasters 14 (1):36–45. doi:10.1111/j.14677717.1990.tb00970x
Scanlon J (2007) Unwelcome irritant or useful ally? The mass media in emergencies. In: Rodriguez H, Quarantelli E, Dynes R (eds) Handbook of disaster research. Springer, New York, pp 413–429
Seid-Aliyeva DE (2006) Role of mass media in the disaster preparedness and sustainable development of society. AIP Conf Proc 825(1)79–83. doi:10.1063/1.2190733
Soltani F (2015) Mass media and its role in increasing society's involvement in disaster management. Int J Health Syst Disaster Manage 3(Suppl S1):12–15. doi:10.4103/2347-9019.168565

## Author Biography

**Joydeb Garai** is currently working as Assistant Professor in the Department of Sociology in Chittagong University, Bangladesh. Besides teaching and supervising students, he is involved in different research works on Climate change adaptation, Disaster management, Community and rural development, Gender and development, Minority and indigenous people and so on. He has completed M.S.S (2008) and B.S.S (2007) Degree from Dhaka University and conducted an advanced thesis on Climate change and Livelihoods of coastal people under the scholastic supervision of Professor Dr. Khondoker Mokaddem Hossain. Several of his research works have been published from high impact factors peer reviewed journals in the world.

# Engaging People with Carbon and Climate Change Using Landscape Scale Conservation and Biodiversity Monitoring

**Adam Moolna, Cheryl Knott, Daveen Wallis, David Crawshaw, Joanne Brierley-Moore, Julia Simons and Anne Selby**

## 1 Introduction

Mitigation and adaptation measures to meet the challenges of changing climate require the engagement now of decision-makers, developers, land users and the general public. Since anthropogenic climate change emerged into public discourse in the 1980s, scientists have been largely successful in their communications persuading publics and governments that human-driven climate change is a real phenomenon. However, we face real current challenges in communicating messages about the actions needed to mitigate climate change due to carbon emissions and to help society and the natural world adapt (Moser 2010). We must modify our behaviour and the economy to minimise future climate change and to mitigate for the changes we are already committed to.

Communicating messages for climate change action is a challenge because climate change and carbon are to a large extent seen as an abstract issue: particularly in the developed world, publics and policy makers struggle to see the detail beyond global carbon budgets and to identify that there are local analogues to rainforest deforestation and melting ice caps (Moser 2010). The irony of making progress in tackling global carbon budgets, with carbon accounting and offsetting, is that carbon is too often treated as a tradable commodity that the developed world can address through trading carbon credits and offsetting elsewhere (Moolna 2012). The global and abstract elements so well communicated for the climate change phenomenon make the very real

A. Moolna · C. Knott (✉) · D. Wallis · D. Crawshaw · J. Brierley-Moore
J. Simons · A. Selby
The Wildlife Trust for Lancashire, Manchester & North Merseyside,
The Barn, Berkeley Drive, Bamber Bridge, Lancashire PR5 6BY, UK
e-mail: cknott@lancswt.org.uk

A. Moolna
School of Geography, Geology and the Environment, Keele University,
Staffordshire ST5 5BG, UK



W. Leal Filho et al. (eds.), *Handbook of Climate Change Communication: Vol. 1*,
Climate Change Management, https://doi.org/10.1007/978-3-319-69838-0_16

local impacts and local action possible seem less apparent and difficult to communicate (Leal Filho 2009; Moser 2010). In the United Kingdom, despite the active lead taken by a few cities such as Manchester, climate adaptation has been largely top down with little overall take up by local government (Tompkins et al. 2010).

We need people to act locally and for policy makers to be working to tackle global climate change through action where they live. Communications is vital for winning public support for wider societal action (Ockwell et al. 2009). To make that happen we must persuade the general public, decision-makers and wider society that climate change is not just abstract figures for carbon emissions and those melting polar ice sheets—it is directly affecting local people and local wildlife everywhere.

Tackling climate change is intertwined closely with biodiversity conservation and the two issues dovetail in their need to be dealt with on a cross-sectoral basis and at a landscape scale. Climate change, landscape scale networks, and the need to facilitate the shifts in species and habitat distributions are amongst the main issues for UK biodiversity conservation, for agriculture, and for environmental management more widely (Gray et al. 2013; Howden et al. 2007). Land use is both a local issue and one with global consequences, moreover land use responds to pressures (and needs to be managed) on scales from local through landscape to global (Foley et al. 2005).

Conservation has made tremendous progress towards the mainstreaming and cross-sectoral working that climate change adaptation requires. In line with the Convention on Biodiversity's latest agreed position with an explicit link between biodiversity and ecosystem services (Kull et al. 2015), UK conservation strategy embodies a shift from a paradigm of individual habitat and species action plans to a focus on managing the environment as a whole (JNCC and Defra 2012). There is a lot in ecosystem-based and landscape scale approaches to conservation that is explicitly climate change relevant. These can be brought to bear in driving wider climate change adaptation, a policy field emerging in England over the last decade and converging with the better established policy field of environmental management (Massey and Huitema 2013).

Responsibility for biodiversity conservation delivery in the UK falls under statutory agencies and obligations are met through a network of stakeholders covering state bodies, private business and non-governmental organisations. Cross-sector working and partnerships join up consideration of biodiversity across agriculture, planning and development, ongoing land use, and explicit conservation management itself. Examples include Local Nature Partnerships and River Basin Management Plans, which came out of the Lawton Review (Lawton et al. 2010) and the UK government's subsequent Natural Environment White Paper (HM Government 2011).

Managing biodiversity has impacts for carbon budgets as has been readily communicated with REDD + (reduced emissions from deforestation and forest degradation) from rainforests to mangroves, and with a less high profile with temperate peatlands (Billet et al. 2010; Drew et al. 2013; Grand-Clement et al. 2013). Changing climate impacts biodiversity by changing environmental parameters of temperature, rainfall and more, with effects including altered biome characteristics and shifting distribution of species and habitats. Ecosystem-based adaptation approaches in conservation are about making ecosystems resilient but also provide opportunities for carbon sequestration (as reviewed by Munang et al. 2013).

### 1.1 This Study

Exploring examples of how climate change is communicated within conservation projects and partnerships identifies learning that can be reflected upon. There are opportunities within conservation for communicating climate change messages for the benefit of both climate change adaptation and biodiversity conservation itself.

The Lancashire Wildlife Trust is leading on two major landscape scale conservation projects that are effectively engaging a wide audience in the importance of ecosystem resilience to, and societal engagement with, climate change. The Carbon Landscape Project that The Lancashire Wildlife Trust is delivering in a cross-sectoral partnership explicitly ties together carbon and climate change with biodiversity conservation. In the north west of England extensive peatlands present a functional carbon environment and a potential carbon sink. The Lancashire Wildlife Trust's Biodiverse Society Project engages communities with local wildlife and the networks, landscapes, and cross-sectoral approaches that are fundamental to climate-sensitive development and ecosystem adaptation. Increasing public awareness of the local impacts of climate change on local wildlife and the need for ecosystem resilience gives us an important additional voice holding politicians, local government and developers to account for a climate-and wildlife-sensitive present and future.

This paper aims to raise awareness of the interconnectivity of climate change and biodiversity conservation and share the lessons learnt from communications and engagement with various settings and audiences to help inform future work elsewhere. The wider environmental and academic communities have expressed much interest in these projects and there has been active discussion of the challenges, future prospects and lessons that can be shared. The Lancashire Wildlife Trust welcomes future links with academic institutions and this paper highlights that formal assessment of partnership working and communications effectiveness would be very useful in improving this and other programmes in the future.

## 2 Analysis

### 2.1 Engaging Policy-Makers, Planners and Publics at the Lancashire Wildlife Trust

The Lancashire Wildlife Trust is part of the Royal Society of Wildlife Trusts—bringing together 47 regional Trusts, 800,000 members, 40,000 volunteers, and 2000 staff across the UK. One of the Wildlife Trusts' core objectives is to engage with key actors in wider society and the general public to work together in partnerships. Engaging the public with their local environment, The Lancashire Wildlife Trust's impacts during 2016 include 50,000 volunteer hours from 1000 regular volunteers, 200 volunteers trained, 20,000 children engaged in schools and at events, and 40 teachers trained in environmental education.

Mainstreaming climate change adaptation and biodiversity conservation at a local level is seen with the local nature partnerships that bring together local government with other stakeholders and non-governmental organisations. Wildlife Trusts across the country are involved in these partnerships. The Greater Manchester local nature partnership, the Natural Capital Group, is chaired by The Lancashire Wildlife Trust, for example, and brings together the ten Greater Manchester councils, non-governmental organisations, universities, and business representatives. The Natural Capital Group reports directly into the Low Carbon Hub, a high level group within the Greater Manchester Combined Authority, showing how the city is aligning its approaches to the twin themes of climate change and biodiversity conservation.

The Wildlife Trusts, through the Living Landscapes and Living Seas programmes, have pioneered landscape scale approaches along with other third sector groups such as the Royal Society for the Protection of Birds (Clarke 2015). The government's Nature Improvement Areas (NIA) programme, which uses an ecosystem services framework to improve ecological connectivity, and the Heritage Lottery Fund's Landscape Partnership programme have supplemented the third sector push substantially (Adams et al. 2013, 2016; Clarke 2015; Fish et al. 2016).

These programmes are about cross-sectoral working and the substantial progress made in developing landscape scale partnerships for biodiversity conservation provides an appropriate existing framework that climate change mainstreaming can be hitched on to.

## 2.2 The Carbon Landscape Project

The Lancashire Wildlife Trust's Carbon Landscape Project is an example of the Living Landscapes approach and also encompasses both a Nature Improvement Area (the Great Manchester Wetlands NIA) and the Heritage Lottery Fund's Landscape Partnership framework (Carbon Landscape Partnership 2016). The Carbon Landscape is a landscape scale initiative with the ambition to make a step change in the restoration of an ancient landscape once devastated by industry. The heritage that this project focuses on is the natural and man-made heritage left behind after the closure of the Lancashire Coalfields and the decline of peat extraction. Engaging people with this tangible heritage of carbon offers a way to make the abstract ideas of globally rising $CO_2$ and climate change immediately relevant.

Nestled between the two cities of Manchester and Liverpool, the area is the only substantial gap in the coast to uplands Merseyside to Manchester urban belt and a vital corridor for the gradual south to north migration of species as the climate becomes warmer. It covers 107 $km^2$ within the areas of three local authorities (Salford, Wigan and Warrington). The Industrial Revolution left behind a physically scarred landscape. Historic conservation efforts have led to a somewhat piecemeal recovery of scattered locations across the landscape and the areas of valuable biodiversity are fragmented. It is only by thinking of the landscape as a whole and on an integrated scale that we can ensure key ecological corridors and stepping stones are created in the right places. In terms of biodiversity, we need to ensure climate change resilience for species to be able to move northward through the only area where this would be possible across the Mersey belt development. This way our landscape can help ensure the impacts of climate change are mitigated.

Why ‘The Carbon Landscape’? The key unifying and distinctive feature of this landscape is that it is based on carbon. Carbon is contained in the area’s mosslands (a local term for these distinctive peat-dominated wetlands) and woodlands and their coal and peat. The importance of carbon in the landscape goes back millions of years to when the coal measures were formed.

This Carbon Landscape, in the very area that fuelled the Industrial Revolution in Manchester and Liverpool, therefore has a powerful story to tell: how fossil carbon fuelled human-driven climate change; how industrial ravages disrupted local landscapes and wildlife; how conservation and climate change mitigation align; and how we need communities and government to work together to build a sustainable future. Communicating this story and understanding carbon in context will shape the buy in of stakeholders and our combined approach to a more sustainable future.

Within this project, The Lancashire Wildlife Trust is using two parallel and complementary approaches to achieve a large size functional ecosystem and climate-resilient landscape whose management is effectively integrated into development planning and local communities’ sense of identity (Aim 1 of the Carbon Landscape Project; see Table 1).

The first approach has been establishing the partnership itself. The funding criteria for the Landscape Partnership programme of the Heritage Lottery Fund have ensured that the Carbon Landscape Project has been explicitly about building cross-sectoral working to reconcile multiple objectives for conservation, agriculture, development and other land uses. Establishing this framework means that institutional and governance concerns, typically the most serious impediments to effective landscape-scale partnerships (Sayer et al. 2013), are addressed from the start. Progress in partner and political engagement has been made through a programme of stakeholder meetings, engagement with the planning system and site visits by senior politicians (Table 2). Over 5 years the partnership will have become firmly established, with partners confident in each other’s abilities, and looking for ways in which they can work jointly on particular schemes. Partners are committed to maintaining landscape improvements and will carry on working together beyond the project to improve the landscape further. Development will continue to threaten but each organisation will be in a stronger position as a partnership to secure compensation funding and to work with developers to incorporate sensitive land management practices into any future developments.

The second approach has been the bottom up engagement of local communities under Aims 2 and 3 of the Carbon Landscape Project (Table 1), again a key aspect of the Landscape Partnership programme (Clarke 2015). The Lancashire Wildlife

**Table 1** Carbon Landscape Project aims and objectives

| Aims | Objectives |
|---|---|
| Aim 1: To restore a derelict landscape, ensuring connectivity and resilience in an area under extreme threats | • To improve 5 flagship sites and create 120 ha of high quality priority habitats<br>• To work in partnership with all landowners and managers to improve a further 380 ha of high quality natural habitats<br>• To enhance connectivity with a focus on watercourses and reducing diffuse pollution<br>• To establish ecological baselines to inform decisions leading to the protection of the natural heritage |
| Aim 2: To reconnect people with their landscape through improved access, and increased learning and volunteering opportunities | • To improve 3 Carbon Landscape gateway sites and improve and promote 20 km of access routes with a focus on the formation of a Carbon Trail and various loops, with signposting, way marking and interpretation<br>• To improve interpretation of the landscape with new facilities at Wigan Flashes and a digital landscape created online<br>• To deliver an events and education programmes that build on Roundview themes that will inspire 2500 adults and 1200 children about their Carbon Landscape<br>• To provide a Volunteer Hub coordinating 500 high quality opportunities and 20 Landscape Champions<br>• Tasks such as monitoring key species will have teams of volunteers and Local Naturalist Groups fully equipped to continue this essential work<br>• Pilot activity will have trialled methodology that we can build on in future years |
| Aim 3: To instil pride and engender community ownership in our Carbon Landscape, upskilling local people, groups and beneficiaries to become custodians of our future | • To engage communities and groups in the Carbon Landscape through 20 projects involving 6000 local people<br>• To deliver a training programme for 200 people per year, giving at least 50 members the confidence to continue activities long term<br>• To employ 9 Landscape Trainees who will help deliver our landscape vision, engaging local people in activity and supporting key local organisations, groups and businesses<br>• To deliver 4 cultural heritage programmes that will involve 500 people and 10 community groups |

Trust, for example, worked closely with partners at the University of Manchester to use a specialist hands-on toolkit for participant-led workshops that let communities take ownership of their landscape through scaffolded learning about climate change and local ecosystems (the RoundView approach; www.roundview.org; Tippett and Connelly 2011). Along with further activities under the Sense of Place and volunteer engagement themes (Table 2), this has informed planning for the second phase delivery of community-focused infrastructure and a 5 year programme of community engagement (Countryscape 2015). Infrastructure such as over 20 km of accessible paths includes health walks, a Carbon Trail heritage and education route, a specialised visitor and information centre at The Lancashire Wildlife Trust's Wigan Flashes nature reserve, and three high profile gateway sites. Community engagement is planned to meet the received demand for training (both to get people into work and for volunteers), for conservation and other projects involving local people, and to build a substantial volunteer movement for local ownership of the landscape.

## 2.3 *The Biodiverse Society Project*

The Lancashire Wildlife Trust's Biodiverse Society Project is an example of what can be achieved by engaging people in wildlife recording across the counties of Lancashire and North Merseyside and sensitising them to the impacts of climate change, raising awareness of the changes in wildlife such as shifting ranges and earlier Spring blooms. Bolstering the spread and skills of the biological recording community will ensure we have the best possible data available to planners and politicians for effective evidence-based management of landscape scale biodiversity.

Local Wildlife Sites (LWSs) are by far the most numerous and comprehensively distributed of designated conservation sites, with 1216 LWSs in Lancashire and 286 LWSs in North Merseyside. They are known by various names across the United Kingdom and based on local selection criteria operated and managed often at a county level. For example, whilst termed Local Wildlife Sites (LWSs) in Merseyside, they are known as Sites of Biological Importance (SBIs) in Greater Manchester and Biological Heritage Sites (BHS) in Lancashire. The landscape scale networks of these sites play a fundamental role for climate change resilience and the long term survival of wildlife by acting as buffers, stepping stones and corridors between statutory designated areas like Sites of Special Scientific Interest (SSSIs).

Importantly, Local Wildlife Sites are non-statutory designated conservation sites, meaning that they have no legal protection and are reliant on the goodwill of the landowner to manage the site appropriately to protect their high biodiversity value. At heart, the LWS system draws attention to those sites that are considered to have significant biodiversity value. It can be used in different applications to achieve the protection, conservation and enhancement of the natural environment. Their main purpose is to flag to local authorities and planners that they need special attention paid in land use planning and in the event of any development proposals. However,

**Table 2** Engagement themes for the Carbon Landscape Project and progress made

| Themes | Progress |
|---|---|
| Sense of Place and community engagement | • Conducted 5 community workshops to help understand public perceptions of the landscape, producing a report with recommendations<br>• Liaised with the local community and groups to develop ideas in more detail, resulting in an Activity Plan for the Sense of Place theme<br>• RoundView workshops have been used to formulate themes that will inspire and motivate people to get actively involved, change perceptions and change their lifestyles through events, the education programme and interpretation along the Carbon Trail<br>• This has allowed the heritage of the Carbon Landscape Way and each of its three character areas to be reinterpreted in a locally meaningful way<br>• Time was allocated for groups to develop funding applications over an extended period within the development phase, some of which have been successful and are now ready to implement<br>• Community engagement has largely focussed on those groups already keen to be involved: a key challenge is widening the audience and engaging other groups that have limited landscape scale links or awareness |
| Volunteer engagement | • Reviewed previous community and volunteer engagement activity identifying what has worked and what has not gone so well<br>• Research for the Outdoor Champions looked at how we can make a step change to existing health walk provision to provide walks that help people understand and appreciate their landscape<br>• Trialled a number of training activities to ensure activity meets the needs of local people<br>• Trialled new methods for a Citizen Science project to ensure the appropriate approaches to volunteer wildlife recording are taken forward from the start of the project<br>• Subsequent production of guidance notes for all of indicator species and appropriate survey methodology<br>• Thorough analysis of current and previous trainee and apprenticeship schemes<br>• Preparatory work assessing accreditation options as part of a Carbon Landscape Traineeship training plan<br>• The 'Carbon Landscape Trialled Training Courses' report describes the trialled training courses that will be expanded through the volunteer and staff training programme<br>• Proposed content for the Training Programme was put together based on the results of a survey sent out to all contacts, groups and volunteers<br>• The challenge has been to adapt activities that suit all and to deliver this approach in a time effective way: volunteers have been happy to take part in providing any feedback but are more keen on taking action |

(continued)

**Table 2** (continued)

| Themes | Progress |
|---|---|
| Partner engagement | • Stakeholder groups and partner meetings discussed the Carbon Trail report and a number of recreational users helped trial certain routes<br>• Regular meetings have agreed communications, community engagement, and biodiversity outcomes<br>• A number of research projects and mapping activities have been created and taken forward by key partners and local universities<br>• With partners having their own agendas and other pressures on time, it has been a slow process to come up with a scheme that is focussed on what we need to do for a landscape change as well as ensuring all partners feel engaged and empowered with the process<br>• The slow ongoing dialogic process of partnership building, however, has resulted in a very strong partnership thanks to the time taken over reflection and the progressing discussions on how different aspirations can be tied together into a coherent and cohesive programme |
| Political engagement | • Formal partners (including local government) joined together for the Carbon Landscape Project: City of Trees, Healthy Rivers Trust, Wigan Council, Inspiring Healthy Lifestyles, Salford City Council, Warrington Borough Council, Natural England, Greater Manchester Ecology Unit, Woolston Eyes Conservation Group, University of Manchester<br>• Engagement with the planning system for formal consideration of landscape-scale networks for biodiversity conservation and ecosystem resilience to climate change impacts (e.g. the Greater Manchester Spatial Strategy)<br>• Invited senior politicians and mayoral candidates to a briefing and site visit to one of the flagship sites in the Carbon Landscape to raise the profile of the project and its context in the wider Great Manchester Wetlands Nature Improvement Area<br>• Further progress will require time and resources to ensure that key politicians are aware of the importance of our landscape for climate change resilience, and that politicians can see how our work can hit a number of additional agendas such as health |

the lack of resources, the lack of a substantial wildlife recording community, and the lack of data available to local authorities means these sites are very vulnerable to inappropriate and poorly informed management and development. That threatens the integrity of our ecosystems at a landscape scale, not just on individual sites, and is a major threat to the climate change resilience of our wildlife and landscape.

The Biodiverse Society Project, in partnership with the local records centres at the Lancashire Environmental Recording Network (LERN) and the Merseyside Biobank, has eight aims and targets for delivering wildlife recording, building skills and capacity, and raising awareness across communities (Table 3). The project is surveying over 200 Local Wildlife Sites across Lancashire and North Merseyside and capacity building with 30 Local Naturalist Groups. Targets include engagement of over 500 volunteers, with nearly 1400 training places available on a variety of recording courses from botanical skills to bird identification (Table 3). By the end of October 2016 the project had surveyed 1277 ha and 128 sites, given 49 different

**Table 3** Aims and objectives for The Lancashire Wildlife Trust's Biodiverse Society project

| Aims | Objectives |
|---|---|
| To undertake wildlife surveys and make recommendations to landowners on how Local Wildlife Sites can be managed and improved | 200 Local Wildlife Sites surveyed |
| To increase the number and quality of records submitted to Local Record Centres and better understanding of the wildlife found on Local Wildlife Sites through regular surveys | An increase in records submitted each year |
| To increase publicly available information about the wildlife that can be found on Local Wildlife Sites through the interpretation materials that will be created | Number of website hits and social media updates recorded |
| To deliver a trainee placement scheme that gives trainees much needed practical training and experience to address skills shortage in surveying: ensuring that early career conservationists are upskilled and able to gain long term employment in the sector | 12 × trainee 12 month placements (4 per year) employed through the project |
| To deliver a training programme targeting a wide range of audiences to develop new skills and gain confidence in the use of these new skills | 1375 training slots provided |
| To increase awareness within local communities of the wildlife value of Local Wildlife Sites and increase respect for these sites, through an events programme that celebrates wildlife found on Local Wildlife Sites that gives people an enjoyable experience | 30 community groups appreciate wildlife that they have recorded on their nearest Local Wildlife Site |
| To establish and train a team of volunteers to continue supporting surveys on Local Wildlife Sites | 500 individuals volunteer on the project |
| To improve the resilience of Local Naturalist Groups that are currently under threat of demise | 30 Local Naturalist Groups become stronger as a result of project support |

landowners management advice, submitted over 9000 biological records via project volunteers, provided 905 training places, used 317 individual volunteers, and accumulated over 3000 hours of volunteer time. All four trainees from the first year have gone on to further employment within the sector and three of the four second year trainees, who finished in December 2016, have already done the same.

By creating a supported network of wildlife recorders within local communities we are empowering communities to address the data deficiency. The various engagement events and taster sessions, helped by the trained recorders who are acting as wildlife champions, are kindling bottom-up community action for wildlife. By including climate change messaging, we strengthen the perceived urgency of the need for landscape scale networks—the climate change issue is here a powerful driver for biodiversity conservation.

## 3 Discussion

Climate change is well behind biodiversity conservation in profile as a local issue and in terms of local groups and local action being taken. The impacts of climate change, however, are amongst the main threats to local biodiversity and conservationists are coming to the fore in communicating to publics and policymakers that we need action now to make our wildlife (and wider society) climate resilient. Climate change communication can benefit by coordinating closely with biodiversity conservation, which is increasingly voicing the threats of climate change to drive progress at a landscape scale. Climate adaptation messaging could, for example, piggy back on biodiversity conservation campaigns to a much greater extent—mimicking the existing model in conservation of using flagship species such as tigers to protect wider ecosystems and less appealing wildlife.

The Wildlife Trusts movement has been at the forefront of the move to landscape scale approaches to conservation in the United Kingdom, in a large part explicitly because landscape scale networks are essential to facilitate the survival and shifts of wildlife in the face of a changing climate. The Great Fen project in south-eastern England, for example, is a celebrated 50 year vision for landscape scale connectivity and anticipatory restoration for ecosystem-based adaptation to climate change (Hughes et al. 2011; Adams et al. 2013).

As The Wildlife Trusts' vision for Living Landscapes sets out: "Imagine a country where… wetlands and peatlands rich in wildlife are soaking up flood water and carbon… our farmland and woodland is producing food and timber but also bursting with wildlife…**wildlife can move freely through countryside, towns and cities, as it adapts to climate change**. A Living Landscape is all this and more." [our emphasis] (The Wildlife Trusts 2010).

The Heritage Lottery Fund's Landscape Partnerships programme has provided a vital driver (and crucially, substantial financial resources) in stepping up the delivery of landscape scale approaches across large swathes of the United Kingdom by stressing the importance of "degree of engagement, commitment and initiative of local residents and businesses, NGOs and statutory bodies, working in partnership to deliver conservation of the natural and cultural heritage, emphasising public access, education, training and community involvement" (Clarke 2015).

The Lancashire Wildlife Trust has brought together carbon emissions and human-driven climate change with the ecosystem approach to biodiversity conservation in the Carbon Landscape Project. The cross-sectoral partnership approach and bottom-up engagement of communities has been driven by the HLF Landscape Partnership funding criteria and is proving crucial to delivering a climate-sensitive development strategy. Dialogic communication bringing stakeholders together (Moser 2010) has been vital for meaningful buy-in. There remain challenges in ensuring work packages are delivered in full and deadlines met, however, and the need for a dynamic chair and lead partner for overall coordination and to set the pace is probably the most important factor.

Sir John Lawton, lead author of the seminal Lawton Report (Lawton et al. 2010), argues the greatest achievement of the Great Manchester Wetlands Nature Improvement Area (NIA) encompassed by the Carbon Landscape Project is the influence exerted through the planning system (for example, engaging with the Greater Manchester Spatial Strategy), surpassing the impact of projects delivered directly on improving and restoring biodiversity (comment at the Great Manchester Wetlands NIA forum, cited in: Great Manchester Wetlands NIA Partnership 2016).

The emphasis on bottom-up approaches from the Heritage Lottery Fund meant the Carbon Landscape Project from the start explicitly sought to build community engagement. The resultant Sense of Place theme has had major importance for taking people on a journey about the local relevance of human-driven climate change and for people feeling empowered to do something positive for their landscape. The Friends of Low Hall partnership with The Lancashire Wildlife Trust and the Wigan Leisure & Culture Trust, for example, has been a shining success in the public taking ownership of a Local Nature Reserve (see the group's website www.friendsoflowhall.co.uk). Reconnecting local people with their landscape through access and learning opportunities in the Carbon Landscape Project empowers communities to take active management and is building up sustainable volunteer resources for the long term. The community voice is also a vital driver for buy in of developers and planners for climate-sensitive development in the coming years of housing growth.

The Biodiverse Society Project has complemented the bottom-up approach of community engagement in the Carbon Landscape Project substantially: mobilising the public to realise ecosystems and wildlife occur in landscapes, to take ownership of those landscapes and understand the role it plays in climate problems and mitigation, and to realise the ecosystem assets and ecosystem services we need to protect. Importantly, both the Biodiverse Society Project and Carbon Landscape Project bring the more formal discussions and academic papers to a "what you can do in your own backyard and local nature reserve" level, they make members of the public and actors at local authorities realise that local iconic species are part of the bigger picture, and they make both climate change and ecosystem approaches relevant. This is a major part of what makes the two projects effective at truly mainstreaming landscape scale climate change adaptation and ecosystem approaches to conservation.

### 3.1 Lessons and Future Prospects

Firstly, an effective coordinator role is central to the degree of success. Within both projects, the community groups and individuals recognise the importance of linking up as a network but they are clear that they cannot do that without help. People want to share their stories, their knowledge and to learn from best practice by others; but without outside support and coordination will not have the resources or outside drivers necessary for it to happen. Most stakeholders remain focused on their immediate local area or sphere of responsibility and the coordinator role is vital in taking separate local groups and multiplying their impact as part of a coordinated landscape scale network.

Secondly, challenges in coordinating timely and full delivery by project partners can be addressed by modified organisation. As lead partner in the Carbon Landscape Project, The Lancashire Wildlife Trust could address varying contributions from different partners by setting out clearer expectations and requirements at each stage to make it easier for a minimum level to be ensured (and relieve pressure on partners exceeding requirements). Ensuring internal project deadlines are met has been a challenge as every partner has work pressures. Using conditional payments to partners dependent on meeting deadlines is one option but is fraught with the danger of damaging relationships and undermining partnership effectiveness.

Thirdly, resources have been vital. The additional budget brought in (£2.2 m from HLF and £0.8 m in additional grants to partners) has mobilised essential resources. The value of in kind support, other staff time contributed, and input of volunteers (skilled and unskilled) mobilised in parallel has proven difficult to assess but is regarded as something substantial that should be monitored and quantified better in future.

This bring us on to lessons about communicating. On initial engagement with "the Carbon Landscape", many people found the name puzzling and did not understand what it meant, they questioned whether it was about carbon capture, and perhaps planting trees to trap $CO_2$? This could be considered positive in that it opens up a discussion and engages people in finding out more. However, it also appears to have put some people off by not referring to biodiversity or wildlife. This was addressed in development by adding the strapline "Restoring Great Manchester Wetlands to the Community". It was found that when words such as "science, politics, policies and climate change" are used people in communities tended to switch off, whereas framing in terms of "local and wildlife" allowed engagement on to which those other issues could be added. In communicating project progress and community engagement, with an overwhelming amount of discussion and information, a major challenge was addressing how that could best be captured and distilled, clarified and communicated. Communication would benefit from better consideration of outputs and messages in planning collection of respondent views. This could perhaps be done in collaboration with appropriate experts or collaborators at universities.

The Biodiverse Society Project finishes in December 2017 and further funding bids are already in the pipeline. Groups are keen to drive forward the agenda but identify the need for external support coordinating groups and their working together. The Wildlife Trust is looking to not only continue the work but to expand by linking with health and young people agendas. The Carbon Landscape Project will finish at the end of December 2021 but it is envisaged that by then the established partnership will be in a position to continue to work together and, further, be looking to explore opportunities in other business areas and geographic locations. Both projects have empowered a highly interested audience, put networks in place, and established landscape scale coordination. It is hoped that the success, and the resources evident, will inspire increased take up elsewhere.

## 3.2 Limitations and Constraints

This study has two key limitations. Firstly, the assessment of the projects' effectiveness in communicating climate change is a rapid review and discussion of existing data and from respondent discussions. It is by no means a systematic review of empirical evidence within pre-defined eligibility criteria to answer a specific research question. Secondly, structured consideration of effectiveness in communicating climate change (or for that matter biodiversity conservation) is not an integral part of either project examined. The aim of the Carbon Landscape Project is to enable a landscape scale approach to biodiversity conservation and of the Biodiverse Society Project to engage a network of communities in wildlife recording and awareness. Climate change was a component theme, albeit a major one, employed for primarily conservation objectives. That means this analysis is dependent on large amounts of qualitative data and feedback that is not structured for a specific assessment of the effectiveness of engagement with climate change messaging. Although time and resource constraints meant it was not possible, the study would have benefited greatly from de novo data collection and semi-structured interviews to enable more robust findings.

## 3.3 Conclusion

We call on the wider climate adaptation and biodiversity conservation communities to come together and use both angles in a combined landscape scale approach that meaningfully engages local people and convinces local government and planners that we can effectively work together. By engaging communities with their local carbon landscape and pushing climate change messaging into development planning and biodiversity conservation, we can both bolster local climate action and help build the wider societal support needed for mainstreaming global action.

**Acknowledgements** With thanks to Carbon Landscape partners: City of Trees; Healthy Rivers Trust; Wigan Council; Inspiring Healthy Lifestyles; Salford City Council; Warrington Borough Council; Natural England; Greater Manchester Ecology Unit; Woolston Eyes Conservation Group; University of Manchester.

With thanks to Neil Hewertson for the original planning of the Carbon Landscape Project at The Lancashire Wildlife Trust.

With thanks to Biodiverse Society partners: Lancashire Environment Record Network (LERN); Merseyside BioBank.

The Heritage Lottery Fund supports both the Carbon Landscape Project (Landscape Partnership award) and the Biodiverse Society Project (Heritage Grant). Further financial support was provided through partner core funds and other restricted grant funding.

## References

Adams W, Hodge I, Sandbrook L (2013) New spaces for nature: the re-territorialisation of biodiversity conservation under neoliberalism in the UK. Trans Inst Br Geogr 39(4):574–588

Adams W, Hodge I, Macgregor N, Sandbrook L (2016) Creating restoration landscapes: partnerships in large-scale conservation in the UK. Ecol Soc 21(3):1

Billet M, Charman D, Clark J, Evans C, Evans M, Ostle N, Worrall F, Burden A, Dinsmore K, Jones T, McNamara N, Parry L, Rowson J, Rose R (2010) Carbon balance of UK peatlands: current state of knowledge and future research challenges. Climate Res 45:13–29

Carbon Landscape Partnership (2016) *The carbon landscape. Landscape conservation action plan part 1*. Preston: The Wildlife Trust for Lancashire, Greater Manchester & North Merseyside

Clarke R (2015) Beyond landscape designation: innovative funding, delivery and governance and the UK protected area system. Manage Environ Qual 26(2):172–194

Countryscape (2015) Carbon landscape. Community workshops report. Manchester: Countryscape

Drew S, Waldron S, Gilvear D, Grieve I, Armstrong A, Bragg O, Harris L (2013) The price of knowledge in the knowledge economy: Should development of peatland in the UK support a research levy? Land Use Policy 32:50–60

Fish R, Church A, Willis C, Winter M, Tratalos J, Haines-Young R, Potschin M (2016) Making space for cultural ecosystem services: insights from a study of the UK nature improvement initiative. Ecosyst Serv 21:329–343

Foley J, DeFries R, Asner G, Barford C, Bonan G, Carpenter S, Chapin F, Coe M, Daily G, Gibbs H, Helkowski J, Holloway T, Howard E, Kucharik C, Monfreda C, Patz J, Prentice I, Ramankutty N, Snyder P (2005) Global consequences of land use. Science 309(5734):570–574

Grand-Clement E, Anderson K, Smith D, Luscombe D, Gatis N, Ross M, Brazier R (2013) Evaluating ecosystem goods and services after restoration of marginal upland peatlands in South-West England. J Appl Ecol 50:324–334

Gray M, Gordon J, Brown E (2013) Geodiversity and the ecosystem approach: the contribution of geoscience in delivering integrated environmental management. Proc Geol Assoc 124(4):659–673

Great Manchester Wetlands NIA Partnership (2016) Great Manchester Wetlands Nature Improvement Area (NIA) strategic plan 2016 to 2025. Great Manchester Wetlands Nature Improvement Area Partnership, Manchester

HM Government (2011) The natural choice: securing the value of nature. HM Government, London

Howden S, Soussana J-F, Tubiello F, Chhetri N, Dunlop M, Meinke H (2007) Adapting agriculture to climate change. Proc Natl Acad Sci 104(50):19691–19696

Hughes F, Stroh P, Adams W, Kirby K, Mountford J, Warrington S (2011) Monitoring and evaluating large-scale, 'open-ended'habitat creation projects: a journey rather than a destination. J Nat Conserv 19(4):245–253

JNCC and Defra (2012) UK Post-2010 Biodiversity framework. JNCC and Defra (on behalf of the Four Countries' Biodiversity Group). London: Defra. Available from: http://jncc.defra.gov.uk/page-6189

Kull C, Arnauld de Sartre X, Castro-Larrañaga M (2015) The political ecology of ecosystem services. Geoforum 61:122–134

Lawton J, Brotherton P, Brown V, Elphick C, Fitter A, Forshaw J, Haddow R, Hilborne S, Leafe R, Mace G, Southgate M, Sutherland W, Tew T, Varley J, Wynne G (2010) Making space for nature: a review of England's wildlife sites and ecological network. Defra, London

Leal Filho W (2009) Communicating climate change: challenges ahead and action needed. Int J Clim Change Strat Manage 1(1):6–18

Massey E, Huitema D (2013) The emergence of climate change adaptation as a policy field: the case of England. Reg Environ Change 13(2):341–352

Moolna A (2012) Making sense of $CO_2$: putting carbon in context. Global Environ Politics 12 (1):1–7

Moser S (2010) Communicating climate change: history, challenges, process and future directions. Wiley Interdisc Rev Clim Change 1:31–53

Munang R, Thiaw I, Alverson K, Mumba M, Liu J, Rivington M (2013) Climate change and ecosystem-based adaptation: a new pragmatic approach to buffering climate change impacts. Curr Opin Environ Sustain 5(1):67–71

Ockwell D, Whitmarsh L, O'Neill S (2009) Reorienting climate change communication for effective mitigation: forcing people to be green or fostering grass-roots engagement? Sci Commun 30(3):305–327

Sayer J, Sunderland T, Ghazoul J, Pfund J, Sheil D, Meijaard E, Venter M, Boedhihartono A, Day M, Garcia C, van Oosten C, Buck L (2013) Ten principles for a landscape approach to reconciling agriculture, conservation, and other competing land uses. Proc Natl Acad Sci 110 (21):8349–8356

The Wildlife Trusts (2010) A Living landscape: play your part in nature's recovery. The Wildlife Trusts, Newark

Tippett J, Connelly A (2011) You want me to do what? Teach a studio class to seventy students? J Educ Built Environ 6(2):26–53

Tompkins E, Adger W, Boyd E, Nicholson-Cole S, Weatherhead K, Arnell N (2010) Observed adaptation to climate change: UK evidence of transition to a well-adapting society. Glob Environ Change 20:627–635

# Knowledge Management as an Enabler of the Paris Agreement Implementation in Africa

Charles Muraya

## 1 Introduction

Besides the Paris Agreement, the post-2015 development agenda in Africa is shaped by regional and global agreements that include Africa's Agenda 2063—*the Africa we want*, the 2030 Agenda for Sustainable Development—*leaving no one behind* and the Sendai Framework for Disaster Risk Reduction. Climate change is integral and central to the success of these compacts. For ease of coordination, informed decisions and better partnerships, enhanced climate knowledge sharing and communications is inevitable.

Knowledge management (KM) provides a link to facilitate harnessing of climate information and knowledge and channeling it to users. As a facilitator and conduit for knowledge delivery therefore, KM dovetails its activities with climate information and knowledge producers such as researchers, scientists, analysts and academics for harnessing and codification, and endeavors to avail the knowledge products to relevant beneficiaries and stakeholders, using the right language and format, and via the most efficient dissemination channels.

In Africa, access to climate knowledge has been hampered by both inadequate production of information and poor Internet penetration. In the realm of research, the level of accessible climate literature from Africa as cited by the IPCC continues to be significantly low (UNISDR 2014) (ClimDev-Africa 2014). According to Mail & Guardian, *"Sub-Saharan Africa only accounts for under 1% of the world's research output, despite having 12% of the world's population"* (Mwiti 2015).

Concerted efforts will be needed in Africa to coordinate partnerships among multiple existing and emerging initiatives like the Africa Adaptation Initiative (AAI) and Africa Renewable Energy Initiative (AREI), which will require infor-

C. Muraya (✉)
African Climate Policy Center, United Nations Economic Commission on Africa, Menelik II Ave, Addis Ababa, Ethiopia
e-mail: murayac@un.org

 
W. Leal Filho et al. (eds.), *Handbook of Climate Change Communication: Vol. 1*, Climate Change Management, https://doi.org/10.1007/978-3-319-69838-0_17

mation and knowledge sharing to ease coordination in support of the Paris Agreement.

At the Paris 21st conference of parties (COP) to the UN Framework Convention on Climate Change (UNFCCC), parties acknowledged that there is "*need to promote universal access to sustainable energy in developing countries, in particular in Africa, through the enhanced deployment of renewable energy*" (UNFCCC, Paris Agreement 2015). This can only be achieved through robust knowledge of options, opportunities, solutions and challenges that affect renewable energy investments in Africa.

This paper seeks to identify opportunities for ramping up the coordination of climate change knowledge to support the Paris Agreement implementation in Africa, through enhanced knowledge partnerships among institutions as well as through increased use of innovative tools and approaches that leverage the rapidly improving cyberspace connectivity.

## 2 The Problem

Climate change increasingly impacts development sectors like agriculture, water, transport and energy. Although awareness of the threats and opportunities created by climate change is on the rise, its concerns are inadequately integrated in decision making by policy makers, practitioners and researchers. According to a 2015 United Nations report "*over the last twenty years, 90% of major disasters have been caused by 6457 recorded floods, storms, heatwaves, droughts and other weather-related events*" (UNISDR 2015). Climate change has been termed as the "greatest challenge of our time" by the G20 (2015).

There is proliferation of information in today's world, climate change included. Besides the colossal amounts of data and information being churned out daily, types of devices and mediums of dissemination have also mushroomed, including cheaply available internet of things (IoT) climate observation devices, tablets, mobile phones and even watches.

Further, multiple initiatives operating in the climate change landscape are generating information and climate knowledge products, with suboptimal coordination and sharing of this knowledge, lessons and practices.

These predicaments, if not addressed, will hamper the realization of the objectives of the Paris Agreement, in particular, and development efforts in general. Strengthened strategies in knowledge management can assist in identifying knowledge gaps among particular user communities, and tailoring, packaging and channeling the information.

## 3 Methods

This paper derives its discussions and assertions from both experiences gained through ClimDev-Africa's engagement with climate information, knowledge brokers and communication experts in Africa, as well as literature review of climate change knowledge published on the internet. ClimDev-Africa convenes an annual climate change and development in Africa conference, which gathers climate change decision makers, researchers, academics, the media, farmers, youth, women and pastoralists to share challenges, lessons and practices in climate change interventions.

ClimDev-Africa has therefore gained significant insights in climate change through partnering with regional climate centers (RCCs) and national hydro-meteorological services (NHMSs) to support the generation of climate data, information and knowledge.

As an information manager at the ClimDev-Africa programme, I felt compelled to share my experience and lessons learnt in regard to climate change knowledge networking across the continent, with a view to advancing the discourse in this increasingly critical area.

## 4 Case Studies

Co-production of climate information products has been shown to yield better results in confronting the vagaries of climate change at the grass-roots level. A number of interventions operating in Africa as outlined below increasingly incorporate this approach, which is underpinned by a strong knowledge and communications framework.

Although there are multiple and diverse interventions currently operating in Africa, the few enumerated below are selected to highlight select players that have a footprint in undertaking climate change projects and similarly generating climate knowledge. This paper does not in any way diminish the role of those initiatives that are not featured here.

The work outlined here is a result of knowledge sharing experience in the context of ClimDev-Africa events and partnerships, augmented by online literature review of institutions active in climate change research, adaptation and mitigation in the continent. No field research was undertaken, nor was any primary data collected to inform the discussion and conclusions.

## 4.1 WISER

The weather and climate information services for Africa (WISER) is a 4 year programme funded by UK's Department for International Development (DfID) whose purpose is "*To enhance the resilience of poor people and of economic development in Africa through improved generation and use of weather and climate information services*". The programme aims to improve the generation and use of weather and climate information across Sub-Saharan Africa. WISER is split into 2 components, one implemented by the UK Met Office with focus on East African projects, while a pan-Africa component is implemented by the African Climate Policy Center (ACPC) in support of RCCs, climate research for development (CR4D), a fellowship programme, climate knowledge management and communications.

The WISER programme kicked off in 2016 with quick start projects geared towards demonstrating both value for money (VfM) and social economic benefits (SEB) of CIS. It emphasizes the last mile delivery of climate information to beneficiaries, and co-generation of climate information products to ensure CIS uptake, and efficiency, economy and effectiveness in the CIS co-production process.

The programme recognizes the role of knowledge management and communications for the success of democratizing the usefulness of climate and weather information, thus firmly integrated these enablers in the programme. For each area of intervention in WISER, the monitoring and evaluation incorporates learning to facilitate feedback for the benefit of subsequent intervention designs.

## 4.2 RCOFs in Africa

Regional climate outlook forums (RCOFs) are organized by RCCs in collaboration with the World Meteorological Organization (WMO), National Meteorological and Hydrological Services (NMHSs), Global Climate Centres (GCC) and development partners. The aim of RCOFs is to provide consensus regional seasonal climate outlooks for applications in climate sensitive socioeconomic sectors for decision support, resilience building and sustainable development (ICPAC 2016).

Africa has 3 RCOFs, one each for West Africa, Southern Africa and Eastern Africa. The IGAD climate prediction and application center (ICPAC) based in Nairobi Kenya coordinates the East Africa greater horn of Africa climate outlook forum (GHACOF), while the SADC Climate Services Centre (SADC-CSC), in Gaborone, Botswana is responsible for the Southern Africa Regional Climate Outlook Forum. The PRESAO: PRÉvisions Saisonnières en Afrique de l'Ouest (West Africa RCOF) is coordinated by the African Centre of Meteorological Application for Development (ACMAD).

RCOFs were originally established to focus on seasonal prediction. However, a number of RCCs, including the GHACOF have extended the platform to address increased capacity to adapt to climate change. The utility of RCOFs is also being extended into developing and disseminating regional climate change information products.

## 4.3 TAHMO

The trans-Africa hydro meteorological observatory (TAHMO) is a project initiated and implemented by a group of American universities with the objective of strengthening Africa's climate observation network through installation of 20,000 weather stations across the continent (one every 30 km). TAHMO distinguishes itself by its innovative sensoring through design of low-cost automatic weather stations.

Using the revamped observation network, TAHMO aims to make high-quality data freely available to governments, scientists and farmers and to "make it possible for Africa to leapfrog to one of the best-monitored continents in the world".

TAHMO in addition runs a School2School initiative that seeks "to foster international school partnerships and science and technology education utilizing on-site climate monitoring to analyze local weather and compare local school environments to partner schools around the world". TAHMO technical staff deliver an in-person training to school staff members about weather station function and minor maintenance aspects (TAHMO 2017).

## 4.4 ClimDev-Africa

The climate change for development in Africa programme (ClimDev-Africa) was established in 2008 and aims at facilitating widely available climate information for informed decision making. With its secretariat at the UN Economic Commission on Africa (ECA), ClimDev-Africa is a joint partnership of the Africa Development Bank (AfDB), ECA and the African Union Commission (AUC). ClimDev-Africa partners with governments, RCCs, regional economic commissions (RECs) and global institutions to build climate resilience of vulnerable communities through both adaptation and mitigation interventions. Its footprint covers a significant proportion of the African continent.

ClimDev-Africa is a key player in convening climate change dialogue spaces, which include the annual climate change and development in Africa (CCDA) conference, which gathers policy makers, researchers and practitioners to deliberate and share lessons, challenges and opportunities encountered in climate change projects. The Africa Pavilion at the COP has in recent years also provided a platform for the continent to reflect on options available to address climate change in line with the global climate change agenda.

The CCDA conference has grown to be a valued space where technical papers are presented, while best practices and lessons are shared. The structure of the conference includes a high level dialogue where African ministers deliberate on the conference theme.

As the secretariat of ClimDev-Africa, the ACPC at ECA acts as a knowledge hub for climate information and knowledge, a role which it fulfils through convening discussion spaces, undertaking climate research, policy analysis and producing climate knowledge products (ClimDev-Africa, African Climate Policy Center 2017).

Using its pan-Africa mandate, Climdev-Africa is repositioning itself to facilitate a wider knowledge partnership that enhances sharing of data, information and knowledge on climate change. It is expected that this network will stimulate the identification of knowledge gaps and avail best practices and lessons to planners, vulnerable communities, scientists, the media and other stakeholders, through a robust information and communications framework.

In the run-up to the Paris conference of parties on climate change, ClimDev-Africa provided technical support to Cameroun, Swaziland, Malawi and Liberia with capacity building to formulate the countries' INDCs. ClimDev-Africa is transitioning its INDCs support to African countries' to monitoring, reporting and verification (MRV) frameworks, data collection, knowledge sharing and capacity building for the NDCs implementation.

## *4.5 UN CC: Learn*

The United Nations climate change learn is a "*One UN Climate Change Learning Partnership*" that coordinates climate change learning within the UN system. UN CC: Learn focus is development of multi-stakeholder driven climate change learning strategies at national level and production of climate change online learning modules. In Africa, UN CC: Learn has undertaken the foregoing interventions in Benin, Uganda, Malawi, Ethiopia, Ghana, Niger and Burkina Faso. UN CC: Learn contributes to the implementation of Article 6 of the UNFCCC on training, education and public awareness-raising, and the Doha Work Programme.

UN CC: Learn is stimulating climate change learning in schools through its partnership with education institutions in the countries where it is operating.

The UN CC: Learn hosts a knowledge platform that acts as a hub for sharing climate knowledge from across the UN system. As of February 2016, UN CC: Learn learning platform hosted 11 introductory learning modules and 6 advanced learning modules.

## 4.6 FCFA

The future climate for Africa (FCFA) is a five-year international research programme jointly funded by DfID and the Natural Environment Research Council (NERC) to stimulate "fundamentally new climate science focused on Africa, and to ensure that this science has an impact on human development across the continent". FCFA is delivered by five independent research consortia, with support from a central coordination unit.

The Coordination, Capacity Development and Knowledge Exchange (CCKE) unit coordinates activities across the different FCFA thematic and geographic research clusters, develops the capacity of key stakeholders, and produces and distributes news, features and reports on FCFA's work (FCFA 2017).

The FCFA project intention is to enable the public to benefit from knowledge and tools made available by the research platforms involved in the project. The programme's success will be measured by the way that its research generates new knowledge which can be used to benefit the poor in a sustainable manner (NERC 2017).

## 4.7 ENACTS

The enhancing access to climate services (ENACTS) is a project implemented by the International Research Institute (IRI) for Climate and Society. ENACTS is a unique, multi-faceted initiative designed to bring climate knowledge into national decision making by improving availability, access to, and use of climate information (IRI 2017).

ENACTS avails an integrated platform aimed at accelerating efforts to simultaneously improve the availability, access and use of climate information at the national level in African countries. The project focuses on the generation of reliable climate information that can be used for national and local decision-making. ENACTS has supported climate observation networks in Rwanda, Ethiopia and The Gambia and the installation of computer hardware for climate data rescue and management in Rwanda and Ethiopia.

Through the support of WISER Programme, ENACTS is undertaking development in 5 countries in East Africa. Funding has been allocated to develop the ENACTS initiative in two new countries (Uganda and Kenya), and to strengthen the initiative in countries where it has already been implemented (Ethiopia, Tanzania and Rwanda) (IRI 2017).

## 4.8 AREI

The Africa Renewable Energy Initiative (AREI) is one of the few initiatives to address climate change mitigation in Africa, an aspect that clearly is central to efforts needed to fulfill the Paris Agreement's goal to keep global temperature below 2°.

AREI prides itself as a transformative, Africa-owned and Africa-led inclusive effort to accelerate and scale up the harnessing of the continent's huge renewable energy potential. Under the mandate of the African Union, and endorsed by African Heads of State and Government on Climate Change (CAHOSCC), the initiative is set to achieve at least 10 GW of new and additional renewable energy generation capacity by 2020, and mobilize the African potential to generate at least 300 GW by 2030 (AREI 2015).

The AREI is firmly anchored in the context of sustainable development and climate change. It shows how low to zero carbon development strategies can be achieved in African countries through climate finance and means of implementation according to the principles of the UNFCCC. It recognizes the critical importance of rapid expansion of energy access for enhanced well-being, economic development and the fulfilment of all Sustainable Development Goals (AREI 2015).

During the establishment phase, AREI seeks to determine what renewable energy efforts exist on the continent and to enhance coordination among them. Actions involve mapping of all existing initiatives and systematization of experiences and lessons learned that are of relevance to all other work areas (AREI 2016).

## 4.9 CIRDA

The climate information for resilient development in Africa (CIRDA) programme is implemented by UNDP in 11 African nations. It is a multi-country programme to strengthen climate information for resilient development and adaptation to climate change in Africa (UNDP 2017).

The CIRDA Programme will strengthen the capacity of the 11 countries and the African region to develop and operate modern climate information and early warning systems by making available technical assistance and provide access to new technologies. Countries will also benefit from regional coordination and a knowledge sharing platform. The Programme provides technical support and capacity building by fostering knowledge sharing between experts, regional organizations and the countries' themselves (UNDP 2017).

### 4.10 ALP

The Adaptation Learning Program (ALP) for Africa aims to increase the capacity of vulnerable households in sub-Saharan Africa to adapt to increasing and uncertain climate change and climate variability. Since 2010, ALP has been working with communities, government institutions and civil society organisations in Ghana, Kenya, Mozambique and Niger with outreach to other African countries.

ALP develops innovative community-based adaptation (CBA) approaches and strategies with farming and pastoralist communities, and promotes their integration into local and national government systems and programmes. Adoption at scale is achieved by generating evidence of impact and good practices for CBA, building capacity and influencing development and adaptation policy and finance nationally and globally. ALP work focuses on gender equality, strengthening adaptive capacity, access to useful climate information, multi-stakeholder decision-making and facilitation of learning for climate resilient agriculture, risk reduction and resilience across Africa (CARE-Africa 2017).

### 4.11 Data Rescue Efforts

Digital climate data systems are a recent feature in Africa's NHMSs. In the past, most of the climate data in the continent was captured only on paper. For many NHMSs therefore, historical data still remains only in hardcopy, making it largely unavailable for climate modelling and research.

Through efforts of players like the WMO and ClimDev-Africa, a number of projects are underway to digitize historical climate data. WMO runs a programme named DARE which is driven by the belief that data is critically important to ensure future generations of scientists and other data users have access to all the information necessary for their studies and projects. The rescued data combined with already available data enables better assessments of projections of the climate into the future that can serve as input for the policy makers (WMO 2017).

ClimDev-Africa has supported Ethiopia, Rwanda and The Gambia in their quest to rescue climate observation data (ClimDev-Africa 2015).

## 5 Contemporary Knowledge Sources

In order to assess top 50 sources of climate change knowledge related to development in Africa, a desk study was undertaken to enumerate accessible publications on the Internet addressing the topic. The methodology used relies on ranking of published climate change knowledge available on Google search engine, and which

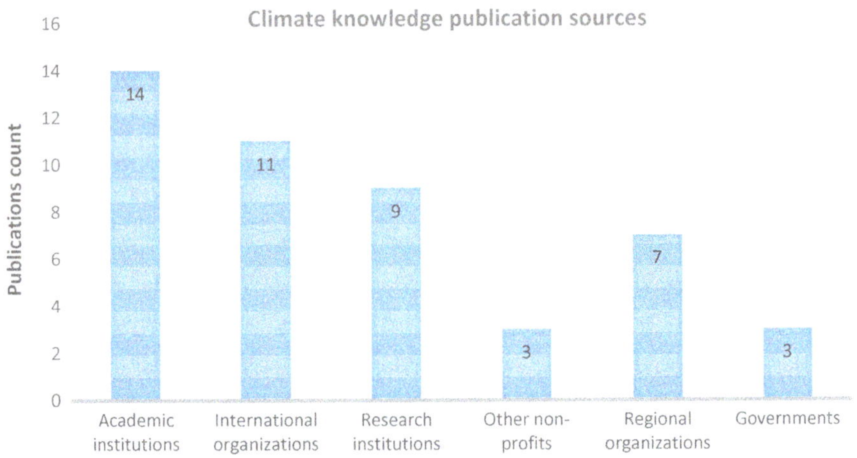


**Fig. 1** Climate change publication sources in Africa—2015–2016

**Table 1** Climate change publications in Africa from 2015–2016

| No. | Category | Institutions | Publications count |
|---|---|---|---|
| 1. | Academic institutions | • Brookings.edu<br>• Mozambique academy of sciences<br>• University of Johannesburg<br>• University of Cape Town<br>• University of Nairobi (UON)<br>• University of the Witwatersrand<br>• North West University<br>• Hamburg University<br>• London School of Economics (LSE), Tufts<br>• Cornell University<br>• Massachusetts Institute of Technology (MIT))<br>• University of Copenhagen<br>• Addis Ababa University | 14 |
| 2. | International organizations | • United Nations Environment Programme (UNEP)<br>• World Bank (WB)<br>• UN Framework Convention on Climate Change (UNFCCC)<br>• UN Climate Change: Learn (UN CC:Learn)<br>• The United Nations Office of the High Representative for the Least Developed Countries, Landlocked Developing Countries and Small Island Developing States (UN-OHRLLS)<br>• Food and Agricultural Organization (FAO)<br>• Inter-governmental Panel on Climate Change (IPCC) | 9 |

(continued)

**Table 1** (continued)

| No. | Category | Institutions | Publications count |
|---|---|---|---|
| 3. | Research institutions | • Western Indian Ocean Marine Science Association (WIOMSA)<br>• European Centre for Development Policy Management Climate Analytics<br>• Researchgate<br>• International Network for Bamboo and Rattan<br>• Procedia Engineering<br>• Council of Scientific and Industrial Research (CSIR)<br>• International Union for Conservation of Nature (IUCN)<br>• National Renewable Energy Laboratory (NREL)<br>• Oversees Development Institute (ODI) | 9 |
| 4. | Other non-profits | • Oxfam and Pan-Africa Climate Justice (PACJA)<br>• Adaptation of African Agriculture (AAA) | 3 |
| 5. | Regional organizations | • UN Economic Commission on Africa (ECA)<br>• Africa Development Bank (AfDB)<br>• African Wildlife Foundation (AWF)<br>• New Partnership for African Development (NEPAD)<br>• Future Climate for Africa (FCFA) | 7 |
| 6. | Governments | • US embassy in Ethiopia<br>• Swiss Agency for Development and Cooperation<br>• Gesellschaft für Internationale Zusammenarbeit (GIZ) | 3 |

was uploaded or updated during the last two years. The search criteria used is "climate change development africa filetype:pdf".

The results of this published literature review are tabulated below (Fig. 1 and Table 1).

# 6 Discussion

## *6.1 Global Climate Framework*

Both the UNFCCC and the Paris Agreement emphasize learning, management of climate knowledge and communications. Article 6 of the UNFCCC addresses the need for collaboration in fostering "education, training and public awareness". The Paris agreement on the other hand extensively calls for sharing of climate knowledge, lessons learnt and best practices. Article 7 outlines measures to tackle climate change adaptation and goes a step further to underscore one of the enablers as:

> Sharing information, good practices, experiences and lessons learned, including, as appropriate, as these relate to science, planning, policies and implementation in relation to adaptation actions;

Article 7 goes on to state that "Capacity-building should be guided by lessons learned, including those from capacity-building activities under the Convention". Article 13 calls for a transparency framework to ease the tracking of progress in implementing the NDCs, and specifically cites the role that "good practices" can play.

On collaboration between countries, Article 7 urges Parties to "Strengthen scientific knowledge on climate, including research, systematic observation of the climate system and early warning systems, in a manner that informs climate services and supports decision-making;", in addition while encouraging adaptation action, the Agreement recognizes the role of indigenous climate knowledge and stipulates that adaptation:

> ...should be based on and guided by the best available science and, as appropriate, traditional knowledge, knowledge of indigenous peoples and local knowledge systems, with a view to integrating adaptation into relevant socioeconomic and environmental policies and actions, where appropriate.

The foregoing thus clearly points to an Agreement whose implementation will require to be supported by robust climate knowledge management and communication strategies and partnerships.

Article 6 of the climate change convention (UNFCCC) on the other hand addresses the need for "education, training and public awareness". It calls on the convention parties to ensure public access to information on climate change and its impacts, and the development of education and public awareness material (UNFCCC, United Nations Framework Convention on Climate Change 1992).

## *6.2 Knowledge Sharing State of Play in Africa*

The case studies presented are evidence that there is no shortage of interventions and initiatives operating in the climate change landscape in Africa. Many of these outfits generate climate data, information and knowledge. In many cases however, knowledge generated by these players is not easily accessible by communities that need it.

A response action for adaptation entails building early warning systems to facilitate mitigation actions prior to climate related disasters. This is why sharing of data from climate observation systems is critical. Although the private sector is beginning to get involved in climate data services, currently the NHMSs at national level and RCCs at sub-regional level are the main players with the mandate to provide solutions in this regard.

For progress to be made on the Paris Agreement in Africa, governments will have to play the central role in terms of ensuring NDCs are not only implemented, but data is collected and made available to all stakeholders in the spirit of the transparency framework stipulated in Article 13 of the Paris Agreement.

In the section on "Knowledge Sources" in this paper, the desk study undertaken exposes the lack of accessible publications on climate change by governments in Africa. If indeed governments through relevant ministries will underpin the success of the Paris Agreement, significant efforts will need to be instituted to ramp up the capacity, skills and resources available to government and sub-regional institutions for knowledge generation, packaging and dissemination to ensure the institutions take their place in the community of climate knowledge purveyors in the continent.

Climate information has increasingly become important and available in the last decade and regional climate outlook forums have enhanced dialogue on seasonal forecasts among producers of information, researchers and different categories of decision-makers (Goddard et al. 2010).

In Africa, the public sector has been slow of the blocks in adopting information technology. This has dogged the capacity to package and share electronic information. It has been shown that basic ICT skills are no longer enough and that public servants will have to be equipped with more advanced ICT skills to function effectively in the public service (Jaarsveldt 2010).

The adoption of e-government has however gathered momentum across the continent. E-government will stimulate the use of the Internet in the public sector, thus creating opportunities for strengthening the packaging and dissemination of climate information and knowledge products.

It is common for public servants to hold workshops and seminars where participants attend in person. In many cases however, outcomes of the events are either not recorded for learning and sharing, or if captured end up in drawers or in personal computers where the information is not accessible. More conscious efforts therefore will be needed to capture, codify, and share both implicit and explicit knowledge.

Despite multiple players generating a wide range of climate change knowledge outputs, the challenge that most users face is difficulty accessing information that is tailored to their needs, in a timely manner, a role that is within the core mandate of RCCs and NMHSs, but that can be optimized through well designed knowledge services.

## 6.3 *Climate Knowledge Partnerships*

In the quest to ensure success of the NDCs and the Paris Agreement implementation in Africa, knowledge partnerships among the key players at all levels will catalyze increased efficiency in climate change interventions being undertaken.

In the past, climate change issues were seen by governments as a climate science community issue. Indeed, in many African countries, the NHMSs had responsibility for only providing weather data for aviation purposes and to a lesser extent undertaking weather forecasting. Gradually, the role of NHMSs has evolved to the provision of seasonal forecasts as well as climate modelling. Products emanating from the latter are increasingly finding a ready community of climate vulnerable users that include farmers, pastoralists, fishermen and planners.

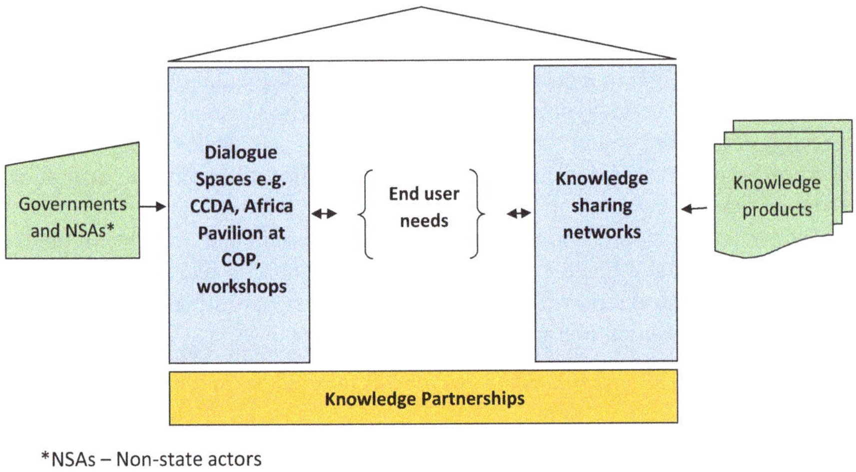


**Fig. 2** Climate knowledge pillars

As shown on Fig. 2, the objectives of climate change learning through transparent implementation, sharing of lessons, opportunities and challenges are best served by strong partnerships, with the involvement of end-users in co-creation and co-generation of knowledge. The key pillars of climate knowledge that are served by robust partnerships include dialogue spaces and knowledge sharing networks.

Because climate change impacts afflict multiple sectors and communities, governments have a key role to play in ensuring enduring and effective solutions are instituted. The chart in Fig. 3 enumerates the players that have the most vital responsibilities in combating climate change. Clearly, the importance of government is underscored by the prevalence of public institutions that must be engaged. This is primarily because responding to climate change is not just one task among others, which can be left to a specialized department or agency: it has to be integrated into the activities of government as a whole across the board (Giddens 2008). National governments have a specific role in establishing the policy and regulatory environment to encourage adaptation by individuals, households and private sector businesses. They can strengthen the knowledge base of climate risk assessments; strengthen the early warning chain of climate change trends, seasonal forecasts and weather alerts, from satellites to national radio stations to local rattles and megaphones and cell phones (IFRC 2009).

Looking at climate knowledge generation as a pipeline or conveyer belt, roles at each stage must be active and effective for constant co-production of knowledge, tailoring, packaging and dissemination. For the Paris Agreement, governments must lead the charge in sharing information and stimulating other partners to come onboard towards creating an environment where all players' intervention iterations continuously become more efficient through monitoring, evaluations and learning (Fig. 3).

**Fig. 3** Climate knowledge partnerships

A category of government affiliated institutions that generate indispensable data and information for adaptation are NHMSs and RCCs. Acting nationally or sub-regionally, these institutions often have the sole responsibility to device mechanisms for provision of weather and climate data on a regular basis. If availed in a timely manner, the data produced has the potential to save lives through enabling communities to take action prior to climate disasters like droughts and floods. Data and information sharing partnerships have been hampered by lack of digitized data, poor internet connectivity, paucity of technical skills and lack of collaboration frameworks.

### 6.4 Open Government

As alluded to earlier in this paper, having adopted the NDCs mechanism as part of the Paris Agreement, governments made legally binding commitments to take action to limit global warming to 2°. In order to have "all hands on deck" by all players and to promote collaboration and effectiveness, the Paris Agreement Article 13 calls for transparency and openness. Traditionally however, governments lean towards withholding rather than availing information. This is one reason why Principle 10 of the Rio Declaration of 1992 underscored that "*Environmental issues are best handled with the participation of all concerned citizens...*"

For meaningful and effective public participation in environmental decision making, governments need to make information available. Participation of the public enhances acceptance and ownership of decisions, improves implementation and sustainability of environmental decisions (UNEP 2012).

According to Wikipedia, governments that enable public viewing of data can help citizens engage within the governmental sectors and "add value to that data" (Robinson 2009). Besides allowing data access to citizens, African inter-governmental climate data sharing can facilitate building of synergies and provide opportunities for strengthening sub-regional responses.

### 6.5 Innovative Information and Knowledge Sharing Approaches

***Knowledge brokers' network***—To mitigate the paucity of readily accessible climate knowledge, establishing a coalition of partners involved in the generation and use of climate knowledge in the continent should be part of the arsenal to combat climate change. These partners include the RCCs, meteorological agencies, government sectors and institutions involved in pan-Africa climate interventions.

Efforts should be made to create a knowledge network that is inclusive and participatory, thus efforts should also be made to bring on board indigenous knowledge practitioners. According to (Jiri, Mafongoya, Mubaya and Mafongoya 2016) given the significant gaps in scientific knowledge, ethno-meteorological knowledge plays a key role in farmers' ability to devise climate variability and change adaptation measures.

***Learning Modules Platforms***—To save on curriculum and climate knowledge delivery costs, increased investments in online learning platforms and related skills will facilitate climate knowledge outreach among climate knowledge beneficiary communities that have access to the internet. In recognition of the large number of

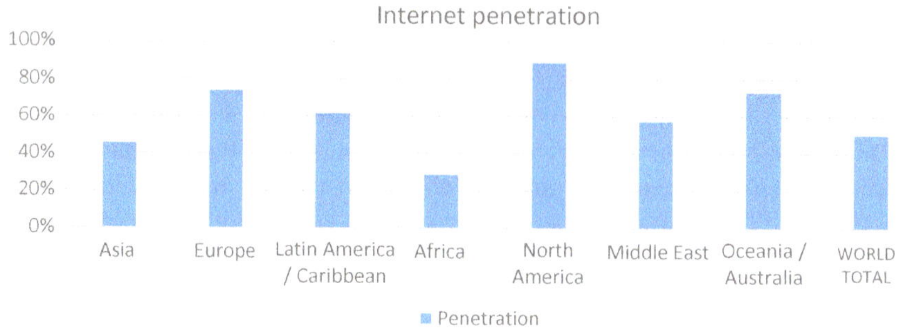


**Fig. 4** Internet penetration by region (data from internet world stats (Internetworldstats 2016)

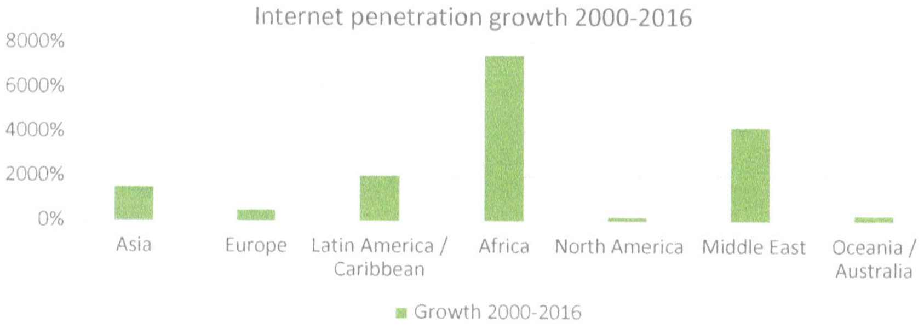


**Fig. 5** Internet penetration growth (data from internet world stats (Internetworldstats 2016))

users whose only mode of access to the internet is mobile devices and considering that broadband is rare, such modules should work well in low bandwidth conditions, thus should minimize use of video, audio, pictures and heavy graphics.

The UN CC:Learn is a pioneer in developing curriculum and online climate change learning modules, and is engaged in 7 Africa countries where it has championed multi-stakeholder efforts to formulate climate change learning strategies.

***Improved Outreach through Webinars and E-Workshops***[1]—To diversify climate knowledge outreach and delivery tools, and to save on costs needed for in person event participation, Africa can leverage dissemination and engagement through webinars and e-workshops respectively. Currently, the use of these knowledge delivery mechanisms is seldom used for climate change events, a state which is likely caused by poor internet penetration and inadequate technical skills. According to the Internet Society, while Africa still lags the rest of the world in

[1]eWorkshops are collaborative and facilitated problem-based online learning events that are run over a number of weeks and typically require 4–6 h commitment per week. Most (if not all) of the work is done asynchronously. eWorkshops are very different to webinars or virtual classrooms, which are run 'real time' (everyone logs in at the same time) (Janssens-Bevernage 2013).

Internet penetration, it is bridging the gap very quickly. For example, in 2005, Internet penetration in Europe was 20 times greater than that of Africa, but by 2014, it was only 4 times greater (Nyirenda-Jere and Biru 2015). Increased use of webinars and e-workshops will enable cost effective outreach and brainstorming on new ideas among the increasingly online user communities (Figs. 4 and 5).

***Last mile climate knowledge dissemination***—At the grassroots level, communities that bear the brunt of climate change impacts are often uninformed on what climate change is, its impacts and what options they have to mitigate its effects.

In this activity therefore, simple to understand flyers can be produced on the "ABCs" of climate change, customized for farmers, fishermen, pastoralists and translated to the most widespread national languages. The leaflets can be distributed through known vulnerable groups, including youth, women organizations, and farming and fisheries cooperatives. Recipients of the information leaflets can have option to send feedback to improve the information.

***Mapping of Climate Change Institutions and Experts***—There are multiple institutions undertaking climate change interventions in Africa today. However, information on availability of expertise is not readily available. To support and enable expediting of implementation of activities, it is prudent to establish a shared, searchable and easy to access compendium of expertise and institutions and their comparative advantage, as an information resource for reference by planners and implementers. The compendium can include a brief narrative of what each institution is doing, its geographic coverage, investment level and partners. Its compilation can be done through a desk study as well as consultation with institutions operating in the climate change space. It would thereafter need to be constantly updated after initial launch.

During its programme implementation and engagements, ClimDev-Africa has built a database of climate change experts in Africa consisting of over 4000 contacts. The targeted metadata for an experts' compendium would include, expert name, organization, area of expertise and a short bio.

## 7 Conclusion

Governments, through development sectors and meteorological agencies, are the main players in the Paris Agreement implementation. In Africa, as seen in the analysis of the most prominent partnerships visible in climate change knowledge activities and interventions space, the public sector has not been sufficiently engaged. Out of 47 climate knowledge publications on Africa in the last 2 years as listed on Google search engine, only a small number of the publications can be directly linked to the public sector.

Article 13 of the Paris Agreement, which emphasis transparency and fine tuning of implementation through capture and sharing of lessons, should be spearheaded by governments as the leading players. As of January 2017, 29 countries in Africa had ratified the Agreement, thus placing on them a legally binding obligation to comply with the Agreement in line with the INDCs they submitted. However, the public sector cannot single handedly satisfy the knowledge needs of the Paris agreement. Partnership with institutions involved in various aspects of climate change will enable building on comparative advantages of every player and avail opportunities for pooling climate knowledge and channeling tailored knowledge products to communities.

With rising Internet coverage, bandwidth and speeds in Africa, there are opportunities to leverage contemporary tools like websites, knowledge platforms, social media, webinars and online learning for increased collaboration, sharing of lessons and transparency by governments, all of which will contribute to the success of the Paris Agreement implementation.

Further studies that can enrich the findings in this paper include research among the key climate change institutions covered in the discussion, to assess levels of hidden knowledge and how to channel it, as well as practices that inhibit the organization, tailoring, packaging and disseminating of climate knowledge.

## References


AREI (2015) Africa renewable energy initiative (AREI). Retrieved from AREI: http://www.arei.org/

AREI (2016, August) Summary: Africa renewable energy initiative. Retrieved from AREI: http://www.arei.org/wp-content/uploads/2016/09/AREI-Summary-english_web.pdf

CARE-Africa (2017) Adaptation learning programme for Africa (ALP). Retrieved from CARE climate change information centre: http://careclimatechange.org/our-work/alp/

ClimDev-Africa (2015) ClimDev-Africa 2014 annual report. Addis Ababa: UNECA. Retrieved from http://repository.uneca.org/bitstream/handle/10855/22785/b11538314.pdf?sequence=1

ClimDev-Africa (2017) African climate policy center. Retrieved from ClimDev-Africa: http://www.climdev-africa.org/afrian-climate-policy-center

ClimDev-Afrrica (2014, November 7) Forum on implications of the IPCC fifth assessment report for Africa. Retrieved from ClimDev-Africa: http://www.climdev-africa.org/content/forum-implications-ipcc-fifth-assessment-report-africa

FCFA (2017) Future climate for Africa. Retrieved from FCFA: http://www.futureclimateafrica.org/

G20 (2015, 11 20) G20 Leaders' Communiqué. Retrieved from http://www.mofa.go.jp/files/000111117.pdf

Giddens A (2008) The politics of climate change. Retrieved from http://www.policy-network.net/uploadedFiles/Publications/Publications/The_politics_of_climate_change_Anthony_Giddens.pdf

Goddard L, Aitchellouche Y, Baethgen W, Dettinger M, Graham R, Hayman P, Kadi M, Martínez R, Meinke H (2010) Providing Seasonal-to-Interannual Climate Information for Risk Management and Decision-making. Procedia Environ Sci 1:81–101

ICPAC (2016, August) GHACOF44 announcement. Retrieved from ICPAC: http://www.icpac.net/wp-content/uploads/GHACOF-44-Announcement.pdf

IFRC (2009, May) Climate change adaptation strategies for local impact. Retrieved from UNFCCC: http://unfccc.int/resource/docs/2009/smsn/igo/054.pdf

Internetworldstats (2016) Internet usage statistics: the big picture. Retrieved from internetworldstats: http://www.internetworldstats.com/stats.htm

IRI (2017) Enhancing access to climate services. Retrieved from IRI: http://iri.columbia.edu/resources/enacts/

Jaarsveldt LV (2010, November 4) Information technology skills for South African public service. Retrieved from Cite Seer X: http://citeseerx.ist.psu.edu/viewdoc/download?doi=10.1.1.1005.8651&rep=rep1&type=pdf

Janssens-Bevernage A (2013, March 12) Why eWorkshops instead of webinars? Retrieved from Dynamind ELearning: http://dynamind-elearning.com/2013/03/why-eworkshops-instead-of-webinars/

Jiri O, Mafongoya PL, Mubaya C, Mafongoya O (2016, April 15) Seasonal climate prediction and adaptation using indigenous knowledge systems in agriculture systems in Southern Africa: a review. J Agric Sci, 160. Retrieved from https://www.researchgate.net/profile/Obert_Jiri/publication/301319794_Seasonal_Climate_Prediction_and_Adaptation_Using_Indigenous_Knowledge_Systems_in_Agriculture_Systems_in_Southern_Africa_A_Review/links/571489cb08ae39beb87cf663.pdf?origin = publication_li

Mwiti L (2015, 1 14) 17 shocking facts about state of science and research in Africa. Retrieved from Mail & Guardian Africa: http://mgafrica.com/article/2015-01-14-17-startling-facts-about-the-state-of-science-and-research-in-africa

NERC (2017) Future climate for Africa (FCFA). Retrieved from NERC: http://www.nerc.ac.uk/research/funded/programmes/fcfa/

Nyirenda-Jere T, Biru T (2015, May 22) Internet development and Internet governance in Africa. Retrieved from Internet Society: http://www.internetsociety.org/sites/default/files/Internet%20development%20and%20Internet%20governance%20in%20Africa.pdf

Robinson (2009) Government data and the invisible hand. Yale J Law Technol, 11: 160. Retrieved from SSRN: https://poseidon01.ssrn.com/delivery.php?ID = 410125074122121067107121096013106106032052042041037011126101084027097125025127089029007042096016033108114100086105103094099076116002063019007095122077122014079024039013045127080119092114088095077006004027098090 1

TAHMO (2017) Trans-Africa hydro meteorological observatory. Retrieved from TAHMO: http://tahmo.org/wp-content/uploads/2014/12/TAHMO_general.pdf

UNDP (2017) Climate information for resilient development and adaptation to climate change in Africa (CIRDA). Retrieved from UNDP: http://undp-cirda.blogspot.com/p/about.html

UNEP (2012) Guidelines for the development of national legislation on access to information, public participation and access to justice in environmental matters. UNEP, Bali

UNFCCC (1992) United Nations framework convention on climate change. United Nations Framework Convention on Climate Change, United Nations

UNFCCC (2015, December 12). Paris agreement. Paris, France

UNISDR (2014, May 16) Climate threat to Africa's resilience. Retrieved from UNISDR: https://www.unisdr.org/archive/37509

UNISDR (2015, 01 01) The human cost of weather related disasters. Retrieved from UNISDR: https://www.unisdr.org/2015/docs/climatechange/COP21_WeatherDisastersReport_2015_FINAL.pdf

WMO (2017, January) Data rescue projects and initiatives. Retrieved from WMO: http://www.wmo.int/pages/prog/wcp/wcdmp/CDM_2.php

# Formulation of an Ethics of Response to Climate Change: The Need for Effective Communication in Higher Education

**P. Castro, A. M. Azul, W. Leal Filho and U. M. Azeiteiro**

## 1 Introduction

Climate change is a key priority area of the *Europe 2020 Strategy*, the EU's strategy for smart, sustainable and inclusive growth, and of the UN Framework Convention on Climate Change (UNFCCC). The handling of climate change is also one of the Sustainable Development Goals (*SGD 13: Take urgent action to combat climate change and its impacts*). Countries around the world are facing climate change impacts and threats to human welfare, biodiversity and ecosystem integrity (IPCLIMATE CHANGE 2014; UN 2015).

Climate change entails the identification of mutually influential relationships between nature, human wellbeing, culture, education and science (Alves et al. 2014; Esteves de Freitas et al. 2016; Lutz and Muttarak 2017; Viegas et al. 2014). The risk perception (Sundblad et al. 2009) and education (Azeiteiro and Leal Filho 2017) can be an important pre-requisites to facilitate behaviour change.

Higher Education Institutions (HEIs) have a critical role in informing and training students (Morgado et al. 2017; Santos et al. 2016; Wachholz et al. 2014; Leal Filho 2010), but also in actively engaging educational institutions and agents (teachers, students, researchers, and other pedagogical agents from formal and

P. Castro (✉)
Centre for Functional Ecology, University of Coimbra, Coimbra, Portugal
e-mail: pcastro@ci.uc.pt

A. M. Azul
Edf Faculdade de Medicina, Center for Neuroscience and Cell Biology,
University of Coimbra, Rua Larga, Polo I, 1º Piso, 3004-504 Coimbra, Portugal

W. Leal Filho
School of Science and the Environment, Manchester Metropolitan University,
Chester Street, M1 5GD Manchester, UK

U. M. Azeiteiro
Departamento de Biologia & CESAM, Universidade de Aveiro, Aveiro, Portugal



W. Leal Filho et al. (eds.), *Handbook of Climate Change Communication: Vol. 1*,
Climate Change Management, https://doi.org/10.1007/978-3-319-69838-0_18

informal education) for local and regional solutions (Azeiteiro and Leal Filho 2017; Morgado et al. 2017; Santos et al. 2016). However, it is not clear the relations between knowledge and attitudes to effectively address adaptation to climate change conditions (Morgado et al. 2017; Santos et al. 2016). Previous studies conducted HEIs in Portugal suggest that there is significant lack of information about the environmental and health impacts of everyday actions, and thus the capacity in making better choices on climate change adaptation options (e.g. Morgado et al. 2017; Santos et al. 2016; Carvalho et al. 2014). Universities are key critical partners for Climate Change adaptation and mitigation (Basanta 2016; Leal Filho 2010). However, the full potential of universities for climate change adaption has not been fully exploited (Basanta 2016; Yoko and Audrey 2015). Curriculum, namely adequate curricula (Climate Education and Literacy) (Azeiteiro and Leal Filho 2017), providing training (Morgado et al. 2017; Santos et al. 2016), Research (for mitigation technologies and adaptation strategies), engaging in Low Carbon Economy Politics, promoting dissemination and valuing communication strategies and effective measures are all urgently needed.

The formulation of an ethics of response to climate change will be a key determinant of the immediate and future wellbeing of those affected by global climate change (COMEST 2010). Ethical values are implemented into practice (individual or group behaviour, corporate or public policies) through principles and rules that together with values, constitute an important part of ethics. The relevance of the topic is shown by the UNESCO intention in producing the "Declaration on Ethical Principles in Relation to Climate Change" which aims to be *a non-binding declaration to promote the moral basis for tackling the threat posed by climate change to people and the planet and to identify and clarify a set of ethical principles to underpin the moral case for the international community to take action to address climate change*. These are issues that need to be addressed by the HEIs so that curricula programs can be adequate to promote better knowledge and attitudes about climate change and an active engagement of future participative professionals and citizens, in order to cope with the environmental crises and climate change challenges and to be part of the solution for climate change problems.

HEIs must be able and pioneering supporting the long-term implementation of global and local climate change policies for HEIs and capacitating professional with the necessary multidisciplinary and intersectoral competencies (Morgado et al. 2017) and participate in the local community (building resilience and enabling local communities to access credible and feasible technologies to help mitigate the impacts of climate change) (Setti et al. 2016; Viegas et al. 2016; Yoko and Audrey 2015). Curriculum development to produce graduates versed with various aspects of climate change adaptation and mitigation needs better definition of strategies. The purpose of this study was to conduct a general assessment of the HEI students' perceptions of graduate and master students on climate change and, how the information/knowledge acquired during their higher education contributed to the formulation of ethical responses on this matter.

## 2 Research Methodology

### 2.1 *Surveyed Institution and Courses*

Research was conducted at the universities of Aveiro and Coimbra in Portugal. The invitations to participate were sent to students of these two Universities. The survey was applied to 1st and 2nd cycle students from Biological and Environmental Sciences.

### 2.2 *Sample*

The respondents were recruited online through an e-mail invitation. The invitation to participate in the study included a short presentation of the study, its aims, responsible researchers and universities, so as to assure an informed consent from participants, followed by a link to the questionnaire itself.

### 2.3 *Questionnaire*

The questionnaire was composed of 21 closed-ended questions and two open questions aiming mainly at characterising the academic provenience of students and ascertain their perceptions, motivations, attitudes, and knowledge relating to climate change as well as on how the University as contributed to their ethical of responses. Although all the presented questions were somewhat transversal to the variables intended to be analysed in the study, seven of them were more targeted to explore their perception, two addressed more specifically ethical concerns and responses, four related with the topic of communication. Their academic education and involvement in the academia and how it contributed to their knowledge and construction of ethical responses were more specifically addressed in five of the questions. The remaining questions contributed to the characterization of the respondents.

The questionnaire was pre-tested among students at the Faculty of Sciences and Technology of the University of Coimbra, not enrolled in its application, and the results were used to refine it. The time required for filling in the questionnaire was about 10 min. The questionnaire was originally adapted from Wachholz et al. (2014) and Manolas et al. (2010) and was already applied by Santos et al. (2016) and Morgado et al. (2017). Google Drive was used to host the questionnaire and allow for the online survey. Students were assured that their responses remained anonymous and took part in the survey willingly.

## 3 Findings

The questionnaire was answered by 32 students, mostly from the Universities of Aveiro (28%) and Coimbra (72%). This survey also received two responses from students studying in the Porto University. Respondents were between 18 and 52 years old, being 72% aged between 20 and 24. The questionnaire was mainly aimed at evaluating the masters' students, but in the end, the respondents were divided equally between masters and students in the course of obtaining their graduation. A common feature was the nature of their graduation or masters' courses, all environmental-related (e.g. master in biodiversity and biotechnology or degree in biology).

All surveyed were interested or very interested about the topic of climate change and from these, only 3% consider it not to be a problem. The majority of respondents (81%), however, perceived the climate change as a serious problem.

Most of the surveyed students stated that climate change is serious problem in Europe and Worldwide (Fig. 1), but at finer spatial scales the respondents seem to change their choice to a "*moderate problem*" or even to "*not a problem*" day-to-day life (Fig. 1).

When asked about the actors responsible for the resolution of climate change, there was a consensus that all identified actors (*International organizations, Governments, Regional governance, Educational institutions, Economic sector, Consumers/individuals, Me*) are "very responsible" in this matter. However, there was a clear indication that the economic sector (72%) and governments (69%) as being "completely responsible" for solving the problems associated with climate change. When considering consumers/individuals and the himself/herself ("me") it was observed that 6% and 9%, respectively, confirmed they had no responsibility at all.

Students were inquired in on how often they adopt specific actions to reduce the emission of greenhouse gases (GHG). Those that were mostly put into practice ("often") were *walk or bike to go to the university* (69%) followed by the

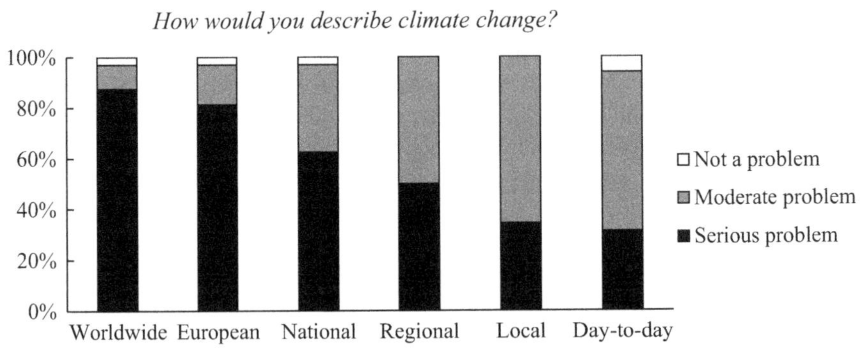


**Fig. 1** How respondents perceive the severity of climate change at different geographical scales

*consumption of meals with more vegetables and less meat* (44%) and the *use public transportation to go to the university* (34%). Surprisingly, only 44% stat they "sometimes" *recycle* and 31% and 25% "rarely" or "never" do that. Students "sometimes" *shop using their own bags* (88%), *buy products without packaging* (50%), *avoid food waste* (69%) and *do not use equipment for heating or cooling* (e.g. air conditioning) (44%). Nearly 25% of respondents "never" *participated in ecological activities with low GHG emissions at the university/community* or "never" opted to *avoid the consumption of meet.*

When asked to select the obstacles that prevent them to adopt lifestyles associated with low GHG emissions, their answers were distributed by the various options available. "Sometimes" and "often" were the main options selected, which together corresponded to: 78%—*requires an additional commitment* and *lack of options* (e.g. eco-facilities, vegetarian restaurants or offer of organic products); 75%—*it is difficult to change lifestyles* (old habits) and 63%—*lack of knowledge and ability to adopt such lifestyles* or *lack of interest.*

The degree of agreement of students with the lines of action listed in Table 1, supposed to mitigate or adapt to climate change was assessed. The majority of students "disagree" "tend to disagree" with the statement *each person, individually, little can do to mitigation and adaptation*, and "tend to agree" or "agree" with all the others (Table 1). They perceive that the level of awareness regarding to the need of finding mitigation/adaptation plans to climate change has risen and tend to agree that *the private sector and policy makers increasingly think on mitigation/adaptation plans*. Still, 79% of them consider that all actions that will be implemented will come too late to solve any problem (Table 1).

When inquired about the ethical characteristics of Human behaviour and its relationship with Nature (Fig. 2), the majority of students not only disagree or tend not to agree that Humans should rule over Nature, but it falls on him the moral duty

**Table 1** Answers of respondents (in %) to the question: how do you analyse the contribution of the following lines of action for mitigation/adaptation to climate change?

| Lines of action for mitigation/adaptation | Disagree | Tend to disagree | Tend to agree | Agree |
|---|---|---|---|---|
| Each person, individually, little can do to mitigation and adaptation | 44 | 28 | 22 | 6 |
| Each person, individually, contributes significantly to mitigation and adaptation | 0 | 25 | 22 | 53 |
| I can influence others to adopt lifestyles associated with low carbon | 0 | 9 | 47 | 44 |
| Technology will have the main role to contribute to mitigation/adaptation | 0 | 19 | 59 | 22 |
| The private sector and policy makers increasingly think on mitigation/adaptation plans | 3 | 41 | 53 | 3 |
| The awareness of need to the mitigation and adaptation increased significantly | 3 | 9 | 66 | 22 |
| All actions will come too late | 0 | 22 | 63 | 16 |

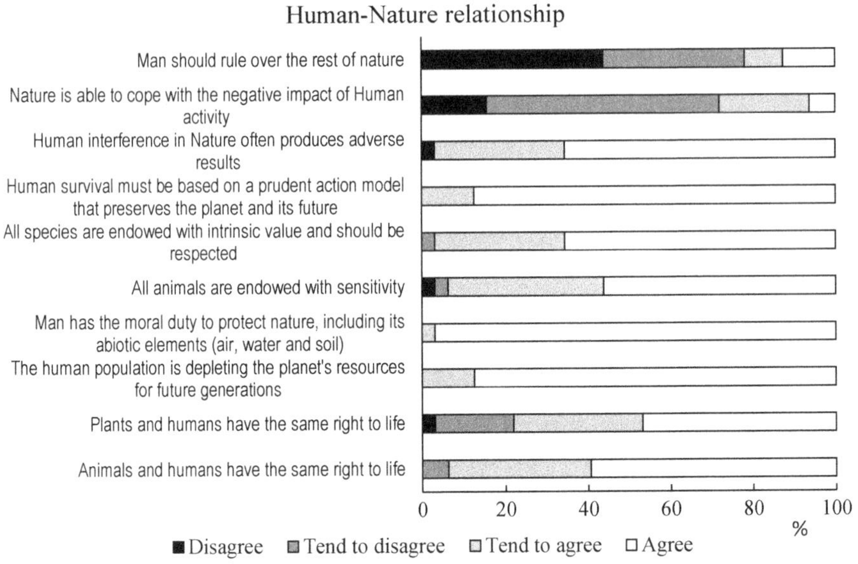


**Fig. 2** Ethical aspects that reflect the Human relationship with Nature

to protect nature, including its abiotic elements (air, water and soil) (97%) and Human survival must be based on a prudent action model that preserves the planet and its future, since Nature doesn't have the to cope with the impacts caused by human practices. The majority of students recognize that animals are endowed with sensitivity and all species are endowed with intrinsic value (Fig. 2).

Regarding the ethical challenges related to the main characteristics of global climate, students seem to attribute high importance (right to life) to other species as well as to human quality of life and health (Fig. 3). Our actions should obey to the principles of responsibility, precaution, polluter pays, safeguarding the interests of future generations (Fig. 3). It is noteworthy that this community share these ethical values/principles and consider "very important" (84%) to *share scientific knowledge and progress and its benefits* (e.g. between more developed and less developed countries).

The majority of students considered to have moderate (84%) or deep (13%) knowledge about the topic, but only 16% of them state not to need more information to form an opinion with regard to climate change. The academy, personal research, internet and the attendance in higher education are the main sources of information, representing each, 72%, 59%, 56% and 41%, respectively.

On climate change and its ethical implications 66% of the respondents agree that their *university develops initiatives on climate change* and it that *contributed significantly to learn more about the subject*, but only 41% *consider sufficient the initiatives developed by my university*. Almost half of the students participate in these initiatives.

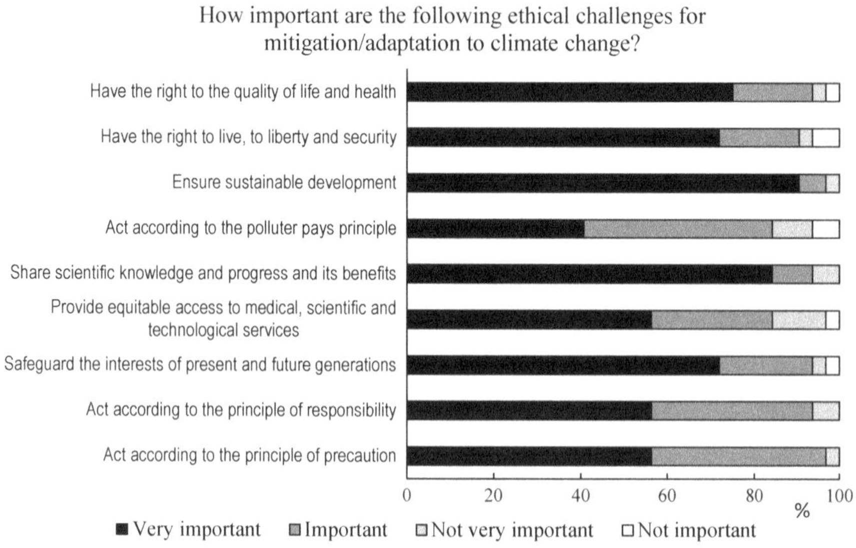


**Fig. 3** Evaluation of ethical challenges posed to Man's actions

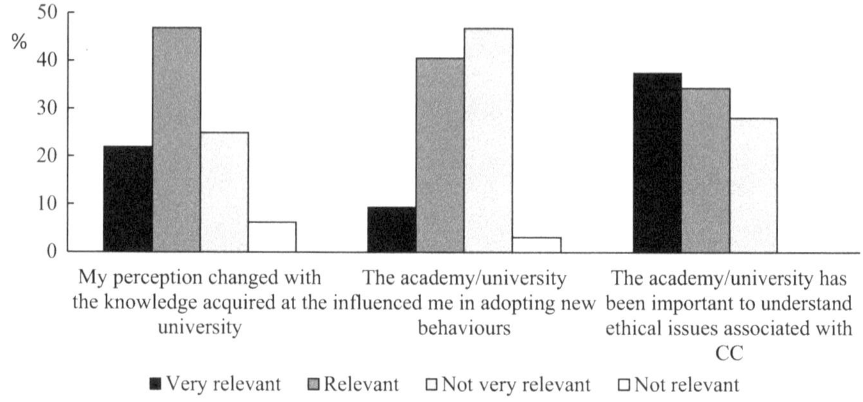


**Fig. 4** Relevance of the academy/university to the understanding of climate change

Considering the area of knowledge of the surveyed students it was expected that almost all (97%) have covered climate change in their course, but only 56% state it has been sufficiently addressed.

When assessing to what extent their higher education had been relevant to understand climate change, 44% considered to be "relevant" the knowledge acquired at the university contributed to change their perception about climate change (Fig. 4), while 25% stated it was "not very relevant". The academy/university life seem to contribute to students' understanding about ethical issues associated with climate change, but a large percentage assert it was "not very relevant" to influence them in adopting new behaviours (Fig. 4).

Students were then asked to give their opinion on what should be implemented in their course or university to improve the knowledge on climate change and its ethical implications. What stands out from all the answers, is that more is necessary. Several examples were given, like organizing discussions, show of documentaries, more communication about experiments/researches, more actions/initiatives to raise awareness, or more meetings must be implemented on the subject. As an example, one student stated *if the university shows that it is interested in reducing environmental impact and actually taking action for it, the interest of the students will certainly increase*. In his/her opinion the university has to be an example to follow. Other suggestions included to be mandatory the subject of climate change and ethical implications in one or more curricular units, the initiatives developed by the university must be better disseminated and the student association should be more involved to promote the dissemination of activities and documentaries that enhance students' knowledge.

As written by two other students:

> Greater emphasis should be placed on the issue of climate change, especially on ethical and social implications, which are less addressed when compared to the biological consequences.
>
> The approach to the subject should, in my opinion, convey the idea that Man is no more than other animals and, therefore, an effort must be made to get everyone to think and act beyond "recycling". We must discover new ways of being human.

## 4 Discussion

Higher education has a very important role to play in training and educating future generations, promoting active citizenship in order to cope with the environmental crisis and the climate change challenges that we face nowadays. Studies assessing higher education students' knowledge and attitudes about climate change and the contribution of their graduate and postgraduate courses to the deepening of this knowledge and to the change of attitudes and behaviours should support the need for change and innovation in climate change education and climate literacy.

The results obtained from this research are aligned with other previous studies (e.g. Santos et al. 2016; Morgado et al. 2017; Keinonen et al. 2017) where surveyed students perceived climate change as a serious problem. Nevertheless, climate change is perceived as a more serious problem outside their local and regional reality. The responsibility seems to be "of others" when addressing measures to cope, mitigate and adapt to this global environmental problem and are not prepared to take individual actions to mitigate it. University education in biological and environmental sciences seems to be reflected in more concrete knowledge as already stated by Morgado et al. (2017) and Santos et al. (2016). Students are familiarized with the climate change phenomenon but this knowledge does not translate necessary into concrete practices and behaviours that mitigate climate change (which is interesting since 97% have covered climate change in their course, but only 56% state it has been sufficiently addressed). It is not always clear the correlation between levels of knowledge and behavioural changes.

When inquired about ethical issues students reveal ethical concerns, can describe ethical issues and even understand what it means to act ethically in this climate change context. 84% of the respondents are aware of sharing science knowledge, progresses between more developed and less developed countries. As revealed by precedent studies (Morgado et al. 2017; Santos et al. 2016; Wachholz et al. 2014) students hold misconceptions about the basic causes and consequences of climate change and most students recognized human responsibility in climate change but demonstrated gaps in their understanding of the phenomenon as well as in the need to adopt behaviours to tackle climate change however an ethical dimension is revealed by this research.

The originality of this research from the previous ones (Santos et al. 2016; Morgado et al. 2017) is that academy/university life seem to contribute to students' understanding about ethical issues associated with climate change, but a large percentage assert it was "not very relevant" to influence them in adopting new behaviours. Clearly positive is that a large number of respondents give their opinion on what should be implemented in their courses or university to improve the knowledge on climate change and its ethical implications namely organizing discussions, show of documentaries, more communication about experiments/researches, more actions/initiatives to raise awareness, or more meetings must be implemented on the subject. Other suggestions included to be mandatory the subject of climate change and ethical implications in one or more curricular units and the initiatives developed by the university must be better disseminated (and the student association should be more involved to promote the dissemination of activities and documentaries that enhance students' knowledge). There is a clear need for better integration, curriculum innovation and change (see Azeiteiro and Leal Filho 2017), participation and communication for the deepening of this knowledge and to the change of attitudes and behaviours.

## 5 Conclusions

As stated by UNESCO *climate change is fundamentally an ethical issue*. If wrong or inadequate responses continue to be provided, disastrous consequences will then continue to occur. It is thus imperative that the ethical dialogue about climate change be extensively promoted. The University represents an entity of excellence in this communication, sharing of information and in the people's formulation of ethical responses.

This study has a limitation in the sense that it was restricted to a small sample of students at two higher education institutions, but the data gathered offer a rough profile as to how climate issues are being handled and perceived at the surveyed universities. The findings suggest that even though the levels of awareness and understanding of students in relation to climate change is rather positive, there are still some information needs to be addressed, especially in respect of the means to translate information and awareness to real action.

Future studies are thus needed, which further explore the links between climate change variability and responses, possibly complemented by studies which may assess earlier effects of curriculum-based climate change information on the development of a stronger awareness to students on the one hand, and a greater commitment to action on the other.

## References


Alves F, Caeiro S, Azeiteiro UM (eds) (2014) Lay rationalities of climate change. Int J Clim Change Strateg Manag 6:1756–8692. http://dx.doi.org/10.1108/IJCCSM-10-2013-0121

Anabela C, Luísa S, Filipe DS, Ana D (2014) Climate change research and policy in Portugal. Wiley Interdisciplinary Reviews: Climate Change 5(2):199–217

Azeiteiro UM, Leal Filho W (2017) Climate literacy and innovations in climate change education (editorial material) Int J Glob Warm

Basanta P (2016) The role of higher education in awareness of climate changes: a case study of north-eastern states. Int J Curr Res 8:2864–2867

Carvalho A, Schmidt L, Santos FD, Delicado A (2014) Climate change research and policy in Portugal. Wiley Interdisciplinary Reviews: Climate Change 5:199–217. http://dx.doi.org/10.1002/wcc.258

COMEST (2010) The ethical implications of climate change. UNESCO, Paris

Esteves de Freitas L, Sato AM, Schottz S, Netto ALC, Lacerda N (2016) Community, university and government interactions for disaster reduction in the mountainous region of Rio de Janeiro, Southeast of Brazil. In: Leal Filho W, Azeiteiro UM, Alves F (eds) (2016) Climate change and health: improving resilience and reducing risks, volume produced as part of the Climate Change Management Series. Springer, New York, pp 313–328. ISBN 978-3-319-24660-4; ISBN 978-3-319-24658-1

IPCLIMATE CHANGE climate change 2014: impacts, adaptation, and vulnerability. In: Field, CB et al. (eds) Cambridge University Press

Keinonen T, Palmberg I, Kukkonen J, Yli-Panula E, Persson C, Vilkonis R (2017) Higher education studentsí perceptions of environmental issues and media coverage. Discourse Commun for Sustain Educ 7(1):5–22, 2016

Leal Filho W (ed) (2010). Universities and climate change—introducing climate change at University programmes. ISBN: 978-3-642-10750-4. Springer, Berlin

Lutz W, Muttarak R (2017) Forecasting societies' adaptive capacities through a demographic metabolism model. Nat Clim Change 7:177–184. https://doi.org/10.1038/nclimate3222

Manolas EI, Tampakis SA, Karanikola PP (2010) Climate change: the views of forestry students in a Greek university. Int J Environ Stud 67:599–609. https://doi.org/10.1080/00207233.2010.499208

Morgado F, Bacelar-Nicolau P, Rendon-Von Osten J, Santos P, Bacelar-Nicolau L, Farooq H, Alves F, Soares A, Azeiteiro UM (2017). Assessing university student perceptions and comprehension of climate change (Portugal, Mexico and Mozambique). Int J Clim Change Strateg Manag 9(3):—IF: 0.708 20154/16 Q2. http://dx.doi.org/10.1108/IJCLIMATECHANGESM-08-2016-0123

Santos P, Bacelar-Nicolau P, Pardal MA, Bacelar-Nicolau L, Azeiteiro UM (2016) Assessing student perceptions and comprehension of climate change in portuguese higher education institutions. In: Leal Filho W, Adamson K, Dunk R, Illingworth S, Azeiteiro UM, Alves F (eds) Implementing climate change adaptation in cities and communities. Climate Change Management. Springer, Cham, pp 221–236. doi: 10.1007/978-3-319-28591-7_12.

Setti AFF, Ribeiro H, Azeiteiro UM, Gallo E (2016) Governance and the promotion of sustainable and healthy territories in the experience of Bocaina, Brazil RGCI—Revista de Gestão Costeira Integrada/ JICZM—J Integr Coast Zon Manag 16. http://www.aprh.pt/rgci/pdf/rgci-612_Setti.pdf. http://dx.doi.org/10.5894/rgci612

Sundblad E-L, Biel A, Gärling T (2009) Knowledge and confidence in knowledge about climate change among experts, journalists, politicians, and laypersons. Environ Behav 41:281–302. https://doi.org/10.1177/0013916508314998

UN (United Nations) (2015) Adoption of the Paris agreement. Conference of the Parties, twenty-first session Paris, 12 Dec 2015. FCLIMATE CHANGEC/CP/2015/L.9

Viegas V, Azeiteiro UM, Alves F (2016) Lay rationalities on climate change: an exploratory study among the artisanal fishers in Peniche (Portugal). In: Filho WL et al. (eds) Climate change adaptation, resilience and hazards. ISBN 978-3-319-39879-2, pp 305–327. DOI 10.1007/978-3-319-39880-8_19. http://www.springer.com/gp/book/9783319398792

Viegas VM, Azeiteiro UM, Dias A, Alves F (2014) Alterações Climáticas, Perceções e Racionalidades Climate Change, Perceptions and rationalities RGCI—Revista de Gestão Costeira Integrada/JICZM—J Int Coast Zone Manag 14(3): 347–363. http://www.aprh.pt/rgci/pdf/rgci-456_Viegas.pdf. http://dx.doi.org/10.5894/rgci456

Wachholz S, Artz N, Chene D (2014) Warming to the idea: university students' knowledge and attitudes about climate change. Int J Sustain High Educ 15(2):128–141. https://doi.org/10.1108/IJSHE-03-2012-0025

Yoko M, Audrey B (2015) climate change education in the context of education for sustainable development: rationale and principles. J Educ Sustain Dev 9:4–26. https://doi.org/10.1177/0973408215569109

# Climate Change: Doing Little Can Change a Lot! Children's Knowledge-Action About Cimate Principles and Effects

Anabela Marisa Azul and Catarina Schreck Reis

## 1 Introduction

### 1.1 The Fact: Climate Change

Climate on Earth has being changing continuously, with ice ages alternated with warmer periods due to natural causes (Maslin 2016). About 55.9 million years ago, a rapid release in natural $CO_2$ conducted to a strong warming causing major ecosystem disruptions, soil degradation, and water-cycle shifts, thereby suggesting that climate is highly sensitive to $CO_2$ rising (Alley 2016). We are now facing a faster release in $CO_2$, as result of human activity, that is distinct from natural phenomena. Climate change is projected to have major impacts on habitats and biodiversity, water availability and supply, food production and security, extreme weather conditions (drought episodes alternated with storms), coastal flooding, new diseases, species extinction (IPCC 2014).

A. M. Azul (✉)
CNC—Center for Neuroscience and Cell Biology, University of Coimbra,
Edf Faculdade de Medicina, Rua Larga, Polo I, 1° Piso, 3004-504 Coimbra, Portugal
e-mail: amjrazul@ci.uc.pt

A. M. Azul
III-UC—Institute for Interdisciplinary Research, University of Coimbra, Casa Costa Alemão —Pólo II, Rua Dom Francisco de Lemos, 3030-789 Coimbra, Portugal

C. S. Reis
CFE—Centre for Functional Ecology, University of Coimbra,
Calçada Martim de Freitas, 3000-456 Coimbra, Portugal
e-mail: catarina.reis@exploratorio.pt

C. S. Reis
Exploratório—Science Center of Coimbra, Parque Verde Do Mondego,
3040-255 Coimbra, Portugal

 
W. Leal Filho et al. (eds.), *Handbook of Climate Change Communication: Vol. 1*,
Climate Change Management, https://doi.org/10.1007/978-3-319-69838-0_19

Climate change mitigation and adaptation are complementary strategies and a major global challenge facing humanity in 21st century and beyond (IPCC 2014), with serious risk of abrupt or irreversible effects for people and ecosystems. Scientific community predict that a large fraction of anthropogenic $CO_2$ emissions is irreversible over several centuries, except if a large amount of $CO_2$ is removed from the atmosphere over a continued period (IPCC 2014).

Scientists have been debating climate change communication and education (CCCE) for over two decades. The need of climate literacy[1] was indeed highlighted by United Nations Educational, Scientific and Cultural Organization (UNESCO 2007; United Nations Literacy Decade, UNLD, 2003–2012) as crucial recommendation for climate change adaptation and future wellbeing. Also the European Commission (Horizon 2020) has recommended new measures to science teaching and learning such as linking environmental issues into the scientific curriculum, thereby helping to make schools a place of science literacy and decision-making skills for acting. Yet, moderately emphasis is still given to climate change communication and education within childhood and primary school curricula. Massive and collaborative investment in climate change communication and education is urgently needed!

## 1.2 *The Challenge: Climate Change Communication and Education to Young Children*

Climate change, although nearly engage children, is a complex topic that covers difficulties in principles understanding, but also some epistemological challenges.

Despite the considerable evidence concerning children's climate change teaching and learning in primary school, less frequently in kindergartens, two key open questions persist: *What nature of knowledge is proper for young children? Who should be involved in the basis of climate change communication and education?* Although science inquiry is considered a fundamental part of school curricula, interdisciplinary work is needed in order to turn scientific knowledge into teaching-learning practices.

The aim of this study was to explore how interdisciplinary participatory research and mutual teaching and learning processes involving academy with school contribute for improving CCCE, with a particular focus on providing a better understanding about the principles and effects behind climate change, but also in developing children's decision-making skills for acting. The participatory research approach is designed in order to involve children actively.

[1]The United Nations Educational, Scientific and Cultural Organization (UNESCO) defines literacy as "*ability to identify, understand, interpret, create, communicate, compute and use printed and written materials associated with varying contexts. Literacy involves a continuum of learning in enabling individuals to achieve their goals to develop their knowledge and potential and to participate fully in their community and wider society.*"

## 2 Theoretical Framework and Participatory Research Approach

Climate change communication and education that combine formal[2] and informal[3] education with active teaching-learning processes open new directions towards participatory knowledge-action models. Researchers may contribute to explore and simplify key climate principles and effects; teachers have a key role in crossing-areas in curriculum and pedagogical shifts; and, most importantly children can be actively involved in cross-examination and construction of their own knowledge based on science principles.

Communication and education to young children should also cover tangible strategies connected to life experiences that can easily promote empathy and engagement. Hands-on and everyday life routines and experiences with an educational focus create empathy connected to active learning processes and individual behaviour.

Remarkable authors that have worked about social aspects of cognitive development of children, namely John Dewey, Jean Piaget, Lev S. Vygotsky, Loris Malaguzzi, or Paulo Freire (Table 1), among others, emphasized the importance of constructive active teaching-learning processes towards decision-making skills from younger ages. Rather than addressing theoretical perspectives in children education research, our interdisciplinary participatory approach followed active teaching-learning processes based on science principles matched with those authors conceptions about working together, involving multiple actors, children in particular, also their families, to a fully accomplish education, communication, and action objectives at societal level.

Our design is based on the assumption that interdisciplinary participatory research involving *academy with school* is central to provide the *effective climate literacy* and *adaptation in new teaching-learning practices for climate change education*. Here, we explore how mutual learning actions of academics with expertise in functional biology and ecology and science communication, teachers with pre-scholar and primary curriculum, and children emotions, perceptions and motivations, contribute for improving a better understanding about climate principles and effects, thereby providing children knowledge and skills for acting.

"*Let's look after the Earth: doing a little can change a lot*!" aimed to explore practices improving climate principles understanding at infancy and primary school, from 3 to 10-years-old, so that collective active learning processes would become effective in individual behaviours. Three main strategies were adopted: (1) hands-on activities emphasizing current climate events; (2) pre-scholar

---

[2]*Formal education* corresponds to the teaching and learning process within a structured education system in which learners are required to demonstrate proficiency.

[3]*Informal education* corresponds to teaching and learning processes externally to formal education system and that meets clear well-defined objectives through certain education actions.

curricular guidelines and crossing-areas in primary school curriculum; and (3) children emotions, perceptions, motivations and life experiences.

The common questions and expressions *"What is climate?"*, *"What is greenhouse effect?"*, *"Why is climate changing?"*, and *"Which problems are caused by greenhouse gas emissions?"*, *"How can each one of us look after the Earth?"*, and *"How doing a little can change a lot?"*, became the baseline for the research approach between children, teachers, and academics. We first explored school textbooks and children books aiming to link scientific concepts to hands-on activities, everyday life routines and the pre-scholar and primary curriculum.

The simplification of climate principles and effects in terms of comprehension and appropriation by young children occurred progressively with the active collaboration of children, the Table 2 presents the glossary of the principles and concepts explored.

The participatory research was designed not only to address climate principles and effects understanding, but also to develop children's empathy and action skills. The Table 3 summarizes the hands-on activities performed in a close collaboration between children, teachers and researchers:

*What do you understand about weather and climate?*—All children
*What is greenhouse effect?*—All children
*How animals react to climate?*—Children 4–5-years-old; children 6–7-years-old
*What is the cause for holes in ozone layer?*—Children 6–7-years-old
*Why don't the oceans freeze?*—Children 6–7-years-old
*Acid-base reaction, how is the rain in your city?*—Children 8–10-years-old
*What are the effects of acid rain on plants and buildings?*—Children 8–10-years-old
*What happens after a forest fire?*—All children
*How much water you need to wash your teeth?*—Children 3–6-years-old
*What are the effects of detergent pollution?*—Children 3–6-years-old; children 6–8-years-old
*How does plastic affect marine life?*—All children
*How long it takes your lunch leftovers to be compost?*—Children 6–7-years-old
*How can you make your trips friendly?*—Children 3–6-years-old
*How often you use a mechanically powered flashlight?*—Children 8–10-years-old

## 2.1 Data Sources

The research design involved the *selection-[re]construction* of 14 activities, was explored *in-and-out-of classroom* according to the age and level of knowledge of children. We used qualitative research methods associated working together, involving multiple actors, children in particular, also their families, following John Dewey, Jean Piaget, Lev S. Vygotsky, Loris Malaguzzi, and Paulo Freire (Table 1) to a fully accomplish active teaching-learning processes towards climate change

**Table 1** Remarks of authors that have worked about social aspects of cognitive development of children

| | |
|---|---|
| John Dewey | Reinforces the importance of education in society (Dewey 1899) real-world experiences through the active learning and teaching (Dewey 1938). "*We cannot overlook the importance for educational purposes of the close and intimate acquaintance got with nature at first hand, with real things and material, with the actual processes of their manipulation, and the knowledge of their social necessities and uses. In all this there was continual training of observation, of ingenuity, constructive imagination, of logical thought, and of the sense of reality acquired through first hand contact with actualities*" [p. 8] "*It is through these occupations that the intellectual and emotional interpretation of nature has been developed*" [p. 19] "*Knowledge is no longer an immobile solid; it has been liquefied. It is actively moving in all the currents of society itself.*" [p. 25] "*When the school introduces and trains each child of society into membership within such a little community, saturating him with the spirit of service, and providing him with the instruments of effective self-direction, we shall have the deepest and best guaranty of a larger society which is worthy, lovely, and harmonious.*" [p. 28] "*Learning? certainly, but living primarily, and learning through and in relation to this living… "drawing it" is excellent, if we mean simply to contrast it with the process of pouring in. But, after all, it is difficult to connect the idea of drawing out with the ordinary doings of the child of three, four, seven, or eight years of age… The child is already intensely active, and the question of education is the question of taking hold of his activities, of giving them direction.*" [p. 37] (Dewey 1899). |
| Jean Piaget | Proposes once children achieve the knowledge, they are capable to act in more complex way while constructing their own knowledge, meaning that children have an active role in their appropriation and construction of knowledge. "*We are now therefore in a position to realize that the fact of being or of not being communicable is not an attribute which can be added to thought from the outside, but is a constitutive feature of profound significance for the shape and structure which reasoning may assume*" [p. 28] "*If we admit that thought in the child depends upon his interests and activities rather than vice versa, then the absence of the desire to argue and collaborate is obviously the initial factor.*" [p. 43] "*Mechanical explanation arouses a more lively interest; both explainer and reproducer try to understand it; the results therefore are better.*" [p. 75] "*child's language does not depend only on the child's development but, as the evidence shows, on the type of relationship which he maintains with the adult.*" [p. 155] (Piaget 1926). |
| Lev Vygotsky | Highlighted the importance of symbolic language and intimate interactions between teachers and children; also figured the importance of older children being actively involved in helping young children in learning processes. "*At preschool age special needs and incentives arise which are highly important to the whole of the child's development and which are spontaneously expressed in play*" [p. 7] "*Play is converted to internal processes at school age, going over to internal speech, logical memory, and abstract thought*" [p.13] "*From the point of view of development, the fact of creating an imaginary situation can be regarded as a means of developing abstract thought.*" [p. 17] (Vygotsky 1966) "*the zone of proximal development. It is the distance between the actual developmental level as determined by independent problem solving and the level of potential development as determined through problem solving under adult guidance or in collaboration with more capable peers.*" [p. 33] "*human* |

(continued)

**Table 1** (continued)

| | |
|---|---|
| | *learning presupposes a specific social nature and a process by which children grow into the intellectual life of those around them.*" [p. 34] (Vygotsky 1978). This suggests that both teachers and children are essential in the efforts to improve climate literacy. |
| Paulo Freire | Strengthens the importance of active constructive collaboration between educator and children (Freire 1996, 1970). "*Through dialogue, the teacher-of-the-students and the students-of-the-teacher cease to exist and a new term emerges: teacher-student with students-teachers. The teacher is no longer merely the-one-who-teaches, but one who is himself taught in dialogue with the students, who in turn while being taught also teach. They become jointly responsible for a process in which all grow... People teach each other, mediated by the world, by the cognizable objects which in banking education are "owned" by the teacher.*" [p. 80] "*As a process of search, of knowledge, and thus of creation, it requires the investigators to discover the interpenetration of problems, in the linking of meaningful themes. The investigation will be most educational when it is most critical, and most critical when it avoids the narrow outlines of partial or "focalized" views of reality, and sticks to the comprehension of total reality. Thus, the process of searching for the meaningful thematic should include a concern for the links between themes, a concern to pose these themes as problems, and a concern for their historical-cultural context.*" [p. 108] "*By stimulating "perception of the previous perception" and "knowledge of the previous knowledge," decoding stimulates the appearance of a new perception and the development of new knowledge. The new perception and knowledge are systematically continued with the inauguration of the educational plan, which transforms the untested feasibility into testing action, as potential consciousness supersedes real consciousness.*" [p. 117] (Freire 1970). |
| Loris Malaguzzi | Support the active participation of children in the construction of knowledge. Malaguzzi brings the Reggio Emilia approach (Smidt 2013) with the constructivism alongside the children perspective and interests. The Reggio approach embraces a progressive programme through mutual learning, with children as co-actor/co-author in planning and implementing the actions. "*Children learn by interacting with their environment and actively transforming their relationships with the world of adults, things, events, and, in original ways, their peers.*" [p. 11] "*Interaction among children affects social, emotional, communicative, and cognitive behaviour and development*" [p. 13] (Malaguzzi 1993). "*It is not an imposition on children or an artificial exercise to work with numbers, quantity, classification, dimensions, forms, measurement, transformation, orientation, conservation and change, or speed and space, because these explorations belong spontaneously to the everyday experiences of living, playing, negotiating, thinking, and speaking by children.*" [p. 53] "*What children learn does not follow as an automatic result from what is taught. Rather, it is in large part due to the children's own doing as a consequence of their activities and our resources.*" [p. 67] "*Value should be placed on contexts, communicative processes, and the construction of a wide network of reciprocal exchanges among children and between children and adults.*" "*Fernando Pessoa (1986) says that the measure of the clock is false. It is certainly false concerning the time of children—for situations in which true teaching and learning take place, for the subjective experience of childhood. One has to respect the time of maturation; of development; of the tools of doing and understanding; of the full, slow, extravagant, lucid, and ever-changing emergence of children's capacities; it is a measure of cultural and biological wisdom.*" [p. 80] (Malaguzzi 1996). |

**Table 2** Glossary

| Glossary | |
|---|---|
| Acid rain | associated to air pollution; it may has natural causes, such as gas emissions from volcanoes and biological processes occurring in soil and water, or result from human activity, for example industry and burning of fossil fuels. |
| Air quality | measured by the purity of invisible (unseen), odorless (tasteless), tasteless (unflavored), and the presence pollutants in the air. The air quality establishes standards to protect public health. |
| Atmosphere | layer of gases surrounding Earth that makes our planet habitable. The main gases include nitrogen and oxygen; the gases carbon dioxide, methane, and ozone account for about onetenth of 1%. |
| Biodegradable | material capable of being decomposed by the activity of fungi or other living organisms therefore incorporated in the natural recycling processes. |
| Biodiversity (Biological diversity) | variety and variability of life on Earth; it may be appointed to species in ecosystems. |
| Calcium carbonate | main component of shells and limestone rocks. |
| Carbon dioxide ($CO_2$) | gas produced during respiration of animals and plants and absorbed by plants. The $CO_2$ release also occurs when fossil fuels (e.g., oil, gas oil and gasoline) are burned. The $CO_2$ is a greenhouse gas and a major cause of global warming. |
| Climate | describes the condition of the weather (temperature, wind velocity, air humidity, precipitation) at a place over a period of many years. |
| Climate change | expresses changes in regular climate events. Climate change can be a natural process or a consequence of human activities, namely through the rising of greenhouse gas emissions ($CO_2$ release, air pollution, deforestation). |
| Chlorofluorocarbons (CFCs) | group of man-made compounds (containing chlorine, fluorine and carbon) used in sprays, refrigerators and air conditioners that causes depletion of atmospheric ozone. |
| Composting | production of humus as result of decomposing of leaves, herbs and remnants of vegetable food. The humus contributes to protect soil, namely moisture and nutrient retention. |
| Conservation biology/ ecosystem | describes plans of action/policies to develop appropriate measures to protect nature and biodiversity, i.e., species, habitats, ecosystems, and protecting. |
| Deforestation | removal of many trees in an extensive area. |
| Ecosystem | corresponds to the system of living organisms interacting with each other and their physical environment. |
| Environment | everything that surrounds and affects a living organism, i.e. its interaction with nonliving thins, such as light, water, air, soil, and other living organisms. |
| Energy | ability to produce work on—or to heat—a thing. Energy is not created or destroyed, only transformed. Our body produces energy. Sun, wind, and water produce energy. Energy can be produced artificially from chemical action in nuclear, wind, solar, hydroelectric and thermoelectric stations. |

(continued)

**Table 2** (continued)

| Glossary | |
|---|---|
| Erosion (in nature) | process of deterioration by wind, water, or other natural agent. |
| Extinction (Biology and Ecology) | disappearance of an organism on Earth, normally a species. Fossils expose many extinct species, such as the dinosaurs. At the present, there are many threatened species resultant from human activity. |
| Fuel fossil | source of energy formed from natural processes, namely from the decomposition of living organisms buried millions of years ago, e.g., coal, oil, and natural gas. Fossil fuels are not renewable. |
| Glacier | huge ice masses formed by accumulation, compaction and recrystallization of snow. Glaciers are maintained by the accumulation of snow and balanced by melting at low in the sea. |
| Global warming | increasing of temperature caused by greenhouse gas emission in the atmosphere. The major contributor to global warming is carbon dioxide ($CO_2$). |
| Greenhouse effect | process that allows the passage of solar radiation through Earth's atmosphere but prevents most of the Earth's surface radiation from escaping into space. The greenhouse effect allows the life on Earth as it exists today. The problem associated to greenhouse effect comes with the rising of greenhouse gases emissions, resulting in an increase of the temperature. |
| Greenhouse gases | gases released to the atmosphere, e.g., water vapor, carbon dioxide ($CO_2$), nitrogen oxide and methane, that contribute to retain solar radiation and rising of temperature. The $CO_2$ is naturally released in cellular respiration and during the burning of coal, oil, and gas to produce energy. Most pollutants resultant from industrial processes (e.g., *mon*oxide, organic carbons, sulfate, soot, some oxides of nitrogen), are not greenhouse gases. |
| Iceberg | block of freshwater ice that has broken off a glacier and floats freely. Only about 10% of the total mass of an iceberg is above the surface of the water. |
| Land use | refers to the total of arrangements, activities and inputs undertaken in a certain land cover type as a result of human actions. |
| Ozone layer | atmosphere layer that absorbs most of the Sun's harmful ultraviolet (UV) radiation. |
| "Ozone hole" | the ozone gas layer in Earth's stratosphere is about 20 miles above Earth's surface. Seasonally occurs a depletion located mainly above the North and South poles and that occur due to human use of ozone-depleting chemicals such as chlorofluorocarbons (CFCs). |
| Pesticide | substance/mixture of substances (chemical or biological agent) used to control pests or weeds in agroforestry. |
| Pollution | associated to harmful substances in air, water or soil; it may cause diseases/death in humans, animals, plants or other living organisms. |
| Recycling | process of converting waste materials into new products. Recyclable materials include many materials that are hardly degraded by nature, e.g., glass, plastic and metal. |
| Reforestation | replanting trees in areas where trees have been cut or burned. |

(continued)

**Table 2** (continued)

| Glossary | |
|---|---|
| Resource non-renewable | resource that can be used only a limit number of times before it run out, such as fossil fuels. |
| Renewable resource | resource that can be used countless times without running out, e.g., energy from the sun, water or wind. |
| Sustainability | utilization of resources without compromising their balance in nature. For example, composting is a sustainable activity. |
| Weather | day-to-day variations in atmospheric conditions, including temperature, precipitation, humidity, air pressure, cloudiness, and wind speed and direction. |

communication and education (CCCE) and decision-making skills from younger ages. Some of the hands-on activities were associated to the experimental method; children were stimulated to formulate hypothesis, perform experiments, describe and explain the results and take their own conclusions. The experimental activities were performed in group, as well as the discussion of results. However, the documentation was individual, which means that after collective explorations and discussions, each child made their own interpretation by drawing or writing. Children were provided with as much time as needed.

We also used some quantitative data, children and teachers responded to a written questionnaire to explore their own actions to *look after the Earth*. In case of kindergartens, the questionnaire was adopted to semi-structured interviews conducted by the teachers, varying in length from 10 to 20 minutes. Children responded to closed questions about their everyday life routines: *How do you come to school?*, *Which materials do you recycle?*, *Do you turn off the light when you leave home?*, *What do you do with clothes and toys that are no longer used?*, *Where do you put the empty containers after a snack?* Children and teachers responded to the open question: *What can you do at school to look after the Earth?* The questionnaire was completed during the lifetime of the project.

The design was accomplished at two kindergartens, with 70 children from 3 to 6-years-old, and at two primary schools, with 250 children from 6 to 10-years-old, in the region of Coimbra, Portugal. A total of 320 children, 16 teachers and 2 researchers, joined the participatory research "*Let's look after the Earth: doing a little can change a lot*!". The selection of kindergarten and primary school has the purpose of diversity sampling in children age and the multiple levels of knowledge in formal education.

## 3 Results and Discussion

The active mutual learning of academics with children and teachers contributed for collective enthusiastic discussions. Researchers were important to introduce and clarify climate change principles among children and teachers during the hands-on activities. Teachers played a key role in suggesting appropriate teaching-learning strategies, pinpointing crossing-areas in curricula and creating awareness. The active collaboration of children was fundamental to guide the scientific topics taking into account their own interests.

Two questions accompanied the very beginning of the participatory research: "*So, can each one of us look after the Earth?*" and "*How doing a little can change a lot?*". Most of the children focused the reflexion following one-side idea of the questions. However it didn't take much time for children understanding that "*Doing a little can change a lot*!" can have two completely opposite meanings. This approach was designed to encourage collective discussions with children to improve their own understanding about climate change and the importance of individual behaviour to action at societal level.

## *3.1 Climate Change and Greenhouse Effect*

Children made daily measurements of weather over a period of two weeks: air temperature, atmospheric pressure, humidity, wind speed and direction, and precipitation. Children from kindergarten and first grade of primary school, learnt about meteorological data, older children from primary school explored concepts previously studied. After children documenting the information with teachers, researchers visited the classes to discuss the topics *weather* and *climate*. Most of the children recognized the terminology *climate* and *weather*, even *greenhouse effect*, in case of primary school, however only few children were able to give an accurate description of their meaning. Surprisingly, both children and teachers revealed misunderstandings about greenhouse effect. The answers to "*It is greenhouse effect good or bad?*" were diverse, but with children and teachers assuming that greenhouse effect is not beneficial for life in Earth. Actually, we assisted to a generalized assumption that greenhouse effect is a result of global warming and it is harmful for life on Earth.

After debating about *climate*, *weather*, and *greenhouse effect*, children, researchers and teachers performed the experimental activity "*What is greenhouse effect?*" (A1, Table 3). Fig. 1 illustrates a diagram of a 7-years-old child that summarizes the Earth atmosphere greenhouse effect, representing the passage of solar radiation through Earth's atmosphere and the prevention of radiation escaping into space (Fig. 1a); children visibly raised to well-established link between *atmosphere* and *greenhouse effect*. A second illustration (Fig. 1b) of a 7-years-old child reveals the researcher and children with the thermometers, collecting data during the experience. The activity "*What is greenhouse effect?*" revealed to be important to children integrate science knowledge from posterior activities but also a better understanding about "carbon cycle" and mitigation and adaptation measures in natural ecosystems and human societies to *remove* $CO_2$ *from the atmosphere* and *reduce the greenhouse effect*.

**Table 3** Active learning practices using cross-areas in curriculum, hands-on activities, and everyday life routines in kindergartens and primary schools

| Activity | Principles concepts | Materials/Time | Procedures |
|---|---|---|---|
| A1. *What is greenhouse effect?* | Atmosphere<br>Greenhouse effect<br>Greenhouse gas | 2 thermometers, 1 glass bottle with lid<br>~ 20 min | *Sunny day*, put one thermometer inside the bottle and close it. Compare the temperature in the 2 thermometers after periods of 10 min. |
| A2. *How react animals to climate?* | Weather<br>Climate | Snails, 1 paper sheet, tray, water<br>2–3 hours | Divide the tray into a wet side and a dry side. Put the snails in the middle. Register the snails' preference.<br>During summer/dry weather, snails retreat into their shells and seal their entrance in to protect their body from drying up. Terrestrial snails are more active at night and on wet days. |
| A3. *What causes the ozone hole?* | Anthropogenic emissions<br>Atmosphere<br>Ozone layer<br>Ozone depletion | Bottle, gum, boiled water, magnifying glass<br>~ 30 min | *Ozone depletion model with hot water and chewing gum*, fill the bottle with the hot water and seal the bottle opening with the masked gum.<br>Observe with a magnifying glass and document!<br>The bottle represents the Earth, the boiled water represent the anthropogenic emissions (chlorofluorocarbons: CFCs), and the gum represent the ozone layer. |
| A4. *Why don't the oceans freeze?* | Melting glaciers<br>Freezing point | 2 glasses, water, salt, spoon, labels, freezer<br>~ 15 min preparation + 10 hours in freeze | Half fill the 2 glasses with water. Add a spoonful of salt, mix, and wait until the water is still. Repeat until the salt begins to collect on the bottom that means water is saturated with salt. Mark the 2 glasses and put both in the freezer. |

(continued)

**Table 3** (continued)

| Activity | Principles concepts | Materials/Time | Procedures |
|---|---|---|---|
| | | | Register what happens every two hours. Salt-water freezes at a lower temperature than freshwater. |
| A5. *Acid or base?* | Acid Base pH | *For the solution test*: red cabbage, distilled water, bowl, strainer. *For testing:* lemon juice, toothpaste, detergents… ~ 15 min preparation + 1 hour waiting | Cut the red cabbage. Heat the distilled water and add it to the cut cabbage. Wait 1 hour and then pass through the strainer. The violet solution will test the pH. Distribute the violet solution equally for several glasses. Keep one glass with the violate solution only; it will be your control. Add the substances you chosen (lemon juice, toothpaste, detergents…) and register what happens. The acid substances will turn your solution to reddish. The base substances will turn your solution to bluish. |
| A6. *How acid is the rain in your region?* | | Rainwater, distilled water, *solution test*, glasses ~ 30 min | Distribute equally *the solution test* in three glasses. Add the distilled water to one glass and your sample of rainwater to the second glass. The third glass will be your control. Register and discuss your results. |
| A7. *How does acid rain affect plants health and buildings?* | Air pollution | 2 similar potted plants, water, vinegar, labels ~ 30 min preparation + 1–2 weeks observations (regular watering) | Prepare two solutions, one with water, a second with equal parts of water and vinegar; identify the solutions and store it. Water the plants with the two different solutions. Register your observations daily, over a period of 2 weeks. Compare what happens to both plants. |

(continued)

**Table 3** (continued)

| Activity | Principles concepts | Materials/Time | Procedures |
|---|---|---|---|
| | | Sea shells, 2 glasses, water, vinegar, labels, spoon<br>~ 30 min preparation + 3–4 days observations | Half fill the 2 glasses with water and add 2 similar shells.<br>Add vinegar to one glass. Identify the glasses and compare what happens to shells. |
| A8. *What happens after a forest fire?* | Deforestation<br>Habitat destruction<br>Erosion | *Performance:* children perform a drama acting as trees, fire and water<br>~ 45 min | Children acting as water run in 2 different scenarios: 1) healthy forest and 2) forest after fire (see the text). |
| A9. *How much water do you use to brush your teeth?* | Water conservation | Water, toothbrush, toothpaste, cup, recipient to collect water<br>~ 30 min | Put the recipient in the washbasin to collect the water from the open tap while you brush your teeth. Brush your teeth using a cup and the tap open. Compare the number of times you can save water by turning of the tap. |
| A10. *What are the effects of detergent pollution?* | Water pollution<br>Ocean/River life | Cooking oil, detergent, spoon, cups, water<br>~ 30 min | Half fill the 2 glasses with water and add 1 spoon of oil to both glasses.<br>Add 1 spoon of detergent to one of the cups.<br>Shake. Register what happen in the 2 glasses. |
| A11. *How does plastic affect marine life?* | Ocean pollution<br>Plastic waste | Rubber bands<br>~ 10 min | Put a rubber band around each finger of one hand. Do the same in the other hand. Try to remove the 2 rubber bands without the help of your hands or friends or family! |
| A12. *How long it takes your lunch leftovers to be compost?* | Decomposition<br>Biodegradable<br>Recycle | Leftover plant-based foods* from lunch (e.g., fruit peel, rice), plastic, aluminum…, soil, labels<br>~ 45 min preparation + 1 month | Bury the lunch leftovers in a shallow hole at the 30cm depth; identify the lunch leftover by using labels written with a pencil and sealed by plastic. Compare what happened to the different materials after 1 month…<br>*Don't use leftover fish/meat-based foods. |

(continued)

**Table 3** (continued)

| Activity | Principles concepts | Materials/Time | Procedures |
|---|---|---|---|
| A13. *How can you make your trips friendly?* | $CO_2$ emissions<br>Pollution<br>Passenger vehicles | Paper, pen<br>~ 20 min | Calculate the annual grams of $CO_2$ emitted from gasoline combusted in different vehicles to transport the same number of passengers. Can you realize how much pollutant is the car comparing with autobus, train or metro? And bicycle? |
| A14. *How often you use a mechanically powered flashlight?* | Non-renewable energy sources<br>Renewable energy | Dynamo flashlight, battery flashlight<br>~ 30 min | Explore the mechanism associated to electricity generated by mechanically powered flashlight and battery flashlights, and realize that we can easily save energy.<br>How often you turn off the electrical appliances when not in use? |

**Fig. 1** *What is greenhouse effect?* **a** Diagram of a 7-years-old child summarizing the Earth atmosphere greenhouse effect with the Sunrays and our planet. **b** Illustration of a 7-years-old child showing the researcher and children with the thermometers, collecting data during the activity

### 3.2 Adaptation to Climate

The *adaptation to climate* can be introduced by the seasonal activity of certain species. When the weather and/or food conditions are somewhat limited, some plants loose their leaves (e.g., apple tree, pear tree, peach tree, cherry tree, birch, oak), and some animals hibernate (e.g., bats, frogs, hedgehogs, snails, squirrels). For example, snails hibernate during summer. The activity "*How animals react to climate?*" (A2, Table 3) was designed to explore the *capacity adaptation to climate*. Children looked for snails in the garden and recorded the places where they were found. The activity involved the observation of snails in wet and dry conditions. After the activity, children put the snails in the garden and explored small insects. Children registered the snails' preference to wet side of the tray and then investigated the results in books. Children could easily extrapolate the behaviours of snails during winter and summer. This activity A2 clearly engaged children to establish links between *climate*, *biodiversity*, and *people*. For younger children, in a sensitive period of their social, emotional, cognitive and proximal development, the first-hand experience with nature, asset the children awareness for the local plants, animals, fungi, weather, climate, water resources, etc., and should be encouraged.

### 3.3 Global Warming and Ozone Hole

Explaining that climate is changing with major impacts on habitats, biodiversity, and people wellbeing, and that adaptation and mitigation measures are needed over a sustainable period of time, is not an easy approach among general public, but is even more complex with young children. Whenever possible, why not play with the most unconventional materials to introduce complex concepts? The debate about *ozone hole, ozone depletion, chlorofluorocarbons (CFCs)* gases, began with the distribution of a gum to all children of the class. Children chewed gums carefully. The gums were then collected by the researchers and used to proceed with the activity *"What causes the ozone hole?"* (A3, Table 3). The apparent disgusting preliminary step was counterbalanced with curiosity and enthusiasm of children to discuss about the experience and the questions, visibly expressed in Fig. 2. The gum symbolized the ozone layer, the plastic water bottles symbolized the Earth, the boiled water symbolized the CFCs, and the cold water symbolized normal air. Children observed the gum holes becoming bigger with time exposition to boiled water. After the activity several notions were discussed about the ozone gas layer in Earth's stratosphere. Children could associate the ozone depletion to human use of chemicals but it is difficult to predict if they would be capable to select products non-CFCs in a supermarket, for example. For many children it was linear the connection between ozone layer depletion and UVB radiation and the importance to protect skin and eyes.

**Fig. 2** *What causes the ozone hole?* Illustration of a 7-years-old child representing a group of children following the ozone depletion model with hot water and chewing gum

## 3.4 Glaciers Melting

The *melting glaciers* are a current topic associated to global warming and climate change but not fully understood by children. The glaciers retreat over the last century represents one of the most iconic pictures for communicating climate change. However, there is not quantitative and consensual information. The activity "*Why don't the oceans freeze?*", with salted water in the refrigerator (A4, Table 3) was particularly important to orient the glaciers dynamics to a different level. *How much and how quickly will glaciers melt?*, promoted enthusiastic debates about the incapacity to form new glaciers after the melting glaciers in seawater. Children were able to reflect about the consequences of global warming but also to discuss the differences between alpine and continental glaciers (Table 3).

## 3.5 Water and Air Pollution

*Air pollution* was explored associated to the topic *acid rain*, namely through the experimental activities "*How acid is the rain in your region?*" and "*How does acid rain affect plants health and buildings?*" (A6, A7; Table 3). Firstly, it was introduced the two categories of chemicals: *acid* and *base* giving examples with current food and products (A5; Table 3). For that, children prepared a solution test with red cabbage and observed the effect in contact with lemon juice, toothpaste and detergents. Afterwards, children collected rainwater near school to evaluate with the solution test (A6; Table 3); the results showed that rainwater was not acid. Children were then investigators and accomplished the experience "*How does acid rain affect plants health and buildings?*" (A7; Table 3). Children conducted the experiences over two weeks with the help of teachers. Children registered the effects of watering the plants with a vinegar solution and the consequences of shells being in

**Fig. 3** *How does acid rain affect plants health and buildings? "The statue of D. Afonso Henriques [The Conqueror], the first king of Portugal after acid rain"*, illustration of an 8-years-old child

well as vinegar solution (A7; Table 3). The researchers visited the school periodically to discuss observations with children and teachers, at the end children made their own conclusions about the effect of acid rain to nature and buildings (Fig. 3).

*Air pollution* was also introduced associated to $CO_2$ emissions. The mathematical problem "*How can you make your trips totally friendly?*" (A8, Table 3). Calculate the total annual $CO_2$ emission by different passenger vehicles, namely bus, car, train and bike. Children naturally realized that car is definitively the passenger vehicle most pollutant. Children also concluded that bikes, skates, or roller skates do not cause air pollution (Fig. 4).

*Water pollution* was explored through the activities "*What are the effects of detergent pollution?*" (A9, Table 3) and "*How does plastic affect marine life?*" (A10, Table 3). Children can identify human activities that cause pollution, but are not really aware to the exact negative impacts for the species and ecosystems. Through the activity A10, children questioned and investigated the effect that detergent pollution has on fresh water birds' ability to float. Children clearly explained the influence of detergent to the protective layer of the birds' feathers and the ability loss of birds to float in polluted water.

**Fig. 4** *How can you make your trips totally friendly?* Illustration of a 4-years-old child

Of the most devastating human activity causing water pollution is large amount of plastics into the ocean; as a result, water and fish are becoming intoxicated (Fig. 5). The danger of plastics in water was easily understood by comparing the difficulty they felt to take off the rubber band around their fingers without the help of anyone. While trying to escape fingers and get the hands free again, children gave several examples of sea fauna affected by plastic (Fig. 6). At the end of both A9 and A10 children could clearly explain the link between the water pollution and human health through the water we use and the food we eat.

**Fig. 5** *What are the effects of water pollution?* Illustration of a 7-years-old child

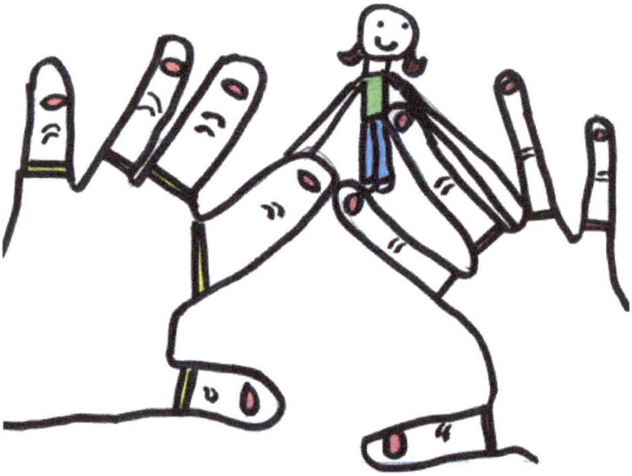

**Fig. 6** *How does plastic affect marine life?* Illustration of a 7-years-old child

### 3.6 Erosion

The *erosion* and the *process of eroding* were introduced through the performance "*What happens after a forest fire?*" (A11, Table 3). Children were divided into four groups: Group 1—Trees; Group 2—Animals; Group 3—Water; and Group 4—Fire. The Group 1 distributed as a forest stayed fixed to soil while the Group 2 played between trees. The Group 3 entered in the forest as raining and try to pass through the branches of the trees, "*is not easy!*" Fire came! Animals runed away! Trees were burning! After the fire, some trees fell down. It was raining again. There were few trunks in the forest, burnt. One animal was present, dead. The water run faster...

After the performance, researchers and children explored the processes involved in soil *erosion* and ecosystem recovery after fire. Some children mentioned the severe wild fires occurring in Portugal by proceeded by high temperatures and scarce rainfall. Many children raised the topic gardening at school, which conducted to interesting debates in preschools and primary schools how to garden and grow their own flowers and vegetables for eating as climate change mitigation and adaptation measures.

### 3.7 Recycling

Recycling is a common practice at kindergarten and primary school. However, children often do not truly understand why they do it. The activity "*How long it takes your lunch leftovers to be compost?*" (A12, Table 3) introduced exciting new questions about the role of soil in recycling materials, together with the re-use awareness of paper and glass materials (Fig. 7).

"*How much water do you use to brush your teeth?*" (A13, Table 3) Some everyday life routines use a little water, and some a lot… After brushing their teeth, with a cup and the water running, children compared the total of water used. Younger children realized the water waste by leaving the water running while older children were able to go through mathematic and calculate the water use in both situations (Fig. 8a). Children discussed several ways to save water; many mentioned taking showers instead of baths, washing the car with a water bucket, and watering plants with unused drinking water (Fig. 8b).

*Saving energy* revealed to be a familiar topic as result of everyday life routines and both kindergarten and primary school curricula. Most of children had a pro-active behaviour towards saving energy, particularly electrical energy. Children articulated clear discourses about saving energy and established the link between energy use and climate change adaptation measures. The perception of the existence of two different types of energy sources and their different effects on pollution and greenhouse gas emissions was almost completely unknown. Nevertheless, children promptly realized the importance of using renewable energy, suggesting several alternative sources of energy, including a new model of solar cars (Fig. 9a, b).

**Fig. 7** *How long it takes your lunch leftovers to be compost?* Illustration of a 5-years-old child

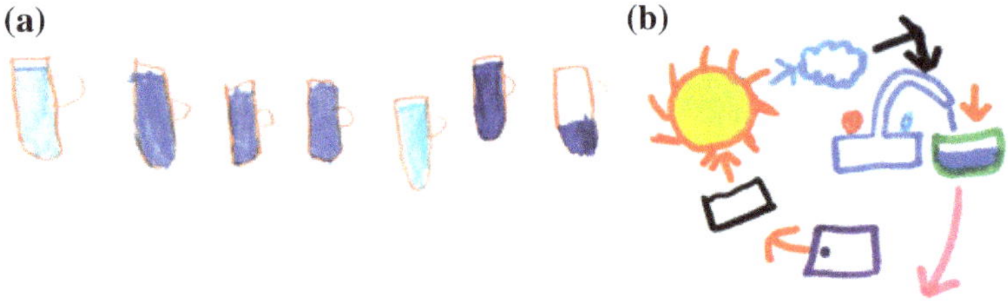


**Fig. 8** **a** *How much water do you use to brush your teeth?* Illustration of a 3-years-old child. **b** Illustration of the water cycle by a 5-years-old child

**Fig. 9** **a** Saving electric energy illustration by a 9-years-old child. **b** Renewable energy, illustration of a solar car by a 9-years-old child. **c** *How often you use a mechanically powered flashlight?* Illustrations by a 9-years-old child

**Table 4** Questionnaire to children and teachers about everyday life routines

| | | | | | |
|---|---|---|---|---|---|
| *How do you come to school?* | | | | | |
| By foot<br>53 | By bicycle<br>1 | By Public transport<br>14 | By car<br>245 | 1) + 4)<br>7 | |
| *Which materials do you recycle?* | | | | | |
| Paper and carton<br>37 | Plastic packaging<br>6 | Glass<br>8 | All<br>249 | None<br>20 | |
| *Showers or baths?* | | | | | |
| Shower<br>271 | Bath<br>32 | Both<br>17 | | | |
| *Do you turn off the light when you leave home?* | | | | | |
| Always<br>212 | Sometimes<br>104 | None<br>4 | | | |
| *What do you do with clothes and toys that are no longer used?* | | | | | |
| In garbage<br>7 (+1) | Family/friend<br>17 | Institution<br>100 | No answer<br>9 | 1) + 2)<br>1 | 2) + 3)<br>24 |
| *Where do you put the empty containers after a snack?* | | | | | |
| Backpack to recycle<br>64 | Dustbin<br>244 | Throw away<br>2 | 1+2<br>6 | No answer<br>4 | |

Children became very enthusiastic about using dynamo flashlight in a dark room when the batteries of the battery flashlight ended: "*How often you use a mechanically powered flashlight?*" (A14; Table 3; Fig. 9c). The drawings reflected their current behaviours about energy awareness as well as knowledge-action concepts associated to the development of renewable sources and safety environmental scenarios of clean (Azul et al. 2009). We undergo the limitations of not going thorough to each drawing or perception of children. However, the diversity in the documentation substantiates the complementarity in improving climate literacy. Besides, children and teachers participated in a written questionnaire; the Tables 4 and 5 summarize the children and teachers answers.

Most of the suggestions made by the younger children reflected individual actions of what should and should not to do, whereas older students revealed critical attitude about the everyday life routines at school and showed autonomy and leadership in developing solutions. Children were able to construct their own knowledge about climate principles and effects how important individual everyday life routines are important for climate changing adaptation. Despite the differences at children ages, it was possible to register an evolution for actions proposals. For example, in case of recycling actions, younger children converged to initiatives to conducted in classroom, while older children were are able to observe that many schools do not have appropriate recycling strategies and proposed to contact the local authorities, or even established a plan to take the separated wastes to the

**Table 5** What can you do at school to look after the Earth?

| Children (age) | Documentation |
|---|---|
| 3–5-years-old | *"Do not throw waste"*<br>*"Recycle school milk packs"*<br>*"Don't waste water"*<br>*"Separate the waste to recycle"*<br>*"Always turn off lights not needed"*<br>*"Save water and light, take care of school textbooks and reuse them for other students"*<br>*"Plant trees"*<br>*"Keep doing what I already do!"* |
| 6-8-years-old | *"Do not throw waste on the floor"*<br>*"Use less paper and water in the bathroom, save food"*<br>*"Use less water and soap in the bathroom"*<br>*"Separate the waste to recycle"*<br>*"Recycle better"*<br>*"Help younger children to recycle"*<br>*"Pick up the trash with friends"*<br>*"Ask for gloves from a school employee, call my friends, and pick up trash with them"*<br>*"Write a letter to the local Mayor asking for recycling bins in school"*<br>*"Go to school on foot"*<br>*"Go to school on foot or by bus"*<br>*"Turn off the lights when leaving the classroom"*<br>*"If I improve at school I will improve in the world"*<br>*"Plant trees"* |
| 9-10-years-old | *"Use solar panels and wind turbines to produce electricity"*<br>*"Reduce noise pollution"*<br>*"Do not throw waste on the floor"*<br>*"Pick up the trash with friends"*<br>*"Since we don't have recycle bins at school, I can save the milk and yogurt packages in my backpack and take them for recycle near my home"*<br>*"Use less water in the bathroom"*<br>*"Save more water in the bathroom"*<br>*"Have more recycling containers"*<br>*"Recycling is giving and receiving"*<br>*"Recycle better"*<br>*"Turn off the electrical appliances when not using"*<br>*"Have a dual flush toilet cistern in the toilet"*<br>*"Plant a vegetable or forest garden"*<br>*"Plant trees"*<br>*"Clean the streets to prevent fires"*<br>*"Go to school on foot or by bus"*<br>*"Come to school by bicycle or by bud (although I doubt that my father allows me to do so!)"*<br>*"We could make a group called: Friends of the Environment"* |

(continued)

**Table 5** (continued)

| Teachers (classroom) | Documentation |
|---|---|
| 3–5-years-old | *"Encouraging problem-solving approach for reducing of water consumption"*<br>*"Promoting waste separation and materials recycling; create children awareness for causes, effects, and legal penalties, with the purpose of reducing pollution; knowledge through research"*<br>*"Perceptions of children about the topic. Create groups to collect waste at school so promoting children awareness as active players in common spaces"*<br>*"Promoting knowledge and behaviours related to recycling and water save"*<br>*"Awareness of natural resources and understanding the importance of saving energy"* |
| 6-8-years-old | *"Encourage group discusses be able to react creatively and responsibly in all environments" Save energy, water and prefer recycled materials*<br>*"Co-operate with waste separation in classroom so that children create awareness by doing"*<br>*"Promote and reinforce the recycling initiatives already on-going in classroom (e.g., paper, plastic, glass, and other specific materials such as piles, stoppers, ink cartridges) to school and community levels."* |
| 9-10-years-old | *"Promote and inspire responsible attitude and awareness about how to contribute climate sustainability, and do it, for example reinforcing the importance of saving water, paper and natural resources, as well as recycling, and use recycled materials."*<br>*"Recycling initiatives at school, namely the proper recollection of paper, plastic, glass, and other specific materials such as piles, stoppers, ink cartridges."* |

closest recycling point. In some classes children decided that while waiting for recycling bins inside the school they could create their own in the classroom and each day one student was responsible to take the recycled bags to the nearest recycling bin outside the school.

The researchers feel very pleased that the *interdisciplinary approach academy-school* reached out other audiences, particularly teachers, in seminars, workshops, and advanced training, but also parents interested in handling climate change education with their children. The researchers feel particularly privileged that outcomes of the participatory research resulted in a book with children as our main contributors, representing the beginning of the collection *Discovering Science* [*Descobrir as Ciências*] (Azul et al. 2009) by Coimbra University Press.

## 4 Final Remarks and Recommendations for Climate Change Communication and Education

Climate change mitigation and adaptation is a major global challenge facing humanity in 21st century and beyond, yet climate change communication and education (CCCE) is still not generally experienced in curricula, and important misunderstanding of climate principles and effects are often overlooked in school. Beyond the climate literacy, and strategies of CCCE, the results indicate that complex concepts about climate principles and effects can be combined continuously with children's interests, everyday life routines, cross-areas in curriculum thereby enabling children to consolidate their knowledge into action. Importantly, regarding the curricula plan for improving the CCCE, the results grow to new directions of interdisciplinary and mutual learning that combine formal and informal education towards participatory knowledge-action models.

**Acknowledgements** Data in this paper were generated as part of a participatory research coordinated by the Center for Functional Ecology, University of Coimbra, entitled "Let's look after the Earth: doing little can change a lot!", co-funded by European Regional Development Fund (FEDER) through the Operational Programme Competitiveness Factors—COMPETE and National Agency Ciência Viva, Lisbon, Portugal. The authors gratefully acknowledge teachers of the kindergartens Serviços de Acção Social da Universidade de Coimbra and Centro de Bem-Estar da Sagrada Família, and the primary schools Ensino Básico do 1º Ciclo da Solum e do Dianteiro, Agrupamento de Escolas de Eugénio de Castro, Coimbra, Portugal.

## References

Alley RB (2016) A heated mirror for future climate. Science 352:151–152
Azul AM, Reis CS, Freitas H (2009) Let's look after the Earth: doing little can a change lot! [Vamos cuidar da Terra: fazer pouco pode mudar muito]. Coimbra University Press, Portugal, 81 pp. ISBN 978-989-26-0568-5 (in Portuguese)
Dewey J (1899) The school and society. The University of Chicago Press, Chicago
Dewey J (1938) Experience and education. The Kappa Delta Pi Lecture Series, Indianapolis
Freire P (1996) Pedagogia da Autonomia—Saberes Necessários à Prática Educativa Editora Paz e Terra. Coleção Saberes
Freire P (1970) Pedagogy of the oppressed. Penguin, Harmondsworth
Intergovernmental Panel on Climate Change (IPPC) (2014) Impacts, adaptation, and vulnerability. In: Field CB et al. (eds). Cambridge University Press, Cambridge
Malaguzzi L (1993) For an education based on relationships. Young Children 49(1):9–12
Malaguzzi L (1996) History, ideas, and basic philosophy: interview with Leila Gandini. In: Edwards C, Gandini L, Forman G (eds) The hundred languages of children: the Reggio Emilia approach—advanced reflections, 2nd edn. London.
Maslin M (2016) In retrospect: forty years of linking orbits to ice ages. Nature:208–210
Piaget, J (1926) The language and thought of the child. English translation, Routledge & Kegan Paul Ltd, London (3rd edition 1959)

Smidt S (2013) Introducing Malaguzzi. Routledge, London, UK
UNESCO (2007) The UNESCO climate change initiative. Available at http://unesdoc.unesco.org/images/0019/001901/190101E.pdf
Vygotsky L (1966) Play and its role in the mental development of the child. Voprosy psikhologii 12(6):62–76
Vygotsky L (1978) Interaction between learning and development. In: Gauvin, Cole (eds) Readings on the development of children.

# Feeling the Heat: The Challenge of Communicating ‘High-End’ Climate Change

Tim Rayner

## 1 Introduction: Unwelcome Messages to Be Communicated

In international climate policy, a global mean temperature rise of 2 °C above pre-industrial levels is widely considered to constitute unacceptably dangerous climate change (Jordan et al. 2013). Exceedance of this threshold within a few decades, however, is now looking extremely difficult to avoid. 2015 was the first year when global temperatures averaged 1 °C above pre-industrial,[1] and 2016 figures—boosted by the periodic *El Nino* effect—exceeded that.[2] On current emission trends, moreover, it is also increasingly likely that warming would not stop at 2 °C, but continue to 4 °C or even beyond if positive feedbacks take effect. Even relatively conservative institutions such as the World Bank (2014) have begun to warn of the dangers of, and urgent need for action to avoid, a rise of 4 °C.

The impacts brought by a greater than 2 °C temperature rise—severe drought, unprecedented heat waves, and major floods in many regions, with serious impacts on ecosystems and associated services—are considered beyond adaptation in many cases. The higher temperatures rise, the more pervasive, systemic, and irreversible impacts are likely to be (Field et al. 2014). Although physical impacts will vary, and some countries may find impacts within their own borders relatively limited, or in some cases even benign, in a highly globalised economy impacts of climate change will spread, affecting inter-dependent supply chains and flows of people and investment (Moser and Hart 2015).

[1]http://www.metoffice.gov.uk/news/releases/2016/2015-global-temperature.
[2]http://www.climatechangenews.com/2017/01/18/the-data-is-in-2016-was-the-hottest-year-on-record/.

T. Rayner (✉)
Tyndall Centre for Climate Change Research, University of East Anglia, Norwich, UK
e-mail: tim.rayner@uea.ac.uk



W. Leal Filho et al. (eds.), *Handbook of Climate Change Communication: Vol. 1*,
Climate Change Management, https://doi.org/10.1007/978-3-319-69838-0_20

Despite these stark risks and growing urgency highlighted by many scientists, citizens globally often report climate change well down their list of priority issues,[3] if they are aware at all—often not the case in developing countries (Ming-Lee et al. 2015). Much of the business community appears relatively unconcerned (Confino 2015), although the 2015 Paris Agreement and election of Donald Trump in 2016 raised the profile of climate change at high-level events such as the Davos summit (Blas and Shankleman 2017). Land-use and infrastructure planners, whose decisions profoundly shape current and future societal resilience to climate impacts, have yet to fully absorb the implications for their work—particularly on long-lasting infrastructure (Smith et al. 2011). So far, governments have not felt the need to commit to the scale of emission reduction needed to avoid crossing the 2 °C threshold; implementing the initial commitments under the Paris Agreement has been projected to put the world on course for a 2.7 °C increase *at best* (UNFCCC 2015).

Recent analyses have highlighted that the likelihood of high levels of warming and greater extremes, potentially occurring sooner than previously thought—e.g. 3 °C in the 2040s (Jackson et al. 2015), 4 °C rise in 2060s (Betts et al. 2011)—may be higher than generally appreciated. Yet arguably these risks are not being communicated adequately, either to political and economic decision makers or to wider publics, all of whom are still regularly informed that holding temperature rise to below 2 °C, with a reasonable probability of success, remains technically and economically feasible. Such a comforting message neglects the extreme economic, political and social challenges associated with rapid decarbonisation, and the technical and wider challenges surrounding the as yet unproven negative emissions technologies that modelling of 2 °C scenarios often assumes—rather heroically—will become available (Geden 2015). Close reading of the latest reports from the Intergovernmental Panel on Climate Change (IPCC), the scientific body charged with advising the world's governments, suggests that the impacts corresponding to higher levels of temperature increase are not only still largely unknown, but also effectively *unstudied* (King et al. 2015).

A further unwelcome message is that scientific uncertainty, often used by those reluctant to invest in responses as an excuse for delay, may not be eliminated or significantly reduced in important respects, but continue to prevent accurate predictions from being made. Moreover, to date the IPCC has failed to provide guidance on the relative likelihood of the main scenarios of future concentrations of greenhouse gases and other drivers of warming on which its climate change projections are based (Sanford et al. 2014). Instead, by treating each 'Representative Concentration Pathway' as equivalently plausible, the '2 °C remains feasible' narrative is reinforced. Since some scenarios are projected to lead to very divergent impacts, effective responses to manage climate risk depend heavily, in some cases, on the scenario actually realized. Sanford et al. (2014: 1) highlight how this leaves

[3]http://data.myworld2015.org/.

policymakers with 'no clear scientific guidance for confronting and managing the growing risk of high-magnitude warming'.

Underestimating the likelihood and consequences of exceeding 2 °C suggests a lower risk future than may in fact unfold, which, in turn, may lead to insufficient or even 'maladaptive' policy and planning responses, i.e. initiatives which may foster adaptation in the short-term but insidiously affect systems' long-term vulnerability and/or adaptive capacity (Field et al. 2014). Rather than proceeding incrementally, seeking to defend what is valued in the present day, adaptation may need to become 'transformational', requiring the abandonment of some currently cherished policy objectives, land uses and practices.

Communicators of climate change thus need to prepare for the likelihood that bad news in the present day will be superseded by worse, and that even when better news emerges—about upward global emission curves beginning to peak, for example (Le Quéré et al. 2016)—the persistent action needed to lower global emissions sufficiently, and prepare for severe impacts, will require that there is no 'let-up' in communication effort (Moser 2016). The scale of the challenge is clearly daunting.

Against this backdrop, the EU-funded HELIX project (High-End cLimate Impacts and eXtremes)[4] brought together individuals from diverse disciplinary and organisational backgrounds for a one-and-a-half day workshop, to discuss how the challenges associated with 'high-end' scenarios can most effectively be communicated to a range of audience types, including infrastructure planners, high-level politicians and civil servants, businesses and lay publics (Rayner and Minns 2015).[5] With this range of audiences in mind, the workshop was informed by presentations summarising lessons from literatures on both climate change communication and the science-policy interface (cf. UCL 2014; Pidgeon and Fischhoff 2011, which cover a similar mix of literatures). This paper takes inspiration from the presentations and facilitated activities that occurred, and the literature that informed them, and reports some key emerging conclusions. It does not claim to be a comprehensive record; nor did the workshop itself attempt to address some key arenas for communication, such as the mass media.

The next section highlights the main risks and opportunities associated with climate communication efforts, as identified in recent literature, in the context of some of the wider lessons of practice to date. The section after that reports on some of the conclusions and recommendations for a new agenda that emerged in the workshop, and elaborates on some of the literature of specific relevance to such discussions. The paper concludes by reflecting on the implications of a recent growth in concern reflected in European polling data, and summarising some of the

---

[4]www.helixproject.eu.

[5]'The Challenge of Communicating Unwelcome Climate Messages' (Cambridge, UK, 16–17 April 2015), included participants from, *inter alia*, UK government departments, consultancies in planning and communication, climate science, psychology, and art.

principles and objectives that should underpin communication in the context of increasingly likely high-end climate change.

## 2 The Risks and Opportunities of Communicating Unwelcome Climate Messages

Communicating the kind of unwelcome climate messages outlined in the introduction carries risks and opportunities. Though literature on psychological responses to the prospect of 'high end' climate impacts is only just beginning to emerge, Hamilton and Kasser (2009) usefully suggest that broadly three types of 'coping strategy' are likely to emerge in response: denial, maladaptation, and adaptive responses. Although Hamilton and Kasser focus on the level of the individual, it can be argued that denial, maladaptation and adaptive strategies are as likely to be followed by organisational and political decision makers as by individual members of the public. Following Hamilton and Kasser (2009: 5), the objective of communication can be framed as being to facilitate and encourage the more adaptive responses.

*Denial strategies* aim primarily at suppressing anxiety associated with predictions of climate disruption by not allowing such information to be accepted in the conscious mind. 'By denying the facts, no emotions need be felt' (Hamilton and Kasser 2009: 2). As expressed by Marshall (2014), we 'don't even think about it'. Commonly associated with climate sceptics with strong ideological objections to the implications that follow accepting high-end predictions—in particular the greater degree of state intervention and shift away from consumerist values needed to avert the worst—it is also a common response from individual citizens without such 'baggage'.

*Maladaptive coping strategies* (at the individual, psychological level) involve acknowledging and accepting unwelcome predictions about climate change up to a point, but blunting some aspects of the science or the associated emotions to reduce emotional impact.[6] These methods of coping can be unhelpful to the individual and at the same time lead to a worsening environmental (and socio-cultural) situation by blocking potential social responses to climate change, perpetuating a social norm that suggests no action is needed, or actively contributing to ecologically damaging behaviours. Recent commentary on the phenomenon of 'soft denial' may also be considered under this heading (Hoexter 2016).

Under the umbrella of maladaptive coping it is possible to differentiate several strategies. These include, *inter alia, distancing strategies* that serve to emphasise the time before the consequences of climate impacts are felt, and the physical distance at which they occur. Climate change is a 'creeping problem' (Moser and

[6]This is thus a more psychological phenomenon than the more concrete type of action referred to in the IPCC definition of the term maladaptation.

Dilling 2004), with the lack of a definite beginning, end, or deadline requiring that respondents create their own timeline. 'Not surprisingly, we do so in ways that remove the compulsion to act' (Marshall 2014: 64). *Diversionary strategies* include minor behaviour changes, or single actions, that allow individuals (or indeed institutional decision makers) to reassure themselves that a response has been made. The uncertainty inherent in predictions allows *unrealistic optimism* (Taylor 1989), or *motivated reasoning* about the dangers of climate change (see e.g. Lench et al. 2014). Rosemary Randall of the group Carbon Conversations identifies the phenomenon of 'regress': the child-like hope that other people, government, a 'superhero' or miracle techno-fix such as geoengineering will come to the rescue (Randall, pers.comm).

The third set of strategies is *adaptive coping*, deployed when individuals accept both the science, the projections, and the accompanying emotions, and then try to act on the basis of both (Hamilton and Kasser 2009: 2). They are adaptive, both in promoting psychological adjustment to new circumstances and in stimulating appropriate actions. Adaptive coping strategies are akin to the later phases of mourning and involve acceptance of, rather than resistance to, some of the pain and distress that follows recognition of the facts of climate science and their meaning.

The available literature suggests that individuals, and arguably whole cultures too, may deploy the above strategies sequentially, having to pass through stages of denial and maladaptation to finally acting adaptively. This is by no means necessarily a smooth, linear sequence. Even those who have deeply accepted climate science and projections go back and forth in and out of denial, grief, anger, engagement, acceptance, etc. (Moser and Berzonsky forthcoming). Following Smith et al. (2011), we can expect that the way individuals cope with a 4 °C world will be influenced by how wider societies and their institutions respond. If only a minority are pursuing adaptive coping while others persist in denial or maladaptive strategies, the former may feel isolated and disempowered, and governments and other institutions will be under less pressure to undertake significant, long-lived climate actions. Thus, facilitating the majority to take at least small steps on an action pathway may overcome this paralysis (Smith et al. 2011: 201).

Survey and other evidence in Western countries (Maibach et al. 2009) suggests that a minority of the population resists or ignores the evidence of climate science, while a majority adopts maladaptive coping strategies. However, a review of 75 studies (Moser 2014a) from across the world indicates how frequently individuals are reporting perceived changes in the environment, or believe they have experienced impacts—despite the documented tendency to 'distancing' noted above. Research has suggested that perceived local warming can influence wider risk perceptions (Zaval et al. 2014). From this, Moser concludes that societies may currently be in transition from a prevalence of threat denial to increasing acceptance. Hamilton and Kasser suggest that the shift from complacency to anxiety is likely to accelerate, and at some point, governments and other organisations will recognise the benefits of promoting and supporting adaptive coping strategies.

Thus, given the gap between the increasingly grim implications of climate science and the persistent human capacity to avoid confronting such realities, the question of how the unwelcome messages from high-end climate research can best be framed for decision-makers and publics alike is an essential precursor to appropriate mitigation and adaptation responses.

## 3 'Emotional Carrots' and Audience Segmentation: Lessons from Past Practice

In general, communications researchers have stressed the need for caution in instilling fear in communicating climate change (e.g. O'Neill and Nicholson Cole 2009; Feinberg and Willer 2011). The consensus has tended to be that 'emotional carrots' (e.g., hope, pride) as opposed to 'emotional sticks' (e.g., guilt, shame, and anxiety) are more likely to increase moral engagement with climate change (Markowitz and Shariff 2012). 'Fear messaging' can be counter-productive, leading to defensive avoidance or 'reactance' (against the suspicion of being manipulated). When people who regard the world as a fundamentally fair and stable place have that belief undermined, they are increasingly likely to ignore warnings and allow events to unfold without intervening (Feinberg and Willer 2011). Even when such messages are initially accepted, they can 'generate a state of prolonged worry and anxiety, over time changing to numbness, desensitisation and disengagement from the issue altogether' (UCL 2014: 60). Such research tends to be qualified by the proviso that such messages may be effective if accompanied by a clear action strategy to reduce risk. But individuals are likely to resent having the onus of responsibility for such action placed on themselves, while government and industrial interest groups apparently do too little (Lorenzoni et al. 2007).

The literature has also highlighted the need to go beyond modes of communication premised on the 'information-deficit model' that has assumed that the same facts are persuasive regardless of the audience. Segmentation analyses have identified a spectrum of audiences ranging from 'alarmed', 'concerned' and 'cautious', to 'disengaged', 'doubtful' and 'dismissive' (e.g. Maibach et al. 2009), each with a unique perspective on climate change and appropriate policy responses, and hence each requiring a tailored approach to communication, including distinct frames, goals, messages and messengers. Individuals accept and trust messages more readily when conveyed by people with similar views, e.g., US Republicans trusting Republican/conservative messengers; Democrats believing Democratic/liberal leaders; business leaders responding to other business leaders (Moser and Dilling 2004). Failure to frame the message according to audience type can be counter-productive, pushing people towards, or reinforcing, denial strategies.

Much of this work has been carried out in the context of climate change mitigation. In terms of adaptation, rather little empirical research exists on how communication may or may not lead to effective responses (Wirth et al. 2014). Recently, however, Hine et al (2016) have presented results running somewhat counter to the conventional wisdom, suggesting that strongly fear-based, negatively emotive messages can actually increase adaptation intentions, even with audience segments previously assumed to react adversely to such messaging (Hine et al. 2016). In Hine et al.'s research, the kind of climate change deniers previously considered immune reported increased adaptation intentions after exposure to fear-based messaging. One possible explanation may be found in the fundamentally different incentives associated with climate change mitigation and adaptation. Mitigation of climate change clearly falls within the class of pro-social behaviour, requiring, in the minds of many, only gradual behavioural change to reduce emissions and thereby provide a global public good. In this case, positive emotions are more effective than negative in driving change (Markowitz and Shariff 2012). Adaptation, on the other hand, is different, often requiring more immediate responses to short-onset threats such as floods and fires. The message is more focused on the need for action to protect one's family and property, rather than wider society and future generations, and more likely to be well received by less 'pro-social' audiences (Hine et al. 2016: 8). If this research can be replicated, it may mean that tailoring and targeting of climate messages, which in some instances can be prohibitively expensive, may not always be required.

## 4 Towards Better Communication of Unwelcome Climate Messages

The workshop discussions centred on a number of key themes, about which there was much, though not total, agreement. The first four in particular are focused on communication from scientists and risk analysts to policy makers, planners and other 'elite' decision makers. Numbers four to six are more widely applicable.

### *4.1 The Full Extent of Climate Risk Needs to Be Framed and Communicated Better*

The currently dominant approach to the framing and communicating of climate risk, as embodied by the work of the IPCC, arguably has important weaknesses. Climate science is too often framed in terms of attempted prediction, or the 'central tendency' (what is deemed 'most likely'). This is despite the fact that it is the 'worst-case', not accurate prediction, that is often most relevant to risk-based decision making (civil servant, UK Foreign Office, workshop presentation). On this 'worst case', however, there is currently relatively little communication. Close reading of the latest IPCC report suggest that the impacts corresponding to high levels of temperature increase are not only relatively uncommunicated, but actually—and astonishingly—relatively unstudied (King et al. 2015).

Also arguably underplayed in current IPCC assessments is that the risks of climate change *tend to increase over time*. It is therefore important to recognize how probabilities change, and avoid using the year 2100 as an arbitrary cut-off point—especially for the higher emission scenarios—because in some cases impacts and the probability of major climate system disruption will still be increasing (King et al. 2015).

Risk assessments tend to differ from predictions/projections in the order in which key questions are asked. Scientific predictions typically ask first what is most likely to happen, and then how that (most likely case) may cause impacts. A risk assessment does the opposite, asking first what would affect society most severely, and then how likely that might be (as a function of time, if it happens to change over time). The two approaches reflect different underlying motivations about what information is considered important and lead to very different answers. Instead of highlighting a range of impacts, the seriousness of which is uncertain, it is more effective to say: 'There is an awful possible future and we can't rule it out' (civil servant, UK Foreign Office, workshop presentation).

Alternative, risk-based assessments are available which, unlike the IPCC, make a judgment about the relative likelihood of different emissions pathways (King et al. 2015). The assessment by King et al. (2015), including consideration of the systemic risks arising from interactions between changes in the physical climate and complex human systems, is in large part a security risk assessment, something that the scientists in the IPCC's working groups are not necessarily competent to perform. Recognizing the depth of uncertainty about the future state of complex systems, the framework uses the tools of scenario exercises and war-gaming to help explore the future.

## 4.2 New Narratives and Analytical Frameworks Are Needed to Inform Robust Scenario Planning and Decision Making

Since some scenarios are projected to lead to very divergent future impacts, effective responses to manage climate risk depend heavily, in some cases, on the scenario that will actually be realised. As Smith et al. (2011) have highlighted, simply presenting the prospect of a 4 °C world as an increasingly likely one is unhelpful and disempowering unless the complexity of dealing with the thousands of decisions that might be affected by climate change can be simplified.

Greater knowledge from climate science is of little use without analytical tools and wherewithal to use that knowledge. Infrastructure and land-use planners need to be reminded of the sophisticated analytical frameworks that are available for robust decision making that can reduce complexity in dealing with the wide range of decisions sensitive to future climate change (e.g. Smith et al. 2011). In situations where near-term decisions have long-term consequences, adaptation planning needs to move from incremental to transformative as a contingency under scenarios of truly disruptive climate change. Pioneering examples include plans for relocation of development from floodplains around London after 2060 and the creation of water-efficient maize for Africa, both of which require planning and investment now to manage impacts in decades to come (Sanford et al. 2014).

Policy makers are often nervous that acknowledging the increasing likelihood of going beyond 2 °C of warming could induce fatalism and lead to a reduction of effort in mitigation (Jordan et al. 2013). But this does not need to be the case. As an over-arching narrative, there is potential in the ABC Framework: **A**im to stay below 2 °C; **B**uild and budget (infrastructure) assuming 3–4 °C; **C**ontingency plan for 5–7 °C of warming (Mabey et al. 2011).[7] This has potential to motivate difficult but much-needed dialogue and planning for the impacts of climate change under high-magnitude warming, addressing the risk that planners and decision-makers may cling to 'unrealistic optimism' as a maladaptive coping strategy. At the same time, building such a science-based dialogue about the costs and challenges of transformative adaptation may even have the effect of motivating greater mitigation effort (Mabey et al. 2011).

## *4.3 Potential High-End Climate Impacts Important to Specific Decision Makers Need to Be Communicated, in Recognisable Language*

Assessments of specific, local, or sectoral risks of climate change should be directed to those with specific, local or sectoral responsibility. Politicians' efforts to communicate the value of climate policy now frequently make use of risk language, such as emission reduction policies acting as an 'insurance policy'. Such language speaks particularly well to the business sector, familiar as it is with assessing investment, insurance and other types of risk. Painter (undated), cites the *Risky Business* initiative, explicitly aimed at business, which uses a risk management frame to report on the risks for agriculture, energy and coastal real estate in the US. The initial report (Houser et al. 2014) received widespread coverage in the US and UK business press (Painter, undated). According to PwC, a company with considerable experience engaging the business community on climate change:

> Science debates have alienated many from the real issues. In reality, when working on climate risk assessments with companies, [we] don't hear much debate about climate sensitivity or the heat of the ocean in 50 years' time. It's not the language of business decisions. [….] Instead, companies talk in terms of operational performance, asset management, business continuity, security of supply of commodities, energy and water, workforce health and well-being. How will investment in resilience and climate mitigation

[7]Contingency planning for 5–7 °C presumably takes decision making into the realm of climate engineering.

> today, show a rate of return in the future? It's okay that there are many uncertainties - businesses are experts at making decisions around uncertainty, and in scenario planning. Bridging this communication divide will be critical for action (Herweijer 2013).

Terms such as 'preparedness' and 'resilience' have been shown to engage audiences such as city-regional governance networks and sustainability professionals more than the somewhat abstract concept of adaptation (see e.g. Kythreotis and Bristow 2016). Such framing, however, brings its own risks. Kythreotis and Bristow (2016) refer to a 'resilience trap': while the concept is not without uses, it could be used to justify short-term strategies, and erroneous re-badging of existing strategies as something new, in ways that may overlook the need for more transformational, long-term approaches. This highlights the importance of inclusive planning processes, and the importance of communicating worst-case scenarios in this context.

### 4.4 More Dialogue, Less Debate and Lecturing

Fuller engagement with unwelcome climate messages and their policy implications requires a more strategic approach than has been witnessed to date. Firstly there is a requirement for more listening and mutual sharing, whereby communication becomes a genuine dialogue (Moser and Berzonsky 2015), and effort is made to understand intended audiences and their decision needs prior to communication design (Wirth et al. 2014). Secondly, in designing dialogical processes sensitive to audience and context, there is a need to draw from a range of skills including natural science, decision science, social science as well as communications specialists.

The kind of approach elaborated at the workshop has much in common with the approach to the communication of climate risk and uncertainty advocated by Pidgeon and Fischhoff (2011). Their vision of a more strategic approach to communication also requires the convening of cross-disciplinary teams that can 'transcend academic norms' and become 'boundary organizations', 'chartered both to conduct basic science and to translate its results into decision-relevant terms'. It 'requires the resources to support research, design and evaluation services for all scientists hoping to communicate their work' (Pidgeon and Fischhoff 2011: 40).

Individuals who are able to generate their own 'unwelcome messages' in dialogic forums informed by scientific input, rather than receiving lectures, are more likely to accept them (Moser and Berzonsky 2015). Such a two-way dialogic approach is inevitably resource- and time-intensive, and may therefore be too expensive for communicating on climate science to the public on a significant scale. On the other hand, if more selectively used (i.e. with key leaders and elites), it may prove to be far more effective than so-called mass communication.

### 4.5 *'Friendly Communicators' Must Recognise Emotions, Loss and the Need for 'Active Hope'*

Knowledge of the implications of high-end climate change scenarios constitutes a burden for those who are generating it, attempting to communicate it, and hearing about it as lay individuals. Communicators must take care to be more than mere 'narrators of doom' (Moser 2014b: 407), and recognise that emotions of anxiety, fear, guilt, loss and helplessness are likely to be aroused. Sensitive handling of such emotions is required by both scientists conducting research and those (including scientists themselves) seeking to communicate the science. If scientists fail to acknowledge the emotional weight of their findings, lay audiences may well become suspicious or simply shut down (Moser 2014b). Scientists must instead act as 'friendly communicators' (Moser and Berzonsky, forthcoming), being emotionally honest, talking openly about their hopes, fears, and anxieties, and willing to tell personal stories. In his 2015 encyclical, *Care for Our Common Home*, Pope Francis (2015: 16) has stated that '[o]ur goal is not to amass information or to satisfy curiosity, but rather to become painfully aware, to dare to turn what is happening to the world into our own personal suffering and thus to discover what each of us can do about it'.

Talking in such personal terms may cause a certain discomfort among scientists, some of whom may fear a loss of professional reputation. But the numbers willing to take this risk, and suffering no apparent loss of reputation, are growing.[8] Lord Stern introduced his two-week old granddaughter at a TEDx talk in New York, asking what kind of century she would be living through.[9] Climate change communication specialist Susanne Moser's recent attempts to engage at a more emotional level have met with gratitude (Moser 2014b).

In order to avoid the tendency for denial or maladaptive responses, audiences themselves need more 'safe spaces' to make emotional connections that can instead open up energy and engagement. According to the developers of the 'active hope' concept, Macy and Johnstone (2012: 70): 'When people are able to tell the truth about what they know, see and feel is happening to their world, a transformation occurs. There is an increased determination to act and a renewed appetite for life'.

Communication regarding the scale of the threat needs to be accompanied with a sense that a response that is in some way effective, allowing a community to maintain its dignity, is possible. To overcome feelings of 'overwhelm', 'people need true hope' (Moser 2014b). Such hope 'can only be constructed from realistic goals, a clear or at least imaginable path, from doable tasks and a meaningful role in addressing the problems at hand' (ibid: 407) (see also Moser and Berzonsky, forthcoming). As noted in the introduction, adaptation may need to become 'transformational', requiring the abandonment of some currently cherished policy

---

[8]See entries on the website http://isthishowyoufeel.weebly.com/this-is-how-scientists-feel.html.

[9]www.ted.com/talks/lord_nicholas_stern_the_state_of_the_climate_and_what_we_might_do_about_it?language=en.

objectives, land uses and practices. But recognition of this, and subsequent decisions, need legitimacy. Decision making frameworks of the kind described by Smith et al. (2011) can have a role in turning this into practical planning. Indeed, '[u]nless the complexity of dealing with the thousands of decisions that might be affected by high-end climate change can be simplified, and the emotional implications are handled sensitively, simply presenting audiences with the prospect of a 4 °C world is more likely to provoke rejection, fatalism and disengagement than adaptive responses' (Smith et al. 2011: 204–5).

## 4.6 Individuals Must Come Together to Learn, Deliberate and Respond as a Collective

Individuals deal with climate change largely in isolation, with little public or political discourse against which personal views may be calibrated (Marshall 2014). That being so, their default assumption is likely to be that few people are concerned enough to contemplate radical solutions. Nurturing 'communities of conviction' may help to overcome the reluctance of individuals to fully engage with unwelcome climate messages, and contemplate such solutions. Marshall (2014) argues that faith groups, which provide support for their members, allowing their personal beliefs to be validated by their peers, hold lessons for communicating and generating action on climate change.

Carbon Conversations groups in the UK offer a supportive group experience that aims to help people halve their personal carbon footprint, dealing with the difficulties of change by connecting to values, emotions and identity. The groups are based on a psychological understanding of how people change, offering:

- space for people to explore what climate change means for themselves, their families and aspirations
- permission to share hopes, doubts and anxieties
- time to work through the conflicts between intention, social pressure and identity
- reliable, well-researched information and practical guidance on what will make a difference
- support in creating a personal plan for change.[10]

Taking action and thereby exerting some control over a threatening situation is known to be an effective response to emotional distress (Hamilton and Kasser 2009). The sense of shared purpose associated with working more collaboratively with others to protect the common interest can reinforce this. Problem-solving as an appropriate coping strategy might impel people to work with others so as to prepare

[10]http://www.carbonconversations.org/what-carbon-conversations.

for a changed climate, through political activism or joining local groups or councils that might develop mitigation and adaptation measures (Moser and Dilling 2007).

## 5 Conclusions

In an era when climate science is indicating that high-end climate change is a more and more likely prospect, the objective of communication efforts can be framed as being to encourage the more adaptive responses to this reality, both on the part of individuals and various elite decision makers. The points outlined above constitute possible elements of a new agenda for communication, but one whose substantial resource implications are obvious. It places significant new demands on scientists and other communicators and the venues in which they communicate, requiring a move beyond traditional formats such as lectures and debates. To build an emotionally literate relationship with *particular audiences* takes more time and *attention to process*; first to hear an audience's concerns, then to offer science and contribute to the development of possible responses, through interdisciplinary teams. Although the scale of this challenge is daunting, given the urgency of the threat of climate change, it should not be regarded as an unreasonable ambition.

At the same time, it is clearly not possible to 'scale up' the kind of dialogical, 'friendly', inter-disciplinary team-based approach outlined here to entire populations. Forms of communication based on information deficit model-based assumptions will remain prevalent. Indeed, it is important to stress that these do have their rightful place, e.g. for correcting error or misunderstanding, for establishing credentials (UCL 2014), or where basic scientific understanding of climate change has yet to develop—as in many developing countries. Findings about communicating climate impacts from Hine et al. (2016) suggest that there may be less need for 'tailoring' of messages in certain cases.

In principle, stark findings about impending impacts societies stand to face should climate change turn out to be at the 'high end' should be honestly delivered, to decision-making elites and wider publics. The critical challenge will be to accompany the communication of troubling findings with a sense that a meaningful collective, societal response, and ultimately a dignified future of some kind, is possible. As research by Hine et al. (2016) has suggested, this may even prompt a readiness to contemplate adaptation in groups previously prone to 'denial'. The transformational implications (for certain, currently cherished policy objectives, land uses and practices that may no longer be viable) will be evident from the science, though scientists themselves should not be expected to determine specific policy implications. Instead, the scientific evidence needs to inform wider deliberative institutions, and should itself ideally be the product of 'co-production' with those who have most at stake (UCL 2014). The organisation of the international research initiative Future Earth, with different regional centres for Europe, Middle East and North Africa, Asia, North America, Latin America and Africa, offers new opportunities for global change research to be both co-produced and communicated in a way that takes account of local and regional contexts.

Though the risk that communicating an increased likelihood of 'worst case' impacts causes fatalism, undermining the continuing mitigation imperative is a real one, this paper has suggested that it can be lessened (Jordan et al. 2013). As the science blogger David Roberts has put it:

> … Even if it does become finally, physically impossible to hit 2 °C, so what? Carbon emissions still need to be driven to zero as quickly as possible to avoid even higher temperatures. The fight remains the same, no matter the temperature outcome (Roberts 2015).

At the time of writing, evidence suggests that more and more citizens are recognising the significant threat posed by climate change. A survey of European attitudes (Pew Research Centre 2016) shows the number of respondents agreeing that 'climate change is a major threat to our country' significantly up from 2013 levels for several countries (see Fig. 1). Even in the Trump era, a majority of Americans (61%) say they are 'very' or 'somewhat' worried about global warming (Leiserowitz et al. 2017a); the same percentage of Trump supporters actually support taxing or regulating greenhouse gas emissions (Leiserowitz et al. 2017b). If Moser is right, and such developments indicate that societies are in transition—

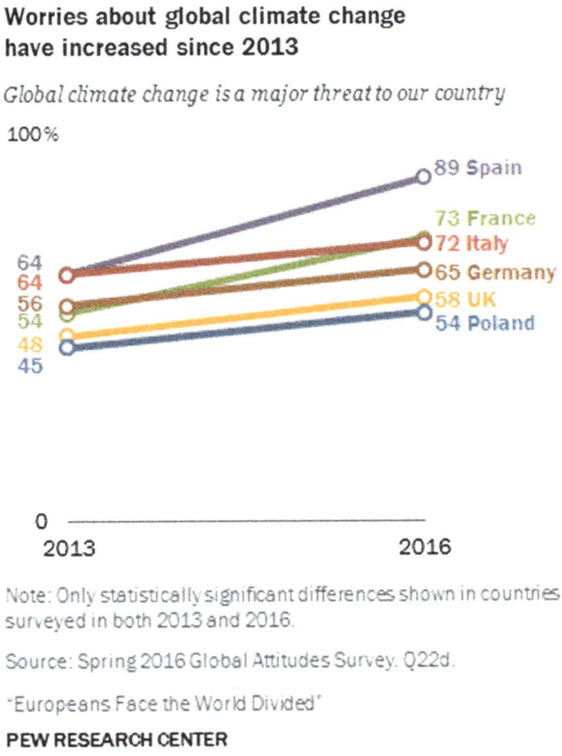


**Fig. 1** Increased concern about the threat of climate change in Europe

albeit a slow and hesitant one—from a prevalence of threat denial to increasing acceptance, and Hamilton and Kasser are right that the shift from complacency to anxiety is likely to accelerate, and eventually prompt governments and other organisations to promote and support more adaptive psychological and policy responses, then communicators need to be ready with appropriate strategies, and policy makers be willing to fund them. Along with others that it has drawn from, this paper begins to offer elements of such a strategy.

**Acknowledgements** The author acknowledges the support of the EU-funded High-End cLimate Impacts and eXtremes (HELIX) project (grant agreement no 603,864) and especially Asher Minns, who initiated the idea of a workshop on 'Unwelcome Messages'. The participation of workshop participants in April 2015, and comments on a working paper by Simon Sharpe, Susanne Moser and Chris Rapley are gratefully acknowledged.

## References

Betts R et al (2011) When could global warming reach 4 °C? Philos Trans R Soc A 369:67–84

Blas J, Shankleman J (2017) Davos Elite focus on climate change, ignoring Trump's skepticism. Available at https://www.bloomberg.com/news/articles/2017-01-15/ignoring-trump-s-skepticism-davos-elite-bets-on-climate-change?cmpid=GP.Markets

Confino J (2015) How concerned are CEOs about climate change? Not at all. The Guardian, 20 Jan. Available at http://www.theguardian.com/sustainable-business/2015/jan/20/global-warming-business-risks-government-regulation-taxes

Field C et al (eds) (2014) Climate change 2014: impacts, adaptation and vulnerability. Contribution of working group II to the fifth assessment report of the intergovernmental panel on climate change. Cambridge University Press, Cambridge

Feinberg M, Willer R (2011) Apocalypse soon? Dire messages reduce belief in global warming by contradicting just-world beliefs. Psychol Sci 22(1):34–38

Francis (2015). Encyclical letter Laudato Si' of the Holy Father Francis on care for our common home. Available at http://w2.vatican.va/content/francesco/en/encyclicals/documents/papa-francesco_20150524_enciclica-laudato-si.html

Geden O (2015) Climate advisers must maintain integrity. Nature 521(May 7):27–28

Hamilton C, Kasser T (2009) Psychological adaptation to the threats and stresses of a four degree world. Paper presented at the four degrees and beyond conference, Oxford University 28–30 Sept

Herweijer C (2013) IPCC/ PwC comment on impact for business, policy, carbon budget, science debate. Available at http://PwC.blogs.com/press_room/2013/09/ipcc-PwC-comment-on-impact-for-business-policy-carbon-budget-science-debate.html

Hine D et al (2016) Preaching to different choirs: how to motivate dismissive, uncommitted, and alarmed audiences to adapt to climate change? Glob Environ Change 36:1–11

Hoexter (2016). A pocket handbook of soft climate denial. Available at http://neweconomicperspectives.org/2016/10/pocket-handbook-soft-climate-denial.html

Houser T, Hsiang S, Kopp R, Larsen K (2014) Economic risks of climate change: an American prospectus. Columbia University Press, New York

Jackson RB, Friedlingstein P, Canadell J, Andrew R (2015) Two or three degrees $CO_2$ emissions and global temperature impacts. Bridge 45(2):16–21

Jordan A, Rayner T, Schroeder H et al (2013) Going beyond two degrees? The risks and opportunities of alternative options. Clim Policy 13(6):751–769

King D et al (2015) Climate change: a risk assessment. Centre for Science and Policy, Cambridge

Kythreotis A, Bristow G (2016) The 'resilience trap'. Reg Stud. doi:10.1080/00343404.2016.1200719
Le Quéré C et al (2016) Global carbon budget 2016. Earth Syst Sci Data. http://www.earth-syst-sci-data.net/8/605/2016/
Leiserowitz A et al (2017a) Climate change in the American mind. Yale University and George Mason University. Yale Program on Climate Change Communication, New Haven, CT
Leiserowitz A et al (2017b) Trump voters & global warming. Yale University and George Mason University. Yale Program on Climate Change Communication, New Haven, CT
Lench HC, Smallman R, Darbor K, Bench S (2014) Motivated perception of probabilistic information. Cognition 133:429–442
Lorenzoni I, Nicholson-Cole S, Whitmarsh L (2007) Barriers perceived to engaging with climate change among the UK public and their policy implications. Glob Environ Change 17:445–459
Mabey N, Gulledge J, Finel B, Silverthorne K (2011) Degrees of risk. E3G, Brussels
Macy J, Johnstone C (2012) Active hope. New World Library, Novato
Maibach E, Roser-Renouf C, Leiserowitz A (2009) Global warming's "Six Americas" 2009. Yale Project on Climate Change and George Mason University Center for Climate Change Communication
Markowitz E, Shariff A (2012) Climate change and moral judgement. Nat Clim Change 2:243–247
Marshall G (2014) Don't even think about it: why our brains are wired to ignore climate change. Bloomsbury, London
Ming-Lee T, Markowitz E, Howe P, Ko C, Leiserowitz A (2015) Predictors of public climate change awareness and risk perception around the world. Nat Clim Change 5:1014–1020
Moser S (2014a) Communicating adaptation to climate change: the art and science of public engagement when climate change comes home. WIREs Clim Change 5:337–358
Moser S (2014b) Whither the heart(-to-heart)? Prospects for a humanistic turn in environmental communication as the world changes darkly. In: Hansen A, Cox R (eds) Handbook on environment and communication. Routledge, London, pp 444–456
Moser S (2016) Reflections on climate change communication research and practice in the second decade of the 21st century. WIREs Clim Change 7:345–369
Moser S, Dilling L (2004) Making climate hot: communicating the urgency and challenge of global climate change. Environment 46(10):32–46
Moser S, Dilling L (eds) (2007) Creating a climate for change: communicating climate change and facilitating social change. Cambridge University Press, Cambridge, UK
Moser S, Hart J (2015) Societal teleconnections and the future of climate change impacts studies. Clim Change 129(1–2):13–26
Moser S, Berzonsky C (2015) There must be more: communication to close the cultural divide. In: O'Brien, Silboe E (eds) The adaptive challenge of climate change. Cambridge University Press, Cambridge
Moser S, Berzonsky C (forthcoming) Hope in the face of climate change. Draft available at http://www.susannemoser.com/publications.communications.php
O'Neill S, Nicholson-Cole S (2009). Fear Won't Do It. Sci Commun 30(3):355–379
Painter J (undated) A new narrative. In: UNA-UK (eds) Climate 2020: facing the future, pp 30–31. Available at https://www.una.org.uk/climate-2020-facing-future
Pew Research Center (2016) Europeans face the world divided. Available at http://assets.pewresearch.org/wp-content/uploads/sites/2/2016/06/Pew-Research-Center-EPW-Report-FINAL-June-13-2016.pdf
Pidgeon N, Fischhoff B (2011) The role of social and decision sciences in communicating uncertain climate risks. Nat Clim Change 1:35–41
Rayner T, Minns A (2015) The challenge of communicating unwelcome climate messages. Working paper 162. Tyndall Centre for Climate Change Research, Norwich
Roberts D (2015) The news on climate is awful. So now what? Updated on May 19, 2015. Available at http://www.vox.com/2015/5/19/8624229/the-news-on-climate-is-awful-so-now-what
Sanford T, Frumhoff PC, Luers A, Gulledge J (2014) The climate policy narrative for a dangerously warming world. Nat Clim Change 4:164–166

Smith MS, Horrocks L, Harvey A, Hamilton C (2011) Rethinking adaptation for a 4 °C world. Philos Trans R Soc A 369:196–216

Taylor S (1989) Positive illusions: creative self-deception and the healthy mind. Basic Books, New York

UCL (2014) Time for change: climate science reconsidered. Report of the UCL Policy Commission on the Communication of Climate Science. University College London, London

UNFCCC (2015) Synthesis report on the aggregate effect of the intended nationally determined contributions. FCCC/CP/2015/7. United Nations Framework Convention on Climate Change, Bonn

Wirth V, Prutsch A, Grothmann T (2014) Communicating climate change adaptation. GAIA 23(1):30–39

World Bank (2014) Turn down the heat: confronting the new climate normal. World Bank, Washington, DC

Zaval L et al (2014) How warm days increase belief in global warming. Nat Clim Change 4: 143–147

# Values as a Route to Widening Public Concern About Climate Change

**Tom Crompton and Shanna Lennon**

## 1 Introduction

Insights from psychology are increasingly being brought to bear in work on climate change (for example: Crompton 2008; Crompton and Kasser 2010; Gifford 2011; Gifford and Nilsson 2014; Stern 2011; Swim et al. 2011). It is hoped that such work will help to deepen practitioners' understanding of how communications about climate change can best contribute to strengthening public acceptance of—and vocal demand for—ambitious policy interventions.

However, it is increasingly apparent that public attitudes towards climate change are importantly shaped by a range of aspects of people's identity. These aspects are, in turn, influenced by factors extending far beyond those explicitly focused on climate change and its communication. Dan Kahan expresses this point:

> [Responding to a survey question by saying] I "do/don't believe in climate change" doesn't convey what you know about science—generally, or in relation to the climate. *It expresses who you are*. Al Gore has described the climate change debate as a "struggle for the soul of America." He's right. But that's exactly the problem. Because in "battles for the soul" of America, the stake that culturally diverse individual[s] have in forming beliefs consistent with their group identity dominates the stake they have in forming beliefs that fit the best available evidence." [Emphasis in original.] (Kahan 2014).

A person's group identity is importantly rooted in the values that a person holds, and perceives themselves as holding, in common with other people in their 'in-group' (Crompton and Kasser 2009). Values are the aspects of people's

T. Crompton (✉)
Common Cause Foundation, Nant y Filltir, Penegoes SY20 8UW, UK
e-mail: tcrompton@commoncausefoundation.org

S. Lennon
Manchester Museum, The University of Manchester, Oxford Road,
Manchester M13 9PL, UK
e-mail: Shanna.Lennon@manchester.ac.uk



W. Leal Filho et al. (eds.), *Handbook of Climate Change Communication: Vol. 1*,
Climate Change Management, https://doi.org/10.1007/978-3-319-69838-0_21

7identities that reflect what they deem to be desirable, important, and worthy of striving for in their lives (Rockeach 1973; Schwartz 1992). Clearly, a person's values are shaped by an extensive range of different influences that extend very far beyond his or her exposure to communications about climate change.

For this reason, it is important that communicators working on climate change take a broad perspective—one that extends far beyond a focus on communicating about climate change. This perspective must encompass an understanding of how people's group identities are developed and sustained, and the role of shared values in these processes.

It is understandable that this insight will prove challenging for many organisations which work to better communicate climate change. In this paper, we do not seek to diminish the importance of such communications, or of bringing an understanding of social psychology to bear in improving the effectiveness of these communications in strengthening public support for action on climate change. Rather, we seek to advance the case that communicating *on climate change* in the most effective ways possible, crucially important as this is, is likely to only ever present a very partial response to the challenge of strengthening public concern about climate change, or public support for policies and practices aimed at mitigating climate change.

This paper is organised as follows: Sect. 2 presents an overview of the social psychology of values. Sect. 3 presents evidence for the importance of incorporating an understanding of values into the design of communications explicitly focused on climate change. This is a necessary, but (for the reasons highlighted above) insufficient response to an understanding of values. Sect. 4 presents evidence for the importance of communications about causes seemingly unrelated to climate change on a person's inclination to voice concern about climate change. This section focuses on a study exploring the interdependencies between the work of two very different non-governmental organisations, each working on seemingly unrelated causes. Sect. 5 takes this discussion further, reflecting on the disparity between people's own values and their perceptions of the values priorities of typical compatriots. Sect. 6 reflects on a specific example of a programme of work in a museum that seeks to respond to the understanding of values outlined in this paper. Sect. 7 reflects on some areas of limitation. Sect. 8 summarises our conclusions.

## 2 Values: An Overview

A wide range of studies has shown that values influence both people's attitudes and behaviours. They affect political persuasions, willingness to participate in political action, career choices, ecological footprints and feelings of personal wellbeing (Schwartz 2011). Social psychologists have identified a set of *intrinsic* and *self-transcendent* values (henceforth, I/ST values) that are consistently found to underpin both concern about social and environmental problems, and action in line with this concern (from decisions impacting on a person's own environmental footprint, to voting choice, to political activism) (Kasser and Ryan 1996; Grouzet et al. 2005). These, then, are the values that organisations setting out to promote public expressions of concern about climate change might reasonably seek to engage and strengthen.

I/ST values include values of *social justice*, *equality*, *unity with nature* and *self-acceptance*. Simply engaging these values—any of them, it seems—leads to increased support for social and environmental causes (Crompton and Kasser 2010 and references therein). I/ST values stand in opposition to *extrinsic* and *self-enhancement values* (henceforth, E/SE values). These include concern about *wealth*, *social status* or *public image*: aims shown to be associated with lower levels of concern about environmental problems, higher environmental footprints, and lower support for environmental policy interventions (Crompton and Kasser 2010). So, just as public expressions of concern about social or environmental problems are promoted through engaging and strengthening I/ST values, influences that serve to engage and strengthen E/SE values are found to undermine such expressions of concern.

This has been demonstrated experimentally. For example, drawing a person's attention, even subtly, to an E/SE value leaves them less likely to offer help to another person (Maio et al. 2009; Vohs et al. 2006) and less likely to express concern about climate change (Chilton et al. 2012). While further research is necessary to fully explain the processes involved, it seems likely that when E/SE values are activated or engaged, people become both more concerned about aims consistent with those values (e.g., power, status, money, competition) and less concerned about aims that are inconsistent with E/SE values, namely the I/ST values that promote greater care, empathy, and environmental concern.

Most experimental investigations of such effects are necessarily short-term. But it seems that repeatedly priming values leads these to become strengthened in a more dispositional way. Research shows that repeatedly exposing people to messages promoting E/SE values tends to lead them to attach relatively greater importance to these values. For instance, studies have found that people express E/SE values more strongly when their peers or parents express these values, when they watch more television or are exposed to other forms of media, and when they live in a more economically competitive social and political context (Bardi et al. 2009; Crompton and Kasser 2009; Kasser et al. 2004). Thus, "altruism, generosity, solidarity, and civic spirit"—constructs closely aligned to I/ST values—have been likened to "muscles that develop and grow stronger with exercise" (Sandel 2012: 130).

## 3 Values in Communications About Climate Change

As discussed in Sect. 2, there is a range of evidence showing that priming I/ST values has the effect of promoting pro-social and pro-environmental attitudes and behaviours, and that priming E/SE values has the opposite effect. But values may be implicit *in communications focused on climate change*, and these values may either promote, or undermine, the commitment of an audience to express concern about climate change. For example, Bolderdijk and colleagues (2013) report that uptake of a behaviour likely to reduce a person's personal carbon dioxide emissions (seeking out information on the benefits of checking one's tyre pressure) was more effectively promoted through appeals to the *environmental* benefits than the *financial savings* that are associated with this behaviour.

Further evidence for this effect is found in a study by Crompton and colleagues (2014). Here participants in a large online panel were asked to read short texts presenting reasons for expressing concern about biodiversity loss. In one condition, these texts were framed to prime I/ST values. For example, part of one text read:

> The importance of environmental protection is still often overlooked and is not adequately reflected in planning and policy. One reason for this is that people's inherent appreciation of, and love for, the natural world is often forgotten. Reminding people of the intrinsic importance that they attach to nature can help to address this problem.

In another set of conditions, participants were asked to read a text about biodiversity conservation written to prime E/SE values:

> Natural assets, and the benefits that they provide, are still often overlooked and are not adequately reflected in planning and policy. One reason for this is that the financial value of the environment, and the commercial benefits that people derive, is often overlooked. Putting a monetary value on nature can help to address this problem.

In a third condition, participants were asked to read 'mixed' texts that combined elements of both the I/ST and E/SE texts.

Participants were then presented with a questionnaire asking about their intentions to support the work of a conservation organisation. This survey explored both intentions to offer non-financial support to a biodiversity conservation organisation (writing to a member of parliament, joining a public meeting or volunteering) and to offer financial support (to donate to an organisation working conservation).

The I/ST text was found to be significantly more effective in leading participants to state an intention to offer non-financial support to an organisation working to conserve biodiversity. Results for a measure of willingness to offer financial support did not reach significance.

Further, the study examined whether this effect was found even among people who attached particular dispositional importance to E/SE values. Three months prior to the study participants were asked to complete a Portrait Values Questionnaire (Schwartz et al. 2001). No interactions were found between a participant's prior values orientation and the type of prime presented to a participant in influencing either attitudinal or intention responses. That is, E/SE texts were no more effective in eliciting expressions of concern from participants who were relatively more E/SE-oriented than from participants who were relatively more I/ST-oriented. Conversely, I/ST texts were no less effective in eliciting expressions of concern about biodiversity conservation from participants who were relatively more E/SE-oriented than from participants who were relatively more I/ST-oriented. In other words, the results showed that I/ST texts were significantly more effective than E/SE (or, indeed, mixed) texts in motivating intention to offer non-financial support *irrespective of a participant's prior values disposition.*

Developing these insights further, Evans and colleagues (2013) examined the whether the values implicit in a communication about one environmental behaviour influence the likelihood of 'spillover' into other pro-environmental behaviours. In this study it was found that recycling rates were influenced by the nature of information that participants were given about a different behaviour—car-sharing. Recycling behaviour was found to be significantly higher among participants who received information about the environmental benefits of car-sharing (i.e. for whom car-sharing was framed in terms of I/ST values) than among participants who received either information about the financial benefits of car-sharing (i.e. for whom car-sharing was framed in terms of E/SE values), or information about both the environmental and financial benefits of car sharing. The authors conclude that:

> Congruent with value theory, positive spillover from one environmental message to another behaviour (car-sharing to recycling) may occur primarily when self-transcending reasons alone are made salient. (Evans et al. 2013: 122)

Such studies suggest that communicators working on climate change should be careful about the values that they prime through their communications. At least in the case of crafting communications intended for public audiences, results such as those outlined above suggest that communicators would do well to consistently present the case for action on climate change in terms of I/ST values, while avoiding appeals to E/SE values.

Today communicators and campaigners working on issues such as climate change appeal to a wide range of potential motivations for people to express concern about climate change, including motivations that seem likely to prime E/SE values. For example, appeal may be made to the projected negative economic impacts of climate change, the contribution that a vibrant renewables sector might make to national economic competitiveness, or the financial savings achievable by making domestic energy-efficiency savings. Yet research such as that presented above suggests that appeals of this nature may inadvertently erode public commitment to support action on climate change.

Other communication strategists advocate a 'values matching' strategy, seeking to align the values implicit in a communication with the values of particular audience segments (Rose 2014). However, empirical studies find that framing communications through I/ST values is likely to be more effective in eliciting expressions of concern even among people for whom E/SE values are of particular importance (Crompton et al. 2014).

Of course, business leaders may have a fiduciary duty to demonstrate to shareholders that steps they take to reduce a company's greenhouse gas emissions are financially prudent, and government ministers may be structurally constrained to demonstrate the economic benefits arising from action to mitigate or adapt to climate change. In engaging audiences of decision-makers in business and government, it may make sense for organisations working on climate change to continue to present the competitiveness or economic case for action. But there is scope for far greater circumspection about the likely impacts of such arguments percolating through into communications aimed at wider public audiences.

## 4 Values in Communications About Unrelated Causes

The last section advanced the case that communications about climate change, at least when targeted at public audiences, would do well to strive for consistency in appealing to I/ST values while avoiding appeals to E/SE values. It also presented evidence that the values implicit in a communication about one pro-environmental behaviour are important in determining spillover into other pro-environmental behaviours.

However, such spillover effects have also been found across seemingly unrelated causes. In another study, short texts presenting reasons for expressing concern about *disability rights* were found to be more effective in leading participants to state an intention to act to help mitigate *loss of biodiversity* where these texts engaged I/ST values as opposed to E/SE values (Crompton et al. 2014). Moreover, texts about *disability rights* (framed to engaged I/ST values) were found to be indistinguishable, in their effectiveness at leading participants to state an intention to act to help mitigate *biodiversity loss*, from texts about *biodiversity loss* (also framed to engage I/ST values). As in the case of the study discussed in Sect. 2, no interactions were found between a participant's prior values orientation and the type of prime presented to a participant in influencing either attitudinal or intention responses.

These results corroborate earlier studies (e.g. Maio et al. 2009; Sheldon et al. 2011) and point to further implications for organisations communicating on climate change: Disability rights and biodiversity conservation are causes which seem to be conceptually distinct. It can be anticipated that if texts focused on biodiversity conservation significantly affect people's motivation to express concern about disability rights (and vice versa) then communications on a wide range of different social or environmental causes will impact on people's concern about climate change.

Studies such as these suggest that those working on climate change communication would do well to consider not just the effects of communications about climate change upon public concern about climate change, but also the effects of communications about a wide range of other causes.

To take just one example of many, communications which highlight the financial value of biodiversity are likely to erode not just public concern about biodiversity loss, but also public concern about climate change. Such a result has important implications for the many environmental organisations that campaign upon both climate change and biodiversity conservation, and it seems probable that analogous results concerning other causes would hold widespread implications for a wide range of other organisations.

## 5 Perceptions of Others' Values

Section 2 highlighted the importance of social modelling in leading people to attach importance to E/SE values. It is known that a person's immediate social network—including their peers and family—has an influence on his or her values. Most people attach greater importance to I/ST values than E/SE values (Common Cause Foundation 2016). People with whom one is in close relationship are therefore likely to promote a person's commitment to I/ST values more strongly than his or her commitment to E/SE values.

But people are also likely to be influenced by their *perceptions* of the values of other people with whom they identify, but with whom they do not have close relationships. For example, a person's perceptions of the values that characterise a typical compatriot—someone with shared national identity—is likely to be important (Sheldon et al. 2011).

Survey data published by Common Cause Foundation (2016) finds that most UK citizens overestimate the importance that a typical fellow citizen attaches to E/SE values, and underestimate the importance that a typical fellow citizen places on I/ST values. A study of a demographically representative sample of a thousand British citizens found that 77% of people held this misconception. Comparable results have been found in the US (Common Cause Foundation, unpublished).

One can imagine how such a situation might arise. Whereas someone may have direct personal experience of the values held to be important by peers and family, he or she can only develop an understanding of the values of a typical fellow citizen indirectly. Here a person is likely to rely heavily on impressions of the value priorities of typical fellow citizens as conveyed by the media, what he or she is are told about the values of typical compatriots by political leaders and other public figures, deductions that can be made from the way in which public policies are shaped, and his or her experience of strangers who might be assumed typical of the population more widely.

Perhaps one of the most common contexts in which people encounter strangers representing diverse demographic backgrounds is in a commercial environment—in supermarkets and shopping centres. Research suggests that people encountered in these environments are likely to be primed to attach particular importance to E/SE values. It is known, for example, that showing people desirable consumer objects tends to lead them to attach greater importance to E/SE values (Bauer et al. 2012). Relative to other contexts, it is therefore likely the case that people in such commercial environments behave in ways that are consistent with E/SE values being primed at that time. One might anticipate therefore that a person's perceptions of the values of a typical fellow citizen are likely to be biased, such that a typical fellow citizen is believed to be more oriented towards E/SE values than is actually the case—something that is indeed found.

Conveying a more accurate perception of a typical fellow citizen's values may prove to be an important route to strengthening commitment to civic participation on causes such as climate change. Indeed, recent research has established a positive association between a person's perceptions of the importance that a typical fellow citizen places on I/ST values and this person's commitment to various forms of civic participation (including voting behaviour, intention to volunteer and intention to become involved in community initiatives) (Common Cause Foundation 2016).

One way to convey a more accurate perception of what typical fellow citizens value is to create opportunities for people from different demographics to meet one another, especially in contexts that are likely to prime I/ST rather than E/SE values. Initiatives to promote street parties represents one such avenue. The non-governmental organisation "Streets Alive" works to promote "community spirit at the street level through street parties and other neighbourly activities". According to the organisation's website, "[w]ith busy, more independent and less localised lives we have fewer chances to bump into people, [but] a street party right outside your door without traffic for a day is a good way to meet them all at once." (Streets Alive 2016).

Such initiatives are likely to have multiple social and environmental benefits. But there are other contributions that organisations with significant communications impact can make to convey a more accurate perception of the importance that most citizens place on I/ST values. In the next section, we discuss steps that a museum in Manchester is taking.

## 6 Applying These Insights: A Practical Example

In the remainder of this paper, we highlight a collaboration between Manchester Museum and Common Cause Foundation to promote I/ST values as a core component of the shared identity of people living in the Greater Manchester city region.

Manchester Museum is a regionally important museum attracting some 450,000 visitors a year. It also attracts the help of 250 volunteers who engage these visitors and facilitate them in exploring the museum's large collection. While the museum has taken extensive steps to communicate directly on climate change, this collaboration sets out to demonstrate the possibility of deepening visitors' commitment to acting on climate change, while communicating on other issues.

The collaboration has several related components:

- Direct communication about the values that most people prioritise
- Creating experiences that provide social proof of the importance of these values
- Drawing on the experience of the museum to support other organisations in the Greater Manchester city region to begin to engage their stakeholders in similar ways.

These components are now each considered in further detail.

## 6.1 Direct Communication About Others' Values

The project is establishing approaches to communicating the insight that most people attach particular importance to I/ST values, but that the widespread importance placed on these values is typically underestimated.

These approaches have not entailed elaborate or costly changes to the museum's exhibition spaces—though such changes would be possible.

For example, the initiative has led to the addition of a short question to the visitors' feedback form:

> Most people care deeply about other people and the places in which they live. We try to reflect these concerns through visitors' overall experience. How did we do?

Interactive boards are being developed upon which visitors are invited to record what matters to them in life, prompted by some possible answers taken from the Schwartz values survey (e.g. "*looking good*", "*being successful*", "*enjoying life*", "*helping others*", "*being creative*", "*respecting tradition*", "*being polite*", etc.). Visitors are then invited to walk around to the other side of the board to view data on the values which others prioritise.

Another approach is to draw attention directly to people's own experience of the kindness of strangers. An interactive board extends the following invitation:

> Most people care deeply about other people and the places in which they live. As part of the wider community, the Museum helps to celebrate this. Please tell us about times when someone you didn't know was kind to you – or draw a picture!

The impact of such interventions seems likely to be very modest. But these steps nonetheless represent an emerging understanding of some of the ways in which a cultural organisation, such as a museum, can help to redress widespread misunderstandings about others' values.

## 6.2 Creating Experiences Which Provide Social Proof of the Importance of I/ST Values

It is likely that a person's perception of a typical compatriot's value priorities will be shaped, in part, by their perception of the assumptions that social institutions make about what people value. An organisation that engages stakeholders (customers, employees or visitors, for example) in the *expectation* that these stakeholders are predominantly dishonest is likely to contribute to perpetuating the perception that people hold the I/ST value 'honesty' to be less important than is actually the case. On the other hand, encountering the expectation of honesty is more likely to convey the understanding that most people are honest, and is also more likely to elicit this value on the part of visitors.

For this reason, one important aspect of the Common Cause Foundation/ Manchester Museum collaboration is to examine a wide range of aspects of a visitor's experience of the Museum, with a view to identifying ways to engage I/ST values and to reduce the engagement of E/SE values.

The Museum has already adopted some such practices. For example, visitors are encouraged to handle original objects from the Museum's collection (as opposed to replica objects, as is the case in many museums). This practice is likely to convey an expectation of trustworthiness and honesty on the part of visitors. Further opportunities include, for example, encouraging volunteers to reflect in conversation with visitors on their personal motivations to give their time freely to help others at the museum, and the sense of purpose that they derive from this.

### *6.3 Building a Wider Network*

Although the museum is a regionally important institution, attracting large numbers of visitors, it is clearly important that many other organisations based in the city region work in analogous ways—contributing to a coherent narrative that roots citizens' understanding of their identity as residents of the city region in shared appreciation of the importance of I/ST values. With this aim, approaches are being made to a range of other organisations in the city region—other arts and heritage organisations, large employers, local media organisations and local government. This is with a view to supporting these organisations to embark upon activities analogous to those outlined in Sects. 5.1 and 5.2 above.

## 7 Challenges

There is extensive research on the way in which values are structured; on the relationship between value orientations and both behavioural and attitudinal responses to social and environmental problems and on the attitudinal and behavioural consequences of priming specific values. However, there is doubtless a need for more research on the attitudinal and behavioural effects of the values implicit in communications aimed at promoting social or environmental concern (Bolderdjik, et al. 2013; Chilton et al. 2012; Crompton et al. 2014; Evans et al. 2013). There is also need for more work on the attitudinal and behavioural effects of the values implicit in communications that prime an understanding of a person's group identity (Sheldon et al. 2011).

However, the authors are aware that it is also necessary to take a reflexive stance at this point. Much as access to information about climate change is known to be of limited effectiveness in strengthening public concern about climate change (Whitmarsh 2011), so one might anticipate that research on climate change communication strategies will be of limited effectiveness in driving changes to such strategies. Extending, and re-presenting, the *evidence* for the likely ineffectiveness, or even counter-productive impact, of many climate change communication strategies seems unlikely to prove sufficient as a route for motivating practitioners to rapidly adopt different strategies. This challenge, however, is beyond the scope of this article.

## 8 Conclusion

An understanding of values highlights some possible pitfalls in current communication strategies, and important new opportunities to build public support and demand for action on climate change.

In the first instance, it seems likely that communications aimed at a public audience, and implicitly engaging E/SE values (such as, for example, financial success, wealth or competitiveness) will inadvertently undermine public concern about climate change and public support for policies to mitigate climate change. Rather, it is argued that communications should seek to engage I/ST values (for example, emphasising the contribution that climate change policies may make to strengthening community, protecting other people and nature, redressing injustices, promoting peace, or as expressions of responsibility towards others). Evidence suggests that communications of this nature will prove to be most effective even among audiences which themselves place particular importance on E/SE values.

The implications of an understanding of values are far broader, however, and point to the potential impact of communications generated by many organisations that have no formal remit to communicate on climate change, and which communicate to public audiences on other issues. The values implicit in such communications are likely to impact on public support for climate policies.

It seems likely that, due to social norming effects, a person's own value disposition is likely to be influenced by his or her perceptions of a typical compatriot's values. Yet research finds that a large majority of people, at least in the UK, believe that a typical fellow citizen places greater importance on E/SE values, and lower importance on I/ST values, than is the case. There are opportunities for a wide range of different organisations to both model I/ST values in engaging their stakeholders, and to help redress the widespread misperception that most people care less about these values than is the case. Simple steps towards this goal, taken by a museum, were discussed.

In sum, we suggest that public support for action on climate change can be promoted through critical reflection on the values that are implicit in communications about climate change, and through the values that are implicit in the communications activities of a wide range of other organisations.

**Acknowledgements** The collaboration between Common Cause Foundation and Manchester Museum, outlined above, is financially supported by The Minor Foundation for Major Challenges.

## References

Bardi A, Lee JA, Hofmann-Towfigh N, Soutar G (2009) The structure of intraindividual value change. J Pers Soc Psychol 97:913–929

Bauer MA, Wilkie JEB, Kim JK, and Bodenhausen GV (2012) Cuing consumerism: situational materialism undermines personal and social well-being, *Psychological Science*, published online, 16 March, Available at: http://pss.sagepub.com/content/early/2012/03/16/0956797611429579. Accessed on: 22.12.2016

Bolderdijk JW, Steg L, Geller ES, Lehman PK, Postmes T (2013) Comparing the effectiveness of monetary versus moral motives in environmental campaigning. Nature Clim Change 3:413–416

Chilton P, Crompton T, Kasser T, Maio G, Nolan A (2012) Communicating bigger-than-self problems to extrinsically-oriented audiences. COIN, CPRE: Godalming, UK, Friends of the Earth, Oxfam, WWF-UK. Available at: www.valuesandframes.org/downloads. Accessed on: 22.12.2016

Common Cause Foundation (2016) Perceptions matter: the common cause UK values survey. Common Cause Foundation: Machynlleth, UK. Available at: www.valuesandframes.org/downloads Accessed on: 22.12.2016

Crompton T (2008) Weathercocks and signposts: the environment movement at a cross-roads. WWF-UK: Godalming, UK. Available at: www.valuesandframes.org/downloads. Accessed on: 22.12.2016

Crompton T, Kasser T (2009) Meeting environmental challenges: the role of human identity, WWF-UK: Godalming, UK. Available at: www.valuesandframes.org/downloads. Accessed on: 22.12.2016

Crompton T, Kasser T (2010) Human identity: a missing link in environmental campaigning. Environment 52:23–33

Crompton T, Weinstein N, Sanderson B, Kasser T, Maio G, Henderson S (2014) No cause is an Island: how people are influenced by values regardless of the cause. Common Cause Foundation: Machynlleth, UK. Available at: www.valuesandframes.org/downloads. Accessed on: 22.12.2016

Evans L, Maio GR, Corner A, Hodgetts CJ, Ahmend S (2013) Self-interest and pro-environmental behaviour. Nature Clim Change 3:122–125

Gifford R (2011) The dragons of inaction: psychological barriers that limit climate change mitigation and adaptation. Am Psychol 66:290–302

Gifford R, Nilsson A (2014) Personal and social factors that influence pro-environmental concern and behaviour: a review. Int J Psychol 49:141–157

Grouzet FME, Kasser T, Ahuvia A, Fernandez-Dols JM, Kim Y, Lau S, Ryan RM, Saunders S, Schmuck P, Sheldon KM (2005) The structure of goal contents across fifteen cultures. J Pers Soc Psychol 89:800–816

Kahan D (2014) What you "believe" about climate change doesn't reflect what you know; it expresses *who you are*, The cultural cognition project at Yale Law School. Available at: http://www.culturalcognition.net/blog/2014/4/23/what-you-believe-about-climate-change-doesnt-reflect-what-yo.html (Accessed on 31 October 2017)

Kasser T, Ryan RM (1996) Further examining the American dream: differential correlates of intrinsic and extrinsic goals. Pers Soc Psychol Bull 22:280–287

Kasser T, Ryan RM, Couchman CE, Sheldon KM (2004) Materialistic values: their causes and consequences. In: Kasser T, Kanner AD (eds) Psychology and consumer culture: the struggle for a good life in a materialistic world. American Psychological Association, Washington DC, pp 11–28

Maio GR, Pakizeh A, Cheung WY, Rees KJ (2009) Changing, priming, and acting on values: effects via motivational relations in a circular model. J Pers Soc Psychol 97:699–715

Rose C (2014) Strategic reasons why values matching is a good idea. Campaign Strategy Newsletter 90. Available at: http://documents.campaignstrategy.org/uploads/campaignstrategy_newsletter_90.pdf. Accessed on: 22.12.2016

Rokeach M (1973) The nature of human values. Free Press, New York

Sandel M (2012) What money can't buy: the moral limits of markets. Allen Lane, London

Schwartz SH (1992) Universals in the content and structure of values: theoretical advances and empirical tests in 20 countries. In: Zanna M (ed) Advances in experimental social psychology, 25. Academic Press, Orlando, FL, pp 1–65

Schwartz SH (2011) Studying values: personal adventure, future directions. J Cross Cult Psychol 42:307–319

Schwartz SH, Melech G, Lehmann A, Burgess S, Harris M, Owens V (2001) Extending the cross-cultural validity of the theory of basic human values with a different method of measurement. J Cross Cult Psychol 32:519–542

Sheldon KM, Nichols CP, Kasser T (2011) Americans recommend smaller ecological footprints when reminded of intrinsic american values of self-expression, family, and generosity. Ecopsychology 3:97–104

Stern PC (2011) Contributions of psychology to limiting climate change. Am Psychol 66:303–314

Streets Alive (2016) About streets alive. Available at: http://www.streetparty.org.uk/about-streets-alive.aspx. Accessed on: 22.12.2016

Swim JK, Stern PC, Doherty TJ, Clayton S, Reser JP, Weber EU, Gifford R, Howard GS (2011) Psychology's contributions to understanding and addressing global climate change. Am Psychol 66:241–250

Vohs KD, Mead NL, Goode MR (2006) The psychological consequences of money. Science 314:1154–1156

Whitmarsh L (2011) Scepticism and uncertainty about climate change: dimensions, determinants and change over time. Glob Environ Change 21:690–700

CPSIA information can be obtained
at www.ICGtesting.com
Printed in the USA
BVHW01*2010120618
518867BV00001B/2/P

9 783319 698373